Reflector and Lens Antennas

Reflector and Lens Antennas

Analysis and Design Using Personal Computers

Carlyle J. Sletten, Editor

Artech House

Library of Congress Cataloging-in-Publication Data

Reflector and lens antennas : analysis and design using personal computers / Carlyle J. Sletten, editor.

p. cm.

Bibliography: p.

Includes index.

ISBN 0-89006-240-4

1. Antennas, Reflector — Design and construction — Data processing.
2. Lens antennas — Design and construction — Data processing.
3. Microcomputers. I. Sletten, Carlyle J., 1922-

TK7871.6.R42 1988

621.38′0283 — dc19 88-10452

ARTECH HOUSE, INC.
685 Canton Street
Norwood, MA 02062

International Standard Book Number: 0-89006-262-X
Library of Congress Catalog Card Number: 88-6324

10 9 8 7 6 5 4 3 2 1

Contents

Preface

This book is intended primarily as an analytical and computer aid in designing reflector and lens antennas. The important equations governing antenna performance are presented, and their limitations and derivations are discussed. Detailed mathematical proofs of unusual expressions are given in Appendices 1A and 1B, and in Chapter 2. Software suitable for the IBM PC-XT or equivalent computer is developed, and printouts in BASIC and FORTRAN are provided. Diskettes with the software are available as an option. The authors' main goal is to lead the student or designer to the physical characteristics of antenna choices, with the geometric configurations and dimensions quickly visible from illustrations and computer graphics. The software provides antenna dimensions and indicates, quantitively, performance trade-offs and the influence of sensitive parameters.

Several books on microwave antennas have appeared in the 1980s, which exploit the ubiquitous capabilities of small computers. The book by Warren Stutzman and Gary Thiele, *Antenna Theory and Design* (John Wiley and Sons, 1981), covers a wide spectrum of antenna topics and leads to the excellent Ohio State University *Numerical EM Code, Reflector Antennas* (1982), by R. C. Rudduck and Y. C. Chang, written to run on main-frame computers. Books by D. M. Pozar, *Antenna Design Using Personal Computers* (Artech House, 1985), and by J. A. Kuecken, *Exploring Antennas and Transmission Lines by Personal Computer* (Van Nostrand, 1986), also cover a wide variety of antenna themes with suggested software options. Recent advances in antenna theory and applications have been documented in the *Handbook of Antenna Design*, volumes I and II, A. W. Rudge, *et al.* (Peregrinus, 1983, 1986) and the *Handbook of Antenna Theory and Design* (Staff of SAM, 1985). Bob Hansen's book, *Microwave Scanning Antennas* (Peninsula, 1986), the volume, *Reflector Antenna Analysis and Design,* by P. J. Wood (Peregrinus, 1980), and *Antennas* by Lamont Blake (Artech House, 1984), contain much new information on reflector and lens antennas.

This textbook fills a niche in the antenna literature as a volume dedicated entirely to the improvement of reflector and lens antennas, which are likely to gain further importance in the millimeter-wave world of tomorrow. The reflector antenna has competed well in the microwave world with phased array technology as reflectors provide the highest gains, widest bandwidths, and best angular resolutions at the lowest costs. Phased arrays are generally assumed to have advantages in attaining lower sidelobes and in generating and scanning multibeams, although the techniques described in this book also offer challenges to these claims.

With its extensive development of computer software, demonstrating with the aid of simple examples the relationships between antenna performance and antenna parameters, this book endeavors to make new and old antenna design methods more readily available to students and engineers in aperture antenna technology. Many small programs are provided that relate such radiation characteristics as pattern sidelobes and gain to dimensions of antenna feeds and structures. This book is written especially for the engineer who is familiar with computer methods and wishes to learn the analytic principles of reflector and lens antennas that lead to the design of high gain antennas used in microwave, millimeter-wave, and infrared systems.

There are some important advantages in using a software approach to antenna design. Large aperture diffraction (and edge diffraction), as well as shaping of reflecting and refracting surfaces, are gigantic numerical problems that are best solved with the power of digital computers. Software can "capture" the arcane sophistication of the analytical and computational procedures that are needed for the synthesis of optimum antenna shapes and patterns. The theory, equations, and objectives of the design procedure are stated in the text, but the computer programs converting input data into printouts of the predicted performance make the antenna "come to life." The time-consuming tasks of programming and debugging have been done, and even the seemingly inevitable printing errors in text are generally absent in the available source code listings. Learning is fastest and surest with concrete examples yielding numbers and geometric configurations. Once the principles are understood, the computer programs become tools to fashion new antennas with fewer experimental trials and iterations.

This book is an advanced text on reflector and lens antenna design containing many new techniques. The tutorial programs listed in the text are but small fragments of the much more complex computer programs available in the optional software package with user's manual and example booklet, also published by Artech House.

There have been some very important advances in recent years in the development of reflector and lens antennas. Without giving credit here to inventors and developers (this is done by the references cited in each chapter), these accomplishments seem to us to be of the most practical importance:

optimized offset subreflector feeding of reflector antennas that reduce aperture blocking and cross-polarization levels;
surface shaping of reflectors and lenses to synthesize radiation patterns;
the development and analysis of conical corrugated horn feeds that give almost rotationally symmetric patterns over wide bandwidths with nearly Huygen source polarization;
wide-angle scanning of dual-reflector antennas and two-point corrected dielectric lenses with multiple sources to improve antenna information-gathering capabilities;
beam waveguide feeding systems to reduce transmission losses, especially at millimeter wavelengths.

Many journal articles and conference papers have been searched to make this textbook on reflector and lens antennas as modern and useful as possible. Some of the work comprises original contributions by the authors, however, and appears in print here for the first time.

The authors are grateful for permission to publish material from original sources. We thank especially AF RADC/EE and GTE-CSD for their help and approval.

Acknowledgments and Dedication

"Microwave optics" was the term Roy Spencer used to describe the electromagnetic techniques based on geometric optics and physical optics for calculating antenna patterns and object scattering. He was one of my teachers at the US Air Force Cambridge Research Center, which was largely staffed by scientists from the MIT Radiation Laboratory after World War II. Roy Spencer, Ralph Hiatt, and Shep Holt were from the Rad Lab and were leaders in the Antenna Laboratory of AFCRC. They were joined soon by talented scientists, including Wade Ellis, Ellen Fine, Kaiser Kunz, and John Ruze at Albany Street in Cambridge, Massachusetts. In 1948, Walter Rotman, Hans Zucker, and I were rookies in this "university," where we learned many techniques that were later developed in this book. The Antenna Laboratory also acquired the Rad Lab Antenna Test Range at Ipswich, Massachusetts, with personnel, Jake Baker, John Andriotakis, Leon Dorr, and Harold Henkle. In Chapters 1 and 3, the author has tried to highlight some of the classic papers that came out of the Rad Lab and the Antenna Lab. Many of these seminal reports were not published, and became unavailable and largely forgotten.

During my 40 years in the field of microwave antennas, I have learned much from many people. Fearful of overlooking important helpers, I will nevertheless list those most indelibly imprinted in my memory. Professor Ronald King and his brother Don were my teachers at Harvard University in the 1940s along with Phillip LeCorbier, Harry Mimno, and L. Brillouin. In those days, we worked closely with Lan Chu at MIT and with Professor Woonton and his staff at McGill University in Canada on topics in Microwave Optics. I recall, too, early collaboration with British antenna experts and I became acquainted with Hubert Gent and his bootlace metal lens antennas. The radio-astronomy antenna experts were always ingenious, and we learned about compound arrays at Cambridge

University. German technology aided us, too, and contributions by Herman Ehrenspeck and Werner Gerbes, who joined AFCRL, were important to us, as were their contacts. In the early years, Professor Lothrope Knutson and the staff of the University of Denmark were strong collaborators. This international scientific aid is reflected by the corrugated horn radiation work of Clarricoats and Olver reported in this volume. My nine months in Madrid under the Fulbright program in 1963–64 brought technical ties with Spain that are apparent from the chapter on feeds by Miguel Calvo. In my work for the United Nations in Brazil in the late 1970s, the capable young faculty at Catholic University in Rio de Janeiro headed by Professors Marcio Rabello and Luiz Costa da Silva were an inspiration to me. We incurred a great debt to Japanese technology when Y. Mizugutch gave a paper at the University of Massachusetts for the IEEE AP-S symposium in 1976 on the proper tilting of offset dual reflectors.

AFCRL was in a strategic place to interact with university research in electromagnetics, much of which was sponsored by the US Air Force. I remember Joe Keller of NYU and his first efforts in the Geometric Theory of Diffraction. A. A. Oliner, N. Marcuvitz, L. Felson, and A. Hessel at Brooklyn Polytechnical Institute taught us about surface waves. Kip Siegel, Tom Senior at the University of Michigan, and Bob Hansen at Conductor, Inc. gave us lessons in diffraction. At the University of Illinois, Ed Jordon and Georges Deschamps were our benefactors. At the Naval Research Laboratory, we knew Bob Adams, Bill Gabriel, and Merrill Skolnik. Closely related to the topics in this book were our conversations with A. Rusch, Vic Galindo, A. Ludwig, Raj Mittra, J. J. Lee, and Bob Kouyoumjian. The papers from Bell Laboratories, especially those of Dave Hogg and C. Dragone, and those of George Goubau of the US Army Laboratories at Fort Monmouth, form the basis for the basic concepts of this book relating to beam waveguides. The scalar horn work of Alan Kay and Al Simons forms a keystone in antenna feed studies.

Since my retirement from AFCRL/RADC in 1976, I have been grateful for my technical association with Irv Goldberg, Len Blaisdell, Ken Rau, Ernie Wilkinson, Barry Watson, George Biernson, Henning Olesen, and others at GTE in Needham Heights, Massachusetts. Perhaps my closest technical collaborator has been Bob Shore of the Electromagnetic Sciences Directorate, RADC, who has taught me much, especially about mathematics and computers. I also have been able to continue communication with the present technical luminaries at the US Air Force RADC (Bedford, Massachusetts) laboratories, especially with Phil Blacksmith, Bob Mailloux, Allan Schell, Jay Schindler, Chuck Drane, and Lee Poirier. You will find the fine technical touch of Shep Holt, formerly of AFCRC, in these pages, where he has written a little and edited some errors.

Special personal thanks are given to the readers of draft chapters of this book and to others for helping in its preparation. Chapter 3 was read by Rick Barnes of MIT Lincoln Laboratory, Lexington, Massachusetts, who also wrote the graphics program LINEPL.BAS. Chapter 5 was reviewed by Adriaan Walther of Worcester Polytechnical Institute. Chapter 7 was read by Tom Kirby of GTE, Needham, Massachusetts. Chapters 3 and 8 were read by Phil Blacksmith of RADC/Hanscom, who also worked long hours on the figures of many chapters. My wife, Ruth, has been my strongest supporter.

I gratefully dedicate this book to these people and many others who have worked with me in the antenna field. Beyond these acknowledgments, this book is written for the young men and women of the twenty-first century who will be using microwave, millimeter-wave, infrared, and optical radiation for instruments and sensors. The four young men writing chapters with me in this book, Bill Shillue, Dan McGrath, Miguel Calvo, and Yueh-Chi Chang, admirably represent this generation. Through our hard work of programming the analytic and geometric structure of electromagnetic radiation, based on the legacy of the twentieth century experience in reflector and lens antennas, we make this design knowledge easily available to modern, computer-literate engineers. The many simple, practical examples in this book make transparent the esoteric art of antenna design. Each computer program with the explanatory text is a tool to deliver numbers and charts that illustrate some facet of microwave radiation phenomena. It has been a tedious task to make these techniques and equations available in MS-DOS PC BASIC and FORTRAN, but those who teach electrical engineering will find this text and software to be an effective gateway into the worlds of radar, communication, satellite sensors, and other systems of electromagnetic applications that treat antenna gains, sidelobes, and aperture efficiencies.

Carl Sletten
January 1988

DISCLAIMER OF WARRANTY

The technical descriptions and procedures and the computer programs printed in this book are provided as is, without warranty of any kind. We make no warranties, expressed or implied, that the equations, programs, and procedures in this book are free of error, or are consistent with any particular standard of merchantability, or that they will meet the user's requirements for any particular application. The programs or procedures should not be employed to solve a problem for which incorrect solution could result in injury to a person or loss of property. Any use of the programs or procedures in such a manner, is at the user's own risk. The authors and the publisher disclaim all liability for direct, incidental, or consequential damages resulting from any use of the programs or procedures in this book.

SPECIAL ACKNOWLEDGMENT

Special mathematical subroutines, copyrighted and sold by Numerical Recipes Software, P. O. Box 243, Cambridge, MA 02238, are used in some of the programs listed in this book and in the optional software package with user's manual and example booklet available from Artech House. These subroutines are used with permission of Numerical Recipes Software.

The Authors

Miguel Calvo, Professor (Catedratico) University of Madrid, Spain, School of Telecommunication.
PhD, University of Madrid.

Yueh-Chi Chang, Electronic Scientist, Bell Aerospace-Textron, Buffalo, New York.
PhD, Ohio State University, Columbus, Ohio.

F. Shepard Holt, Professor of Mathematics (retired) Tufts University, Medford, Massachusetts.
PhD, Massachusetts Institute of Technology, Cambridge, Massachusetts.

Daniel T. McGrath, Electronic Scientist, Electromagnetics Sciences Division, RADC, Bedford, Massachusetts.
MS, United States Air Force Institute of Technology, Wright-Patterson AFB, Ohio.

William P. Shillue, Electronic Engineer, GTE, Needham Heights, Massachusetts.
BSEE, Cornell University, Ithaca, New York.

Robert A. Shore, Research Scientist, Electromagnetic Science Division, RADC, Bedford, Massachusetts.
PhD, Harvard University, Cambridge, Massachusetts.

Carlyle J. Sletten, (Editor), Senior Member Technical Staff, GTE-Communication Systems Division, Needham Heights, Massachusetts.
MA, Harvard University, Cambridge, Massachusetts.

Chapter 1
INTRODUCTION TO BASIC TECHNIQUES FOR ANALYSIS OF APERTURE ANTENNAS

R. A. Shore and C. J. Sletten

1.1 INTRODUCTION

This chapter summarizes important factors in the design of high performance aperture antennas. Emphasis is on radiation patterns with low sidelobes and high gain, although other design goals are identified. These fundamental considerations apply to either lens or reflector aperture antennas. Lens antennas usually do not have feed-blocking problems. Offset feeding techniques are developed for reflectors to avoid aperture blockage and reduce cross polarization. Reflector antennas generally have lower weight and cost less to fabricate than lenses, making them the designer's choice for many applications.

The shape of the antenna aperture edge contour—for example, circular, rectangular, or elliptical shapes—has great influence on the antenna's near-in sidelobe characteristics and the shape of the main radiation lobe. Elliptically shaped apertures give rise to oval-shaped main beams with the narrow beamwidths being in the plane of the largest aperture dimension. Dr. R. C. Spencer [1], writing in 1942, discussed the use of antenna aperture shape to control sidelobe level. He writes,

> We have noted that the sidelobes for the rectangular aperture are especially low in the directions along the diagonals. For example, [Figure 1.1(a)] suggests a method of eliminating sidelobes in the vertical direction by cutting the paraboloid so that the amplitude distribution in this direction approximates a Gauss error curve.

> The diagonal aperture shape in [Figure 1.1(b)] tends to reduce lobes in both the horizontal and vertical directions over those of the circular (aperture) case, but they will be increased in the 45 degree directions.

Circular antenna apertures, however, are the usual choice for high gain antennas in the microwave and millimeter-wave bands because they produce relatively low and equal first sidelobes in all ϕ-cuts through the (usually) rotationally symmetric main beam. Also, conical corrugated horns generate nearly circularly symmetric illumination patterns, approximating the assumption of rotationally symmetric distributions on circular apertures. The advantage in analysis of treating radial amplitude functions on circular apertures for the study of optimum gain *versus* sidelobe solutions also leads us to circular apertures. Elliptical or rectangular shaped apertures can be substituted for circular apertures when beam or coverage patterns suggest such geometries, and their radiation patterns may be computed by numerical methods.

Primarily, we will discuss paraboloids with circular apertures as main reflectors, although most of the results apply equally well to spherical or parabolic torus antennas, or for that matter to lenses with circular apertures. We present frequently used design equations and analysis tools with some discussion of their origins and applications. We also examine the in-phase amplitude radiation patterns of circular apertures when excited by "ideal" distributions that yield the best antenna (aperture) gain for given sidelobe levels. Then, we discuss practical limitations that prevent realization of optimum pattern performance. These "spoiling" effects are principally feed spillover, cross polarization, feed and strut blocking, and aperture phase errors.

The chapter includes two important analytical and computer program developments that are used frequently in later chapters: (1) the optimum tilt configurations for offset dual-reflector antennas and (2) a technique for computing the radiation pattern of conical corrugated horns. Derivations of expressions used in these important "computational utilities" are given in the Appendices 1A and 1B. A list of program names for the computer algorithms developed for Chapter 1 is given in Appendix 1C with printouts of selected source code to illustrate the computer techniques. All computer programs are available on diskettes in the optional software package with user's manual and example booklet. The computer programs are written in Microsoft BASIC and FORTRAN for the IBM PC or compatible computers.

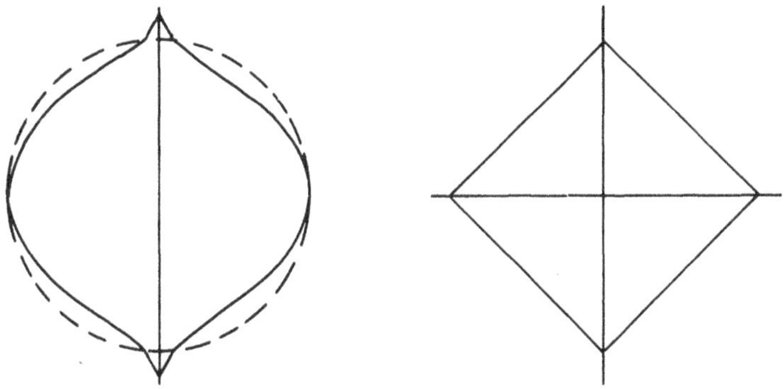

Figure 1.1 Aperture Edge Contour Shaping to Reduce Sidelobes: (a) Elimination of Vertical Lobes; (b) Reduction of Vertical and Horizontal Lobes

1.2 BASIC EQUATIONS FOR COMPUTING APERTURE ANTENNA RADIATION PATTERNS

To calculate the radiation patterns of reflector or lens antennas we need analytical expressions for the feed pattern and the fields across the aperture of the antenna for a given reflector feed configuration. Then a procedure for computing the antenna far-field patterns from the aperture distributions is required.

The feed is the element that connects the source of electromagnetic radiation, such as a transistor oscillator, by way of a transmission line, such as a waveguide, to the antenna. In Chapter 2 important feeds for illuminating reflectors and lenses are described. Additional feed pattern information is given in Chapter 3. The half-wave electric dipole radiator, an important and basic feed, has a far-field radiation pattern given approximately by [2]

$$E_\theta = j60I_0 \cos[(\pi/2)\cos(\theta)]/\sin(\theta)e^{-j\beta r}/r$$
$$E_\phi = 0 \tag{1.1}$$

where I_0 is the current at the center of the dipole, and $\beta = 2\pi/\lambda$, with λ being the wavelength.

The spherical angles θ and ϕ are defined in a rectangular coordinate frame x,y,z with the half-wave dipole lying along the polar z-axis. The radiated spherical wave decreases like $1/r$ as a function of distance from the radiator with the electrical phase given by the term $e^{-j\beta r}$. The feed pattern is rotationally symmetric with no variations in ϕ. The radiated vector far field is assumed to be TEM (no field components in the r direction) with H_ϕ everywhere orthogonal to and proportional to E_θ. Because this pattern is quite broad, dipole radiators usually have a reflector element to direct the radiation toward the aperture antenna.

Half-wavelength slot radiators [3] in waveguides [with small ground planes, see Figure 1.2(a)] are almost unidirectional with the pattern peak in the $\phi = \pi/2$ plane (toward the reflector antenna) given by [Figure 1.2(b)].

$$E_\phi = -j/\pi V_0 \cos[(\pi/2)\sin\theta]/\cos(\theta)e^{-j\beta r}/r$$
$$E_\theta = 0 \tag{1.2}$$

where V_0 is the voltage across the slot terminals.

However for the half-wavelength slot antenna $E_\theta(\theta,\phi)$ is in general nonzero. For the slot antenna shown in Figure 1.2(a), where the slot axis lies along y and the slot aperture is perpendicular to the z-axis, there are both E_θ and E_ϕ fields except in the $x = 0$ plane.

More typically, aperture antennas are fed by horn radiators. Horn patterns can often be approximated by Gauss error functions or by a $\cos^N\theta$ function so that, for a horn pattern linearly polarized in the y direction at $\theta = 0$, we can write

$$g(\theta)_H \approx A \cos^N\theta(\sin\phi\boldsymbol{\theta} + \cos\phi\boldsymbol{\phi})e^{-j\beta r}/r \tag{1.3}$$

where N is a constant (ranging from one to perhaps 500) and the $\cos^N\theta$ pattern is assumed to be zero for $\theta > 90°$. A is the power density per unit solid angle at $\theta = 0$; it is often normalized and the horn pattern expressed in dB. The horn pattern is assumed to be a function of θ only, independent of ϕ, and the polarization that of a Huygens source.

These simple examples of antenna feeds illustrate the vector fields of the feed radiation patterns with complex variables used for computing electrical phase from path-lengths. For details about feed radiation patterns, including relations between linear polarization and circular polarization, the reader is referred to Chapter 2. Radiation patterns of dipoles and slots are developed further in Chapter 3.

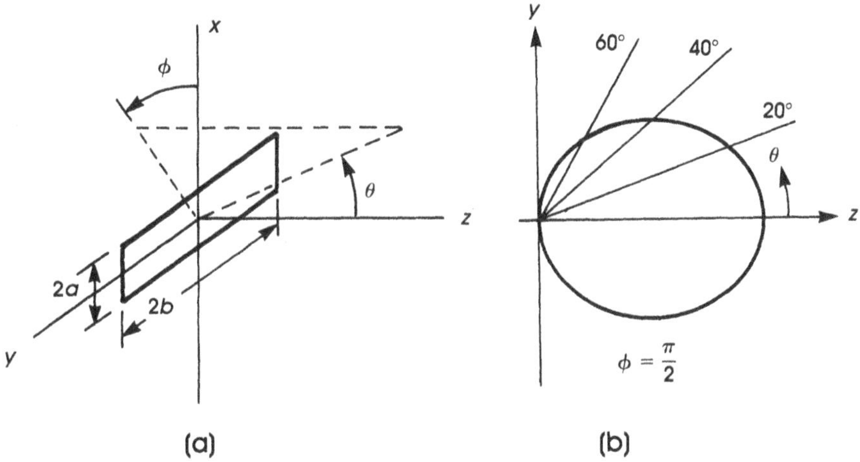

Figure 1.2 (a) Resonant Slot in Ground Plane; (b) Electric Field E_ϕ versus θ in $\phi = \pi/2$ Plane
(From Collin and Zucker, Part I [3]. Reprinted with permission.)

1.2.1 Computing Aperture Fields

The usual tool for calculating the aperture fields from the feed patterns reflected or refracted by surfaces or structures on the aperture antennas is *geometric optics* (GO). According to GO, electromagnetic energy flows along straight ray paths in TEM modes with phase determined by path lengths, βr, and amplitude calculated by the convergence or divergence of ray tubes. For spherical, diverging rays energy decreases like $1/r$. Fortunately GO has been extended by the *geometric theory of diffraction* (GTD) [4,5] to handle radiation from reflector edges and other discontinuities by methods which define diffracted rays that have characteristics like the GO rays.

Snell's laws determine quite accurately the bending of rays by reflecting and refracting surfaces. The most useful form of Snell's equations for perfectly conducting reflector surfaces are listed here [6]. For a GO ray in the direction given by its unit vector $\mathbf{S}_i$ incident on a perfectly conducting surface with normal $\mathbf{n}$, its unit reflected vector, $\mathbf{S}_r$, is given by

$$\mathbf{S}_r = \mathbf{S}_i - 2(\mathbf{n} \cdot \mathbf{S}_i)\mathbf{n} \tag{1.4}$$

The electric fields at a perfect conductor obey the relation

$$\mathbf{E}_r = 2(\mathbf{n} \cdot \mathbf{E}_i)\mathbf{n} - \mathbf{E}_i \tag{1.5}$$

with the magnitudes $E_r = E_i$.

Using the condition that rays are TEM waves, the magnetic fields can be expressed as

$$\mathbf{H}_i = (\varepsilon_0/\mu_0)^{1/2}(\mathbf{S}_i \times \mathbf{E}_i) \tag{1.6}$$

$$\mathbf{H}_r = (\varepsilon_0/\mu_0)^{1/2}(\mathbf{S}_r \times \mathbf{E}_r) \tag{1.7}$$

where ε_0 is the permittivity of free space and μ_0 is the permeability of free space.

Surface currents, $\mathbf{J}_s$, can be computed from

$$\mathbf{J}_s = 2(\mathbf{n} \times \mathbf{H}_i) = 2(\mathbf{n} \times \mathbf{H}_r) \tag{1.8}$$

Snell's law for lenses separating homogeneous dielectrics with indices of refraction, N_0 and N_1, at a refracting boundary, where subscripts 0 and 1 identify media 0 and 1, respectively, can be written as

$$\mathbf{n} \times (N_0\mathbf{S}_i - N_1\mathbf{S}_t) = 0$$

where $\mathbf{S}_t$ is the transmitted ray in dielectric medium 1. If the relative index of refraction in medium 1 is N_1 and $N_0 = 1$, a more familiar form of Snell's law is written:

$$\sin\theta_0 = N_1 \sin\theta_1$$

where θ_0 is the angle that the incident ray S_i makes with the normal to the dielectric interface, and θ_1 is the angle of the transmitted ray with this normal.

We can write the E-field component [7], normal to the plane of incidence, of the transmitted ray as

$$E_t = 2E_i \cos\theta_0 \sin\theta_1/\sin(\theta_1 + \theta_0) \tag{1.9}$$

and of the reflected ray as

$$E_r = E_i \sin(\theta_1 - \theta_0)/\sin(\theta_1 + \theta_0) \tag{1.10}$$

For the components of E lying in the plane of incidence, we can write for the transmitted component

$$E_t = E_i 2 \cos \theta_0 \sin \theta_1 / [\sin(\theta_0 + \theta_1) \cos(\theta_0 - \theta_1)] \tag{1.11}$$

and for the reflected component

$$E_r = E_i \tan(\theta_0 - \theta_1)/\tan(\theta_0 + \theta_1) \tag{1.12}$$

Given a feed pattern $g(\theta, \phi)$, the ray optics expressions above, when applied to points on the surfaces with normals **n**, can be used to trace rays out to a reflector or lens surface and then to an aperture plane. The antenna aperture has a given edge boundary and the aperture fields are in general complex vectors.

In Chapter 5 these expressions for reflection and refraction at dielectric boundaries are developed in computer programs.

1.2.2 Radiation from Reflector Edges

To calculate radiation fields due to edges or rims of metal reflectors, the GTD or *uniform geometric theory of diffraction* (UGTD) [8–10] can be used. According to this extension of GO, any point, P, surrounding the reflector antenna receives radiation from a small number of edge points, usually two. The E-field at any point due to these rim points is calculated using diffraction coefficients, D, which are functions defined for both parallel and perpendicular E-components illuminating the two special points on the rim. Using ray-fixed coordinates these functions can be evaluated from Fresnel integrals or Cornu spirals (which are incorporated into the diffraction coefficients) to find rim-caused radiation patterns. The fields of special interest are in the shadow regions. Unlike for GO, the total radiation field at the shadow and reflection boundaries is continuous. The two rim-point locations contributing to any point P can be shown to be on minimum or maximum ray paths connecting the feed phase center to the rim points and thence to the point P (*cf.* Fermat's principle [3,7]). We can indicate formally the vector (or dyadic) expressions [10] for computing UGTD fields from two rim points Q_1 and Q_2 illuminated by the feed by the edge fields $E(Q_{1,2})$ as:

$$E^d_{1,2} = E(Q_{1,2})D[L(r),\phi,\phi_1]A(r)e^{-j\beta r} \tag{1.13}$$

The angles ϕ and ϕ_1 are measured with reference to the tangent to the reflector surface at the edge points. The functions L and A are simple geometric expressions involving r. E-field components parallel and perpendicular to the edge surface at Q_1 and Q_2 are calculated for similar components of D. Details needed for computer evaluation including multiple diffracted rays and effects of field gradients are discussed in Chapter 4.

The rim of the reflector also causes a backlobe or caustic lobe due to the sum contributions of all the rim points. This lobe can be computed by a separate GTD procedure [9].

1.2.3 Aperture Diffraction Patterns

Far-field patterns for aperture antennas are derived in [7,11] beginning with Maxwell's equations. Radiation patterns are calculated for simplified assumptions such as in-phase aperture fields, with simple, closed-form integrable amplitude variations such as cosine functions of the radius of a circular aperture. Tables (from Reference [11], pp. 187 and 195) are helpful for estimating levels of first sidelobes and antenna gain of rectangular and circular apertures. The Fourier transform method [11,12] for computing the aperture antenna far-field pattern, $g(\theta,\phi)$, is given by the following expressions when the aperture field distribution is given by $f(x,y)$:

$$g(\theta,\phi) \approx \int_A f(x,y)e^{j(k_x x + k_y y)}dxdy \tag{1.14}$$

where

$$k_x = k \sin \theta \cos \phi \text{ and } k_y = k \sin \theta \sin \phi$$

with the radiation patterns expressed in the spherical angles θ and ϕ. This rectangular coordinate formulation is used as a starting point for the efficient and rapid radiation pattern methods developed in Chapter 4.

For circular apertures it is convenient to write the relation of the aperture fields to far-field patterns in polar coordinates ρ,ϕ_1 in the aperture as:

$$g(\theta,\phi) \approx \int_A f(\rho,\phi_1)e^{ju\rho \cos(\phi-\phi_1)}\rho d\rho d\phi_1 \tag{1.15}$$

where ρ is normalized to the aperture radius, $\mathbf{A}$, and $u = (2\pi/\lambda) \sin \theta$. Because circular apertures are used frequently in this book, this polar coordinate formulation is coded in the program PATT.FOR to compute far-field radiation patterns numerically. The double integral can be evaluated numerically from two integrals of the form,

$$I = \int_{-\pi}^{\pi} \left\{ \int_0^1 A(\rho,\phi) \begin{bmatrix} \cos(v(\phi)\rho) \\ \sin(v(\phi)\rho) \end{bmatrix} \rho d\rho \right\} d\phi \qquad (1.16)$$

The integration [13,14] with respect to ρ is by Filon's formula and the outer integration with respect to the angular variable ϕ is performed by the extended Simpson's rule. The number of radial and angular points or cuts depends of course on the number of wavelengths in the aperture. To avoid grating lobes in the near-in sidelobe region spacings, spacings of up to two wavelengths on the aperture rim can be tolerated. Typically 21 to 51 radial points and from 75 to 301 angular cuts can be computed by a microcomputer.

Although the program PATT FOR tends to be somewhat inefficient because the aperture points are much closer together near the center of the aperture than on the edges, this program is the "workhorse" in several chapters of this book where circular apertures are the simple examples examined. Usually only aperture phase and amplitude distributions for one polarization component are entered into PATT by the file PHAMP.DAT computed for the particular antenna of interest. When the aperture contours are not circular, zeros can be used to fill an oversized circular aperture using computer conditionals that conform to the antenna's edge contours.

The source code for PATT. is listed in Appendix 1C. The θ- and ϕ-intervals for the far-field pattern cuts are entered interactively. The θ prompts, THETAI,THETAF,DTHETA, appear on the screen asking for the initial θ, final θ, and the delta θ intervals. For ϕ, values in degrees from 0 to ± 180 may be entered. The program first prints out the antenna gain in dBi and files the points on the radiation pattern in the file CJS.DAT. The radiation patterns can be computer plotted using the versatile graphics program LINEPL.BAS written by Rick Barnes of MIT Lincoln Laboratory. The program CHRN.FOR with CORH.DAT provides dimensions for constructing conical corrugated horns.

1.3 "IDEAL" CIRCULAR APERTURE DISTRIBUTIONS

What is the best power distribution on a circular antenna aperture for the highest gain for a given sidelobe level? Unlike for linear arrays of discrete radiators, there are no continuous distributions across apertures that can be shown analytically to give the optimum gain versus sidelobe performance. However, we can be guided toward good gain versus sidelobe performance by the results of several workers [15–19].

It is of interest to compare the performance of antennas with imperfect feed patterns and phase errors with ideal antennas that have the best known aperture distribution free from phase errors. The results presented here illustrate that realizable radiation patterns do approach the better known model aperture distributions. In general, "ideal" aperture distributions have radial functions that are akin to a Gauss error function in shape, without discontinuities (except at the aperture edge) or discontinuous first derivatives. Sidelobes can be related to Gibb's phenomena in Fourier transform theory, where discontinuities are shown to increase lobing levels.

For the rotationally symmetric circular aperture several radial functions have appeared in the literature that give good performance for first sidelobes at about the −30 dB level from pattern peak. These radial distribution functions [13] are listed in Table 1.1 and the first five sidelobe levels and angular locations are summarized in Table 1.2. These computations were done using the program PATT. Case 1 shows the sidelobe peaks for uniform aperture distributions.

The radial functions, Case 2, due to Kouznetsov [15], have vanishing derivatives at the aperture center and edges and are plotted in Figure 1.3 which also presents a profile of the first sidelobes. Case 3, Minimum Spreading Factor Functions, are taken from optical diffraction literature [16] and numerical integration of these dome-shaped distributions based on Equations (3A and 3B) of Table 1.1 give the results as tabulated.

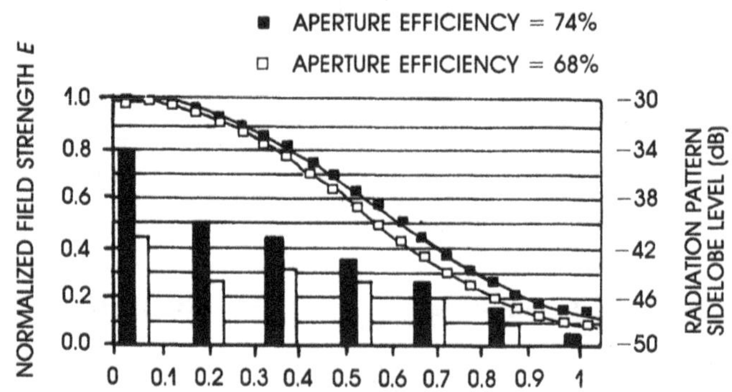

Figure 1.3 Radial Kouznetsov Amplitude Distribution on Circular Aperture, with Bar Chart Representation of Resulting Sidelobes (From Sletten, *Microwave Journal* [22]. Reprinted with permission)

Table 1.1 Circular Aperture Illumination Functions

---

1 Uniform

2 Kouznetsov Low Sidelobe Functions

A $G(r) = 1/7 + 6/7\,(1 - r^2)^2$
B $G(r) = 3/29 + 18/29\,(1 - r^2)^2 + 8/29\,(1 - r^2)^3$

3 Minimum Spreading Factor Functions

A $G(r) = 0.202 + 0.417\,(1 - r^2) + 0.249\,(1 - r^2)^2 + 0.131(1 - r^2)^3$
B $G(r) = 0.074 + 0.302\,(1 - r^2) + 0.233\,(1 - r^2)^2 + 0.390(1 - r^2)^3$

4 $G(r) = b + (1 - r^2)^3$

A $b = 0.5$
B $b = 0.4$
C $b = 0.32$

5 $G(r) = \cos^{n/2}\theta_0 \cos^2\theta_0/2$
where a = aperture radius = 48
and B = transform constant = 515.64
A $n = 198.29$ (−15 dB taper)
B $n = 264.72$ (−20 dB taper)

6 $G(r) = 1,\ 0 \leqslant r \leqslant r_1$

$$G(r) = 1 + \frac{1-b}{2}\left[\cos\frac{\pi(r - r_1)}{1 - r_1} - 1\right],\ r_1 \leqslant r \leqslant 1$$

A $r_1 = 0.15,\ b = 0.23$
B $r_1 = 0.10,\ b = 0.23$
C $r_1 = 0.05,\ b = 0.23$

7 76.2 cms, 12° Corrugated Horn*

A No Phase Error
B Phase Error Included

---

* Type of feed: Gregorian.

Table 1.2 Aperture Efficiency, Half-Power Point, and Sidelobes for Different Circular Aperture Illumination Functions*

Case	*Ap. Eff.*	*BW Half-Power*	u_1	S_1	u_2	S_2	u_3	S_3	u_4	S_4	u_5	S_5
(1)	1.00	1.62	5.12	−17.57	8.42	−23.81	11.62	−27.96	14.80	−31.08	17.96	−33.60
(2A)	0.74	2.00	6.51	−34.02	9.06	−39.78	11.86	−41.02	14.89	−42.72	18.02	−44.52
(2B)	0.68	2.08	6.95	−41.31	9.13	−44.99	11.81	−43.46	14.87	−44.67	18.00	−46.35
(3A)	0.81	1.88	5.97	−26.71	8.91	−32.45	11.97	−36.42	15.07	−39.48	18.19	−41.96
(3B)	0.71	2.06	6.72	−34.68	9.35	−39.79	12.26	−43.01	15.32	−45.83	18.38	−48.25
(4A)	0.88	1.82	5.71	−30.77	8.35	−28.38	11.58	−31.75	14.77	−34.71	18.00	−37.18
(4B)	0.84	1.86	5.85	−35.57	8.34	−29.35	11.57	−32.51	14.77	−35.43	17.94	−37.87
(4C)	0.80	1.90	6.03	−45.48	8.31	−30.50	11.55	−33.39	14.76	−36.26	17.94	−38.68
(5A)	0.81	1.91	6.10	−30.01	8.89	−33.18	11.93	−36.92	15.03	−39.88	18.15	−42.31
(5B)	0.71	2.03	6.63	−40.93	8.96	−36.94	12.03	−40.52	15.11	−43.36	18.22	−45.73
(6A)	0.80	1.92	6.18	−30.28	8.79	−38.42	11.65	−35.72	14.84	−39.58	17.97	−40.96
(6B)	0.80	1.92	6.14	−31.83	8.70	−35.95	11.65	−35.49	14.83	−39.03	17.97	−40.55
(6C)	0.80	1.91	6.11	−33.21	8.62	−34.06	11.66	−35.35	14.82	−38.40	17.97	−40.29
(7A)	0.71	2.04	6.66	−33.22	9.33	−41.35	12.08	−43.67	15.07	−44.88	18.17	−46.79
(7B)	0.69	2.06	6.97	−30.96	—	—	11.66	−37.59	15.12	−43.71	18.21	−46.48

$u = \frac{2\pi}{\lambda} a \sin\theta$ S = Sidelobe dB from beam peak

* Cases correspond to the illumination functions of Table 1.1.

Case 4 is taken from Sherman's study [17] and offers the constant b as an edge taper variable. Case 5 represents aperture distributions due to a conical horn pattern approximated by a $\cos^N(\theta_0)$ power pattern which yields a radial aperture function of the form,

$$\cos^N(\theta_0)\cos^4(\theta_0/2)$$

where the feed horn pattern angle, θ_0, is related to the radius r_c of its circular image on the main reflector aperture by $r_c = B\tan(\theta_0/2)$ (*cf.* Appendix 1A, Equations (1A.11, 1A.12, 1A.13)). The aperture radius, a, of Case 5 equals 48 and $B = 515.64$.

Case 6, due to F. S. Holt, fits a cosine function with two parameters to a small sector of uniform field near the center of the aperture.

Case 7 indicates that with a 2.44 meter diameter offset paraboloidal antenna illuminated via an ellipsoidal subreflector at 12 GHz by a conical corrugated horn 76.2 cms long and semiflare angle of 12°, patterns similar to the "ideal" aperture distributions can be obtained. The effects of horn phase errors do not degrade the gain (aperture efficiency) and sidelobe levels appreciably (Case 7B). The actual horn illumination pattern when reflected from the paraboloid section has an amplitude shape much like the *ad hoc* functions in Cases 2–6.

1.3.1 Hansen's One-Parameter Distribution

R. C. Hansen [18,19] has developed a symmetric circular aperture distribution following the procedure of Taylor for the line source. The distribution provides an optimum compromise between narrow beamwidth, low sidelobes, and high efficiency. His one parameter H is related to the first sidelobe to mainbeam peak ratio, SLR, by the equation

$$\text{SLR} = 17.57 + 20\log[2I_1(\pi H)]/(\pi H)] \tag{1.17}$$

where I_1 is the modified Bessel function of order 1 and the constant 17.57 is in dB. The aperture distribution is given by

$$g(\rho) = I_0[H(\pi^2 - \rho^2)^{1/2}] \tag{1.18}$$

where I_0 is the modified Bessel function of order 0 and ρ is a radial variable equal to π at the aperture edge. The radiation patterns corresponding to different aperture distributions can be computed from

$$F(u) = \begin{cases} 2J_1[\pi(u^2 - H^2)^{1/2}]/[\pi(u^2 - H^2)^{1/2}], & u \geq H \\ 2I_1[\pi(H^2 - u^2)^{1/2}]/[\pi(H^2 - u^2)^{1/2}], & u \leq H \end{cases} \tag{1.19}$$

where J_1 is the Bessel function of the first kind of order 1 and $u = (D/\lambda) \sin \theta$ with D the aperture diameter. The 3-dB value, u_3, is found from the solution of the equation

$$\frac{I_1(\pi H)}{\sqrt{2H}} = \frac{I_1[\pi(H^2 - u_3^2)^{1/2}]}{(H^2 - u_3^2)^{1/2}} \tag{1.20}$$

Pattern performance and aperture efficiency are given in Table 1.3. Values of SLR are plotted in Figure 1.4. The program HANSEN.FOR can be used for calculating the aperture efficiency, edge taper, aperture distribution, and 3-dB value.

Table 1.3 Characteristics of Hansen One-Parameter Distribution

SLR *dB*	*H*	u_3 *rad*	η	*Edge Taper dB*
17.57	0	0.5145	1	0
20	0.4872	0.5393	0.9786	4.5
25	0.8899	0.5869	0.8711	12.4
30	1.1977	0.6304	0.7595	19.3
35	1.4708	0.6701	0.6683	25.8
40	1.7254	0.7070	0.5964	32.0
45	1.9681	0.7413	0.5390	38.0
50	2.2026	0.7737	0.4923	43.9

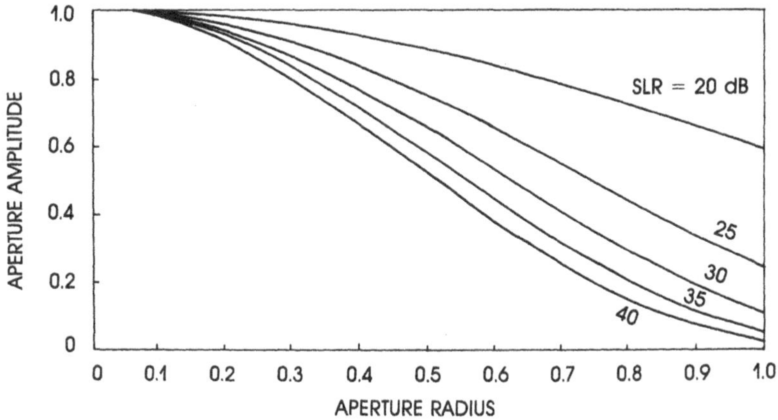

Figure 1.4 One-Parameter Circular Aperture Distribution, with Sidelobe Ratio as Parameter
(From Hansen, *IEEE Trans.* [18]. Reprinted with permission.)

1.3.2 Using Dolph-Chebyschev or Taylor Linear Array Distributions

The optimum linear array functions using Chebyschev polynomials can be useful guides for continuous apertures. In this approach the linear array coefficients corresponding to a given beamwidth and (uniform) sidelobe distribution are first computed. This linear array amplitude distribution is then used as a continuous radial distribution across the reflector aperture, assumed to have a rotationally symmetric aperture distribution. To find linear array coefficients for reflector antennas many wavelengths in diameter it is easier to use approximate methods [20,21] which give a profile of the discrete array excitation coefficients. The anomolous end array coefficient is usually ignored in the process of representing a continuous radial function with discrete array coefficients. This approach leads to pattern sidelobe peaks that decrease as theta increases rather than uniform sidelobe peaks for all theta as obtained for the corresponding linear arrays. More details are given in [22] and the example patterns and distributions from that report are shown in Figures 1.5 and 1.6. Those wishing to find approximate array coefficients for large linear arrays, and to explore this approach to gain versus sidelobe optimization on continuous aperture antennas, may find the program DRANE.FOR helpful.

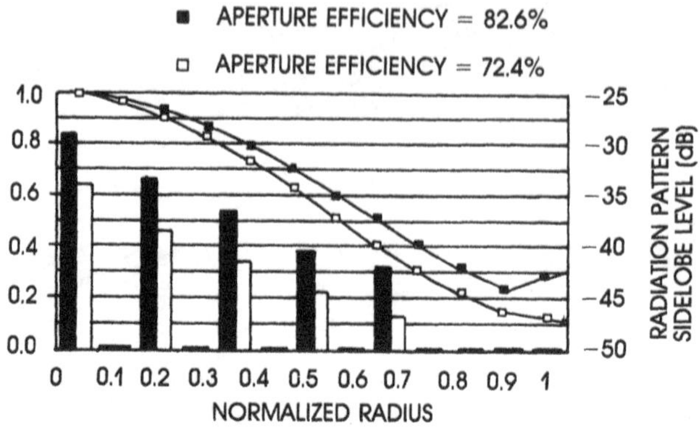

Figure 1.5 Radial Dolph-Chebyschev Amplitude Distribution on Circular Aperture, with Bar Chart Representation of Resulting Sidelobes (From Sletten, *Microwave Journal* [22]. Reprinted with permission)

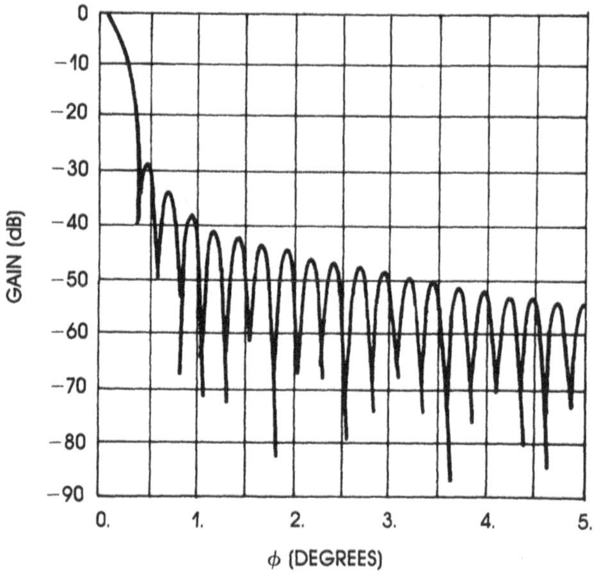

Figure 1.6 Radiation Patterns for 30 Unit Radius (Continuous) Aperture (From Sletten, *Microwave Journal* [22]. Reprinted with permission)

1.4 OFFSET DUAL-REFLECTOR FEEDING

Satisfactory offset dual-reflector configurations for feeding sections of paraboloids were not developed until the mid-1970s, although earlier ideas had been reported for eliminating aperture blocking and cross polarization by use of offset subreflectors illuminated with conical feed patterns. Japanese progress was reported by Mizugutch [23] in 1976 who presented basic expressions for the optimum tilt angles of offset dual-reflector antennas. GTE in 1979 published a report on an experimental antenna [24] using this optimally oriented feeding for an offset Gregorian antenna.

The derivation of the mathematical expressions for the confocal feeding of an offset paraboloidal section by another rotationally symmetric conic section such that the resulting antenna patterns are symmetrical and free of reflector-induced cross polarization is an involved task [25–29]. The analysis is given in Appendix 1A. It is not generally available in English textbooks.

From an engineering design point of view we need to know the equations that govern the tilt orientations of the feed source, usually a conical corrugated horn, in the subreflector and the angle that the subreflector axis is tilted with respect to the paraboloid (main dish) axis. We also need the subreflector coordinates and edge contours to construct these mirrors. These equations [29] are presented now. Software is listed in Appendix 1C to select the best configuration for required main dish radiation patterns or feed horns.

Referring to Figure 1.7, the main reflector, usually an offset section of paraboloid with circular aperture of radius R, is defined by the focal length f and the height of the midpoint of the aperture which can be located at a distance y_c perpendicular to the z-axis. Then in the rectangular coordinates x,y,z of Figure 1.7 (with origin at the focus F_0, z-axis along the axis of the parent paraboloid directed outward from the paraboloid, y-axis directed downward, x-axis directed out of the plane of the paper) the equation of the paraboloid is

$$x^2 + y^2 = 4f(f - z) \tag{1.21}$$

with the edge contour given by $x^2 + (y - y_c)^2 = R^2$.

The offset subreflector, an ellipsoid or hyperboloid of revolution, can be defined by an eccentricity e and the interfocal distance $2c$. In dual-reflector feeding on focus the conic section subreflector is confocal with the paraboloidal (main) reflector focus. For optimum orientation of the subreflector in the main reflector, the subreflector axis (connecting the two foci) is tilted by an angle β with respect to the z-axis of the paraboloid. Referring to the derivations in Appendix 1A, Equation (1A.10),

$$y_c = -4fe \sin \beta/(1 + e^2 - 2e \cos \beta) \tag{1.22}$$

The feed, usually a horn with phase center at F_1, is tilted by an angle α with the subreflector axis so that the feed axis points toward M_s, the midpoint of the subreflector. We find the angle α from (1A.5) of Appendix 1A,

$$\tan \alpha = (1 - e^2) \sin \beta/[(1 + e^2) \cos \beta - 2e] \tag{1.23}$$

The subreflector surface may be written in x_0,y_0,z_0 coordinates (with origin at F_1, z_0-axis directed from F_1 along the feed axis, x_0-axis parallel to the x-axis) as

$$x_0^2 + y_0^2 + z_0^2 = e^2(z_0 \cos \alpha - y_0 \sin \alpha + d)^2 \tag{1.24}$$

and in x,y,z coordinates about F_0 as

$$x^2 + y^2 + z^2 = e^2(z \cos \beta - y \sin \beta + d)^2 \tag{1.25}$$

with

$$d = c(1 - e^2)/(2e^2) \tag{1.26}$$

It is useful to write the ellipsoidal or hyperboloidal surface in spherical coordinates about F_1 (when the surface furthest from F_1 is needed) as

$$r = ed/(1 - e \cos \alpha \cos \theta + e \sin \alpha \sin \theta \sin \phi) \tag{1.27}$$

θ_H is the cone angle from F_1 that defines the rays that illuminate the edge of the subreflector which reflects them on to the edge of the main reflector. The angle θ_H can be computed from

$$\tan(\theta_H/2) = R(1 + e^2 - 2e \cos \beta)/[2f(1 - e^2)] \tag{1.28}$$

θ_H is often the semiflare angle of a conical corrugated horn where its pattern is about -17 dB from pattern peak on the horn axis. The amplitude of the edge rays are attenuated due to traversing a longer path than

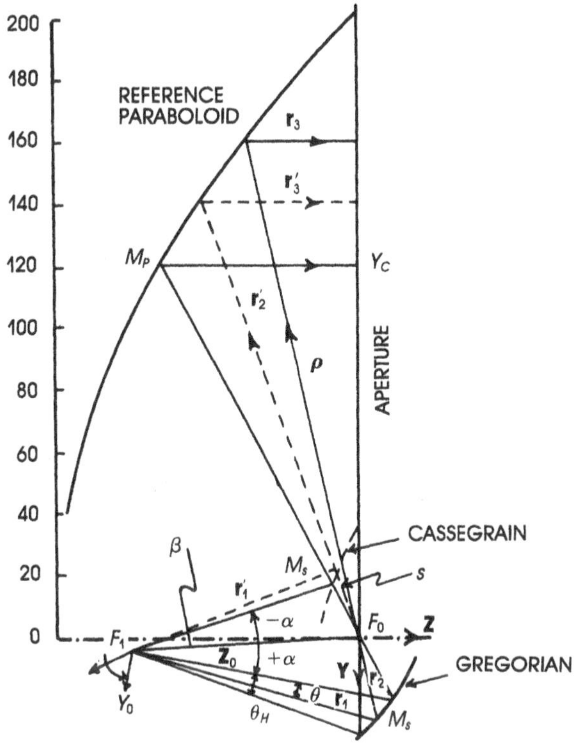

Figure 1.7 Offset Dual-Reflector Geometry (in Profile) for Cassegrain and Gregorian Antennas
(From Sletten, *Proc. IEEE* [29]. Reprinted with permission)

the central ray connecting points M_s and M_p. This path attenuation can be shown to be a drop in power of $\cos^4(\theta/2)$.

The folded optics of the dual reflector system has a magnification, M, of

$$M = (1 - e^2)/(1 + e^2 - 2e \cos \beta) \tag{1.29}$$

and the coordinate transformation for relating the spherical coordinates of the horn radiation pattern to points x,y,z on the paraboloid surface (or to x,y on the aperture as the rays are reflected parallel to the z-axis) are given by:

$$x = B \tan(\theta/2) \cos \phi \tag{1.30}$$

$$y = y_c - B \tan(\theta/2) \sin \phi \tag{1.31}$$

where $B = 2fM$ [*cf.* Appendix 1A, Equations (1A.10–1A.14)]. B is the transform constant for calculating the aperture distribution from the angular pattern of the feed.

The derivations in Appendix 1A are carried out for the z vector pointing away from the main reflector. In general the same equations are valid for the Cassegrain and Gregorian subreflectors with d negative and $e > 1$ for the hyperboloid. Because the quartic equations are double valued and the angles α and β can be positive or negative the configuration of each dual-reflector antenna must be carefully stated.

Configurations should be chosen to avoid aperture blocking by feed or subreflector. The usual procedure for configuring the antenna is to make an initial choice of the main reflector parameters, f, R, and y_c. Then the subreflector eccentricity e is varied to make θ_H approximately equal to the semiflare angle of the feed horn. The parameter c does not change the angles α and β which are always made to obey the design Equations (1.22) and (1.23); c can be used as a sizing parameter to shrink or expand the subreflector size. The subreflector should be at least 15 wavelengths in diameter to avoid increasing the diffraction-caused cross polarization.

Although the main beam of the paraboloid will be rotationally symmetric and cross-polarization levels will be -35 dB or lower from the main lobe peak when the antenna is fed from F_1, these conditions may not be true when beam scanning is done by feeds at points other than F_1. For a multibeam offset dual-reflector system the configuration chosen determines the off-focus aberrations.

Three feed configurations are programmed corresponding to applications based on the equations developed above and in Appendix 1A. The offset Gregorian antenna and the very wide-angle offset Gregorian are very useful for very low sidelobe antennas. The offset Cassegrain (hyperbola of revolution) and the front-fed offset Cassegrain [30] are programmed separately. Both of these Cassegrainian systems can be configured for very wide-angle multibeams or for wide-angular scanning by feed motions. The three configuration programs GREG.BAS, CASS.BAS, and FFOC.BAS are available in the optional software package with user's manual and example booklet.

The software is developed to allow variation of the design parameters such as distance above axis of the center, y_c, of the main paraboloid section of a focal length f. The configuration program GREG.BAS listed in Appendix 1C for the offset Cassegrain antenna is typical of these programs. To operate GREG.BAS for an offset section of paraboloid with focal length f, radius of its circular aperture, R, and midpoint of the aperture y_c = YC, these values are entered interactively. The parameters of the offset hyperboloidal subreflector E, the eccentricity, and, C, the distance between the foci of the subreflector, are also required inputs. The program computes and prints the configuration parameters, β, the angle between the paraboloid axis and the subreflector axis, α, the angle between the horn axis and the subreflector axis, the horn semiflare angle, the hyperboloid constant, D, the transform constant B, and other parameters. Generally E and C are varied to obtain the size subreflector desired. C can be varied without changing the angles α or β. When acceptable configuration parameters are obtained the operator can print out a central profile of the subreflector (and its center point, M) in X_1,Y_1,Z_1 coordinates with origin at the paraboloid focal point, Z_1 directed toward the paraboloid vertex, and Y_1 directed upward. Figure 1.8 shows the central profile of a Gregorian antenna with X_1,Y_1,Z_1 coordinates having their origin at F_0. Also the central profile of the main reflector and the edge points of the subreflector are printed by GREG.BAS. Example input parameters are: $E = 0.538$, YC = 62, $F = 70$, $C = 35$, $R = 48$, where length units in either inches or centimeters can be used.

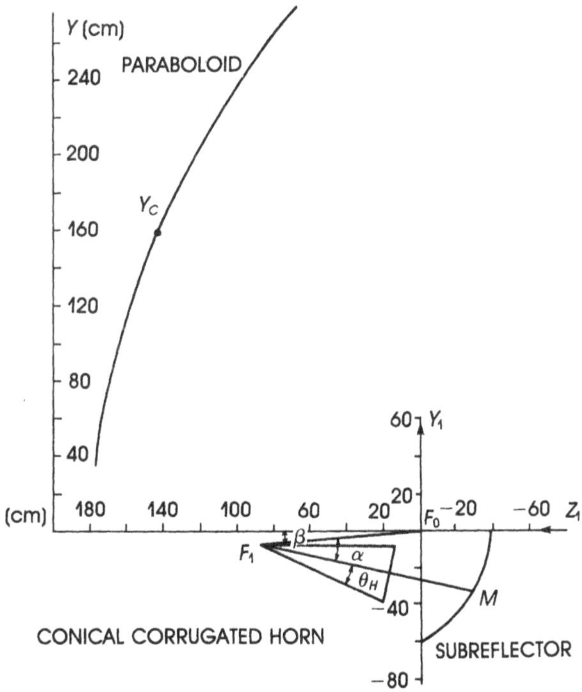

Figure 1.8 Cross-Sectional View of a Gregorian Antenna (From Sletten and Shore, *IEE Proc.*, Vol. 129, Pt. H, No. 3, June 1982. Reprinted with permission)

1.5 COMPUTING THE NEAR-FIELD RADIATION PATTERNS OF CONICAL CORRUGATED HORNS

The conical corrugated horn feed operated in the dominant hybrid mode is the "workhorse" of the feeds for reflector and lens antennas. The radiation pattern programs developed here are used frequently in other chapters. In this book the conical corrugated horn feed is treated by two different analytical methods. In Chapter 2 its patterns are computed by methods similar to those used for pyramidal and smooth-wall type horns. Here in Chapter 1 we give the results obtained by Clarricoats and co-workers.

Using a spherical mode expansion method and assuming that only the dominant HE_{11} is present in the horn under balanced hybrid conditions, Clarricoats, *et al.* [31], have derived a series expansion for the copolar electric field on a spherical surface of radius R_1 from the apex of a conical corrugated horn of length R_0:

$$E(R_1,\theta) = \sum_{n=1}^{N} C_n h_n^2(kR_1) f_n^1(\theta) \tag{1.32}$$

where

$$C_n = \frac{e^{(-jkR_0)}}{R_0} \frac{2n+1}{2n^2(n+1)^2 h_n^2(kR_0)} \int_0^{\theta_1} f_\nu^1(\theta) f_n^1(\theta) \sin\theta \, d\theta \tag{1.33}$$

Here,

$$f_\nu^1(\theta) = \frac{P^1(\cos\theta)}{\sin\theta} + \frac{dP_\nu^1(\cos\theta)}{d\theta}$$

$P_\nu^1(\cos\theta)$ is the associated Legendre function of the first kind of degree and order 1; $h_n^2(x)$ is the spherical Hankel function of the second kind of order n; k is the wavenumber, $2\pi/\lambda$; θ is the angle between the horn axis and the line from the horn apex to the field point; θ_1 is the horn semiflare angle; ν is given by the smallest solution of the equation:

$$f_\nu^1(\theta) = 0$$

and N is sufficiently large so that further terms in the sum can be neglected.

FORTRAN computer programs HORN1, HORN2, HORN3, HORN4, and HORN5 have been written to obtain the amplitude and phase radiation patterns and to fit polynomial curves to a set of discrete values of the patterns. Details of the computational method are given in Appendix 1B. Program CHRN.FOR is included for calculating the dimensions of corrugated conical horns. Source code for these programs is available in the optional software package and instructions on running the programs are given in the optional user's manual and example booklet.

1.6 "SPOILING" EFFECTS

Tables 1.1 to 1.3 provide design criteria for shaping aperture amplitude contours and summarize, for the circular aperture, the influence of amplitude distribution and edge taper. These tables indicate the sidelobe levels for given aperture efficiencies that are ideally possible. In all realistic antenna designs there are "spoiling" effects that tend to increase sidelobes and to reduce antenna gain.

1.6.1 Aperture Spillover Losses and Sidelobes

Illumination of reflector antennas is usually accomplished by radiating feeds such as EM horns or dipole radiators or indirectly by subreflector radiation patterns. Such patterns do not fall to zero beyond the main reflector edges. The illumination fall-off (for feed patterns) is governed by diffraction principles that relate power gradients in the feed patterns to the feed aperture size (D) in wavelengths, with sharper fall-offs possible when the ratio (D/λ) is large. Feed energy that misses the aperture represents loss in gain as the reflector cannot reflect it along the beam axis. The feed spillover energy also causes broad lobing in an angular region around the main dish edges. The feed pattern in the sectors beyond the main reflector becomes the antenna pattern in the far-out lobing region, except for the influence of main dish edge diffraction.

When the radiation pattern of the feed is known and can be represented in the angular sector that intercepts the main reflector and beyond by integrable functions then to good approximation the spillover loss can be computed [19] from the following expressions:

$$P(\text{total}) = \int_0^{2\pi} \int_0^{\pi} P_H(\theta,\phi) \sin \theta d\theta d\phi$$

$$P(\text{apert}) = 2\pi \int_0^{\theta_h} P_H(\theta) \sin \theta d\theta$$

$$P(\text{spill}) \approx 2\pi \int_{\theta_h}^{\pi/2} P_H(\theta) \sin \theta d\theta \tag{1.34}$$

$$\%(\text{spillover}) = \frac{P(\text{spill}) \times 100}{P(\text{apert}) + P(\text{spill})}$$

where θ_h is the angle subtended at the feed by points on the reflector rim.

We are again assuming radiation patterns that are rotationally symmetric, i.e., no ϕ dependence in the feed pattern. Because corrugated horn patterns can often be well approximated with $\cos^N\theta$ functions, the BASIC program SPILL gives the general behavior of the gain loss in dB resulting from the feed pattern energy not intercepted by the main reflector. In general this loss in gain must be subtracted from the aperture gain calculated in the above section on ideal aperture distributions. (There are other gain losses such as ohmic losses in the feed, cross-polarization losses, *et cetera,* that also need to be subtracted to compute the antenna gain of the main beam.)

To run SPILL.BAS, a simple example is suggested that demonstrates the kind of data it provides. Enter $N = 12.6$ and the "Range of Theta" angles from the feed axis to the reflector edge as: 36°, 40°. The program prints out the edge taper in dB and the percentage (and dB) loss due to spillover. The program is useful for estimating spillover losses for symmetric feed patterns well approximated by a $\cos^N\theta$ function.

To obtain an accurate pattern behind the main reflector and the modification of the feed spillover pattern in and about the directions subtended by the main reflector edges, we must compute the effect of edge diffraction. These fields can be found using the uniform geometric theory of diffraction technique discussed earlier. There is in addition a caustic lobe that depends on the integration of contributions from all the rim points. This caustic lobe or lobes can be considered separately and the main modification of the feed pattern behind and around the reflector computed by UGTD.

1.6.2 Cross-Polarization Effects

Polarization purity is often highly desirable in antenna patterns, especially when orthogonal polarizations are transmitted to double the channel capacity in communication by a technique known as frequency reuse. Polarization is defined as a state of the vector polarizations, with linear (usually "horizontal or vertical") and right- or left-handed circular polarizations being the polarizations most often used. The relationships between polarization states are treated in Chapter 2. Polarization of the antenna radiation pattern depends primarily on the purity of polarization in the feed pattern and on the cross polarization introduced by the reflector or lens surfaces.

It is convenient to use the Huygens source as the primary definition of polarization [32] because it corresponds best to most measurement situations and is nearly the polarization of a "linearly" polarized corrugated horn. Jones [33] has shown that when a paraboloid is fed on-axis with a Huygens source, the radiated polarization of the reflector antenna is purely linear. However when the feeding axis of the source, such as a corrugated horn with Huygens source polarization, is tilted off the paraboloid axis as in offset feeding, the cross polarization increases in the aperture fields and radiated fields of the antenna.

It is helpful to write out the expressions for an x-polarized and y-polarized (at $\theta = 0$) Huygens source in spherical and rectangular coordinates.

For the x-polarized reference Huygens source

$$\begin{aligned} \mathbf{i}_x &= \cos\phi\boldsymbol{\theta} - \sin\phi\boldsymbol{\phi} \\ &= \{1 - \cos 2\phi(1 - \cos\theta)\}\mathbf{x} - (1 - \cos\theta)\sin\phi\cos\phi\mathbf{y} \\ &\quad - \sin\theta\cos\phi\mathbf{z} \end{aligned} \tag{1.35}$$

For the y-polarized reference Huygens source

$$\begin{aligned} \mathbf{i}_y &= \sin\phi\boldsymbol{\theta} + \cos\phi\boldsymbol{\phi} \\ &= -(1 - \cos\theta)(\sin\phi\cos\phi\mathbf{x} + \{1 - \sin^2\phi(1 - \cos\theta)\}\mathbf{y} \\ &\quad - \sin\theta\sin\phi\mathbf{z} \end{aligned} \tag{1.36}$$

with cross-polarized states given by $\mathbf{i}_y$ for the x-polarized source and $\mathbf{i}_x$ for the y-polarized source.

The first equation in rectangular coordinates has been entered as the feeding source polarization in the BASIC program CROSS. The tilt angle of the source can be varied from the paraboloid's z-axis and the aperture polarization of rays reflected from any point of the paraboloid surface printed. It is seen that when the tilt angle is zero the cross-polarization level is also zero everywhere and when the tilt angle is increased the cross-polarization components also increase except in the y-z and x-z planes where they remain zero due to symmetry.

The program CROSS.BAS demonstrates the cross-polarization components generated by a y-polarized Huygens source illuminating an offset section of paraboloid. To run the program it is suggested that the first θ entered in the program be zero to see that then the x-polarized aperture fields are all zero. Then enter for example $\theta = 10$ degrees and note the cross-polarized field levels in the aperture for various aperture points X, Y.

We can still use offset feeding however for low cross-polarized radiation patterns because when two conic section mirrors are suitably oriented the cross-polarization level in the aperture fields again are cancelled [25]. This cross-polarization elimination technique is demonstrated using POLARB.BAS (in Chapter 8) to display the cross-polarization levels at the feed of a second offset paraboloid facing the first mirror with the same focal length and parallel axes and with both feed horns tilted by the same angle toward the centers of both mirrors. This facing offset paraboloid configuration is sometimes used in beam wave guides. But for feeding a large aperture paraboloid the polarization correction is usually done with an offset subreflector with ellipsoidal or hyperboloidal shapes. Dual-offset reflector feeding was discussed in Section 1.4 and expressions are derived in Appendix 1A for obtaining beam symmetry with proper configurations of subreflectors and feeds. Although not proven in Appendix 1A, the optimum tilt angles for beam symmetry also eliminate the reflector induced cross-polarization excitations. For more information on polarization suppression in multiple reflectors, see References [23,25,27,34].

1.6.3 Aperture Blocking

For symmetric, axially fed reflectors (but not lenses) radiated energy from the feed and fields reflected from the reflector are scattered by the feed itself and from struts or supports holding the feed in the focal region of the main dish. By using offset feeding the feed and usually the struts can be positioned so that radiated energy is not blocked or intercepted, at least in a geometrical optical sense. It is important, however, to be able to compute the losses and lobing introduced by scatterers in or near the antenna aperture.

To a first approximation the feed scattering can be treated as if it created a hole in the aperture field of the main reflector which has the shape of the largest projected area of the feed. This approach is discussed in Silver ([11], p. 192), where the effect of aperture blocking is shown to cause a loss in peak gain approximately equivalent to the power in the blocked area. The sidelobes caused by aperture blocking are closely related to the lobing produced by the geometry of the blocked zone, producing broad sidelobes with beamwidths in radians about $(\lambda/2) \times$ (hole diameter). Blocking then causes an increase in the near-in sidelobe levels as the unblocked radiation pattern lobes are superimposed on the blocking pattern lobes.

Radial struts when illuminated by a plane wave in front of the main reflector cause a cone of radiation which produces lobes depending on the geometry of the strut supports. There are many helpful references to handle most situations [9,35,36].

Usually the strut is treated as a right circular metal rod illuminated by the tapered plane wave reflected from the aperture, with feed illumination of the struts neglected. Because the struts are often at different angles in relation to the aperture plane wave excitation the geometry requires coordinate transformations that are best handled by special computer software which finds the positions and magnitudes of the several conical lobes. For the paraboloid antenna, the OSU software [9,36] treats dielectric struts scattering and feed blocking for most of the important geometries. Programs for computing blockage effects are developed for PCs in Chapter 4 using approaches similar to those used at Ohio State University.

1.6.4 Aperture Phase Errors

The antenna sidelobe and gain performance can be degraded by aperture phase errors, which is another way to state that the reflector or lens does not collimate the energy into plane waves. For beam shaping the aperture distributions may sometimes be designed to have nonplanar aperture phase fronts but, generally, phase errors on beam antennas increase the sidelobe levels and reduce the antenna peak gain.

There are two main sources of phase errors. First the feed itself, the corrugated horn, for example, does not produce a true spherical phase front as it would if the energy radiated from a point source. Then mechanical distortions such as feed misalignment or mechanical tolerances in the construction of the conic section reflectors cause aperture phase errors because optical path lengths of rays from the feed to the aperture depart from the ideal model.

Horn phase errors can be computed using programs of Chapter 2 and the programs HORN1-5 discussed earlier. Methods for repositioning the horn to partially correct for these errors or shaping the subreflector for horn phase correction are presented in Chapters 2 and 3.

1.6.5 Effects of Surface Tolerance Errors and Distortions

When reflector surfaces are not perfectly constructed to form true paraboloidal or subreflector shapes phase errors appear in the antenna aperture distribution due to path length variations on ray paths. If δR is the (normal) distance error from the true surface then the pathlength distance error δL is

$$\delta L = \delta R(\mathbf{S}_i \cdot \mathbf{S}_r)$$

where $\mathbf{S}_i$ and $\mathbf{S}_r$ are incident and reflected rays from the reflector surface.

When the surface errors are introduced randomly due to imperfect machining, for example, it is possible to specify the antenna surface by two parameters, the rms error for the surface, and C which is the correlation interval characteristic of the size of the departure zones. Ruze [37,38] has developed a theoretical model that quite accurately gives the statistical distribution of sidelobes and the loss in gain due to such a stochastic description of surface errors on antennas. This model leads to a very useful set of equations that are programmed in STOCH.BAS. In particular the sharp increase in first sidelobes and the loss in gain are well documented by experiments. The model also accurately predicts that the far-out sidelobes rise very little. There is a dependence, too, on aperture size with small reflectors affected more than large ones due to a given root mean square (rms) error. The dominant influences of random errors are contained in these equations:

$$P(\theta,\phi)_{sl} = 16S(\theta,\phi)(C\pi\delta)^2/(G_0\lambda)^2 e^{-(\pi u C/\lambda)^2}$$
$$S = \cos\theta[\cos^2\theta\cos^2\phi + \sin^2\phi] \tag{1.37}$$

where $u = \sin\theta$, $\delta =$ the rms error in radians, G_0 is the antenna gain without surface errors, and λ is the wavelength.

It is of course better to measure the actual deviations of the reflector surface from a true paraboloid and then to use numerical algorithms to compute the rise in sidelobes and loss in gain due to surface errors. This procedure is presented in Chapter 3.

References

[1] Spencer, R.C., "Paraboloid Diffraction Patterns from Standpoint of Physical Optics," MIT Rad. Lab. Rep. T-7, Oct. 21, 1942.

[2] Jasik, H., *Antenna Engineering Handbook,* New York: McGraw-Hill, 1961, Chapter 2.

[3] Collin, R.E., and F.J. Zucker, *Antenna Theory, Part I,* New York: McGraw-Hill, 1969, Chapter 14.

[4] Keller, J.B., "Geometrical Theory of Diffraction," Journal of Optical Society of America, 52, Feb. 1962, pp. 116–130.

[5] Ratnasiri, R.A.J., R.G. Kouyoumjian, and P.H. Pathak, "The Wide Angle Sidelobes of Reflector Antennas," 23 March 1970, Ohio State Univ. Electro-Science Lab. Rep. 2183-1 and also AFCRL 69-0413.

[6] Collin, R.E., and F.J. Zucker, *Antenna Theory, Part II,* New York: McGraw Hill, 1969, Chapters 16 and 17.

[7] Stratton, J., *Electromagnetic Theory,* New York: McGraw-Hill, 1941, Section 9.

[8] Kouyoumjian, R.G., and P.H. Pathak, "A Uniform Geometrical Theory of Diffraction for an Edge in a Perfectly Conducting Surface," Proc. IEEE, 62, Nov. 1974, pp. 1448–1461.

[9] Lee, S.H., R.C. Rudduck, C.A. Klein, and R.G. Kouyoumjian, "A GTD Analysis of the Circular Reflector Antenna including Feed and Strut Scatter," Ohio State Univ., RADC-TR-77-259, 1977.

[10] Stutzman, W.L., and G.A. Thiele, *Antenna Theory and Design,* New York: John Wiley and Sons, 1981.

[11] Silver, S., *Microwave Antenna Theory and Design,* New York: McGraw-Hill, 1949, pp. 147–199.

[12] Spencer, R.C., "Fourier Integral Methods of Pattern Analysis," Rad. Lab. Rep. No. 762-1, Jan. 21, 1946, and Rep. No. 762-2, March 30, 1946.

[13] Faigen, I.M., C.F. Reichert, C.J. Sletten, and R. A. Shore, "Reflector Antennas with Low Sidelobes, Low Cross Polarization and High Aperture Efficiency," NASA CR-174670, March 1984.

[14] Davis, P.J., and P. Rabinowitz, *Methods of Numerical Integration,* New York: Academic Press, 1975, p. 46, pp. 122–126, 369–370.

[15] Kouznetsov, V., "Sidelobe Reduction in Circular Aperture Antennas," 1978 European Antenna Conf. Digest, London.

[16] Jacquinot, P., and B. Roizin-Dossier, in *Progress in Optics,* Emil Wolf (ed.), Vol. III, Amsterdam: North Holland, 1964.

[17] Sherman, J.W., "Aperture Antenna Analysis," in *Radar Handbook*, M. Skolnik (ed.), New York: McGraw-Hill, 1970, Chapter 9.

[18] Hansen, R.C., "A One-Parameter Circular Aperture Distribution with Narrow Beamwidth and Low Sidelobes," IEEE Trans. Antennas and Propagation, pp. 477–480, July 1976.

[19] Hansen, R.C., "Planar Arrays." Chapter 10 of *The Handbook of Antenna Design*, Vol. 2, A.W. Rudge *et al.* (eds.), London: Peter Peregrinus, 1983, pp. 159–161.

[20] Van der Maas, G.J., "A Simplified Calculation for Dolph-Tchebycheff Arrays," J. Applied Physics, Vol. 25, pp. 121–124, Jan. 1954.

[21] Drane, C.J., "Useful approximations for the Directivity and Beamwidth of Large Scanning Dolph-Chebyschev Arrays," Proc. IEEE, 56, 1968, pp. 1779–1787.

[22] Sletten, C.J., W. Payne, and W. Shillue, "Offset Dual Reflector Antennas for Very Low Sidelobes," Microwave Journal, May 1986, pp. 221–240.

[23] Mizugutch, Y., M. Akagawa, and H. Yokoi, "Offset Dual Reflector Antenna," IEEE AP-S Int. Symp. Digest, Oct. 11, 15, 1976, pp. 1–5.

[24] Bardine, B.H., and E.J. Wilkinson, "A Low Sidelobe Earth Station Antenna for the 4/6 GHz Band," Microwave Journal, Nov. 1980.

[25] Tanaka, H., and M. Mizusawa, "Elimination of Cross Polarization in Offset Dual Reflector Antennas," (Japanese) Trans. IECE, Vol. 58-B, No. 12, UOC, 1975.

[26] Shore, R.A., and C.J. Sletten, "Dual Offset Reflector Antenna Systems with Rotationally Symmetric Aperture Distributions," RADC-TR-84-140, AD-A150734, June, 1984.

[27] Mizusawa, M., and T. Kitsuregawa, "A Beam-waveguide Feed Having a Symmetric Beam for Cassegrain Antennas," IEEE AP-S Int. Symp. Digest, Nov. 1973.

[28] Dragone, C., "Offset Multireflector Antennas with Perfect Pattern Symmetry and Polarization Discrimination," Bell System Technical J., Vol. 57, No. 7, pp. 2663–2683, 1978.

[29] Sletten, C.J., "Numerical Technique for Shaping Reflecting Surfaces to Synthesize Antenna Patterns," Proc. IEEE, Vol. 69, No. 3, pp. 392–394, March 1981.

[30] Makino, S., Y. Kobayashi, and T. Katagi, "Front Fed Offset Cassegrain Type Antenna," IEEE AP-S Int. Symp. Digest, June 17–21, 1985, pp. 341–344.

[31] Clarricoats, P.J.B., and A.D. Olver, "Near-field Radiation Characteristics of Corrugated Horns," Electonics Letters, Vol. 7, Aug. 12, 1971, pp. 446–448.

[32] Ludwig, A.W., "The Definition of Cross Polarization," IEEE Trans. Antennas and Propagation, Vol. AP-21, pp. 116–119, Jan. 1973.
[33] Jones, E.M.T., "Paraboloid Reflector and Hyperboloid Lens Antennas," IRE Trans. Antennas and Propagation, AP-2, pp. 119–127, July 1954.
[34] Gans, M.J., "Cross Polarization in Reflector Type Beam Waveguides and Antennas," Bell System Technical J., March 1976, pp. 289–316.
[35] Ruze, J., "Feed Support Blockage Loss in Parabolic Antennas," Microwave Journal, Vol. 11, No. 12, 1968, pp. 76–80.
[36] Rudduck, R.C., and Y.C. Chang, "Numerical EM Code, Reflector Antennas," NEC-REF (Ver.2), Columbus, OH: Ohio State Univ., Dec. 1982, TR 712242-16(713742) for US Dept. of the Navy.
[37] Ruze, J., "The Effect of Aperture Errors on the Antenna Radiation Pattern," AFCRC TR 57-105, also "Antenna Tolerance Theory—A Review," Proc. IEEE, Vol. 54, April 1966, pp. 633–642.
[38] Skolnik, M.I., *Introduction to Radar Systems,* New York: McGraw-Hill, 1962, pp. 336–343.
[39] Shore, R.A., and C. Sletten, *op. cit.* [26], Appendix 1.
[40] Dragone, C., *op. cit.* [28].
[41] Stegun, Irene A., "Legendre Functions," Chapter 8 in *Handbook of Mathematical Functions and Formulas, Graphs, and Mathematical Tables,* Abramowitz, M., and I.A. Stegun (eds.), National Bureau of Standards, US Government Printing Office, Washington, DC, 1964. See Eq. 8.5.4.
[42] Stegun, I.A., *op. cit.* [41]; Eq. 8.5.3.
[43] Arfken, G., *Mathematical Methods for Physicists,* Third Edition, Orlando, Florida: Academic Press, 1985, p. 668.
[44] Pearson, C.E. (ed.) *Handbook of Applied Mathematics,* Second Edition, New York: Van Nostrand Reinhold, 1983, p. 371 and p. 362.
[45] Lebedev, W.N., and R.A. Silverman (trans.), *Special Functions and Their Applications,* New York: Dover, 1972. See Eq. 7.3.6.
[46] Mahamud, S.F., and P.J.B. Clarricoats, "Radiation from Wide Flare-angle Corrugated Conical Horns," IEE Proc., Vol. 129, Pt. H, 1982, pp. 222–228.
[47] Antosiewicz, H.A., "Bessel Functions of Fractional Order," Chapter 10 in Abramowitz, M., and Stegun, I.A., *op. cit.* [41]. See Section 10.5.
[48] Antosiewicz, H.A., *loc. cit.* [47]. Eq. 10.1.19.

Appendix 1A
Dual Offset Reflector Antenna Systems with Rotationally Symmetric Aperture Distributions

The purpose of this appendix is to give a condensed derivation of the equations to design an offset dual-reflector antenna system with a rotationally symmetric aperture distribution. A detailed analysis may be found in Shore and Sletten [39].

We consider a dual-reflector system as shown in Figures 1A.1 and 1A.2. The subreflector is a section of either an ellipsoid (Gregorian system) or a hyperboloid (Cassegrainian system) with eccentricity e and interfocal distance $2c$. The focal point F_0 of the main reflector is also one of the foci of the parent subreflector surface, and the feed phase center is located at the other subreflector focal point, F_1. A Cartesian coordinate system is established with origin at F_0, z-axis along the axis of the parent paraboloid in the plane of the paper and directed outward from the reflector, y-axis in the plane of the paper and directed downward, and x-axis directed out from the plane of the paper. The feed pattern is described with reference to the spherical coordinates θ_0, ϕ_0 related in the usual way to the Cartesian x_0, y_0, z_0-system with origin F_1, z_0-axis directed from F_1 along the axis of the feed in the plane of the paper, and x_0-axis parallel to the x-axis.

The entire dual-reflector system is assumed to be symmetric with respect to the x, y-plane. The orientation of the subreflector axis with respect to that of the main reflector is specified by the angle β through which the subreflector axis must be rotated around F_0 to coincide with the z-axis. The orientation of the feed axis with respect to the subreflector axis is specified by the angle α through which the subreflector axis must be rotated around F_1 to coincide with the feed axis. Positive α and β are associated with a counterclockwise rotation direction.

The equations in programs GREG.BAS and CASS.BAS (and in most of the programs in later chapters) use a coordinate frame in which z is directed toward the vertex of the paraboloid and the y-axis is upward with the x-axis out of the paper. These differences in coordinates cause differences in the signs of coordinates and in certain expressions.

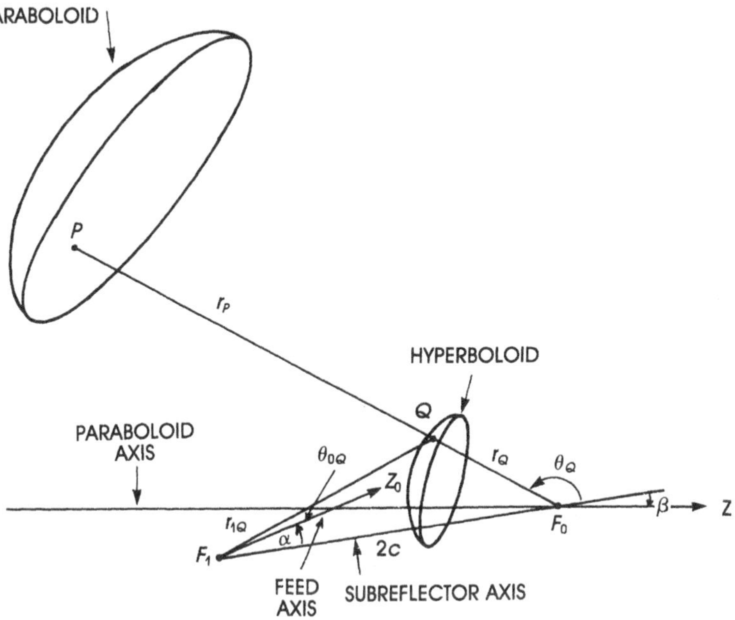

Figure 1A.1 Cassegrain Reflector System Geometry

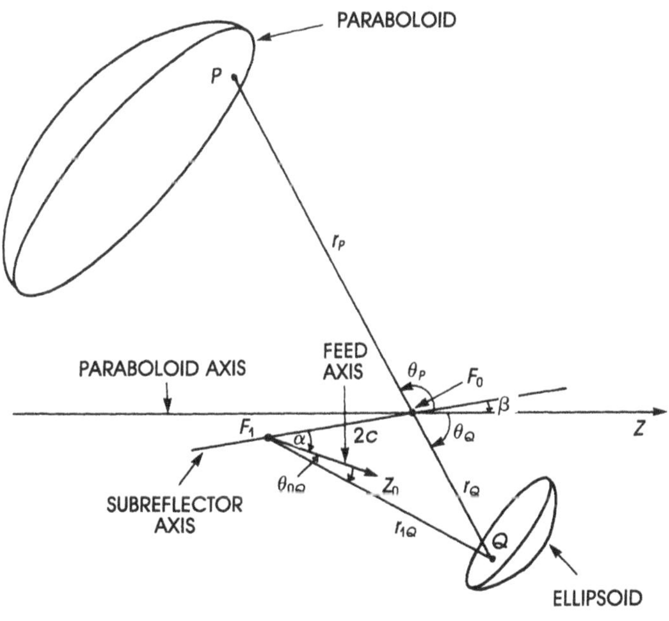

Figure 1A.2 Gregorian Reflector System Geometry

By tracing through the dual-reflector system a ray emanating from the feed phase center at F_1 in the direction specified by θ_0, ϕ_0 and expressing the image point P of the ray on the main reflector aperture as a function of θ_0, ϕ_0, and the system parameters e, f, α, and β, we obtain

$$x_P = -2f\frac{E\cos(\phi_0)}{A + B\sin(\phi_0)} \tag{1A.1}$$

$$y_P = -2f\frac{C + D\sin(\phi_0)}{A + B\sin(\phi_0)} \tag{1A.2}$$

where

$$A = u_1 + u_2\cos(\theta_0)$$

$$B = u_3\sin(\theta_0)$$

$$C = u_4 + u_5\cos(\theta_0)$$

$$D = u_6\sin(\theta_0)$$

$$E = (1 - e^2)\sin(\theta_0)$$

and

$$u_1 = \cos(\alpha - \beta) + e^2\cos(\alpha + \beta) - 2e\cos(\alpha)$$

$$u_2 = 1 + e^2 - 2e\cos(\beta)$$

$$u_3 = 2e\sin(\alpha) - e^2\sin(\alpha + \beta) - \sin(\alpha - \beta)$$

$$u_4 = \sin(\alpha - \beta) - e^2\sin(\alpha + \beta)$$

$$u_5 = 2e\sin(\beta)$$

$$u_6 = \cos(\alpha - \beta) - e^2\cos(\alpha + \beta)$$

Note that the interfocal distance, $2c$, of the parent subreflector surface does not appear in this result. Thus, the image point on the paraboloid aperture depends only on the subreflector eccentricity, e; the paraboloid focal length, f; the angles, α and β, specifying the relative orientations of the axes of the feed, subreflector, and main reflectors; and the ray direction. The dependence on the paraboloid focal length is that of a scale factor only.

The image on the paraboloid aperture of the circular cone of rays, θ_0 = constant, can be shown to be the circle with center at (O,y_c) and radius r_c,

$$x_P^2 + (y_P - y_c)^2 = r_c^2$$

where

$$y_c = -2f\frac{u_4 + u_5\cos(\theta_0)}{u_1 + u_2\cos(\theta_0)} \tag{1A.3}$$

and

$$r_c = \left|\frac{2f(1 - e^2)\sin(\theta_0)}{u_1 + u_2\cos(\theta_0)}\right| \tag{1A.4}$$

The center of the circle is in general a function of θ_0 so that the circles corresponding to different values of θ_0 are not concentric. However, differentiating (1A.3) with respect to θ_0 and equating the derivative to zero to make y_c independent of θ_0 leads to the equation

$$\tan(\alpha) = \frac{(1 - e^2)\sin(\beta)}{(1 + e^2)\cos(\beta) - 2e} \tag{1A.5}$$

which is equivalent to

$$2e\sin(\alpha) - e^2\sin(\alpha + \beta) - \sin(\alpha - \beta) = 0 \tag{1A.6}$$

Equation (1A.5) is the central design equation. It gives the relation between the angle α (between the subreflector axis and the feed axis), the angle β (between the subreflector axis and the main reflector axis), and the subreflector eccentricity, e, that must be satisfied for the images of circular cones of rays from the feed phase center, θ_0 = constant, to be concentric circles on the main reflector aperture.

It is also possible to express the angle β in terms of e and the angle α by the equation

$$\tan(\beta) = \frac{(1 - e^2)\sin(\alpha)}{(1 + e^2)\cos(\alpha) + 2e} \tag{1A.7}$$

which is equivalent to

$$2e\sin(\beta) + e^2\sin(\alpha + \beta) - \sin(\alpha - \beta) = 0 \tag{1A.8}$$

Using (1A.7) and (1A.8) the useful relation

$$\tan(\alpha/2) = \frac{1 + e}{1 - e}\tan(\beta/2) \tag{1A.9}$$

can be derived. (See also Reference [28] for an alternative derivation.)

Equations (1A.3) and (1A.4) for the y-coordinate of the center of the circular image on the main reflector aperture of the circular cone of rays θ_0 = constant, and for the radius of the image circle become

$$y_c = \frac{-4f\,e\sin(\beta)}{1 + e^2 - 2e\cos(\beta)} \tag{1A.10}$$

and

$$r_c = \frac{2f\,|1 - e^2|\tan(\theta_0/2)}{1 + e^2 - 2e\cos(\beta)} \tag{1A.11}$$

while Equations (1A.1) and (1A.2) for the x- and y-coordinates of the image point of the ray in the direction (θ_0,ϕ_0) become

$$x_P = -2f\frac{(1 - e^2)}{1 + e^2 - 2e\cos(\beta)}\tan(\theta_0/2)\cos(\phi_0) \tag{1A.12}$$

and

$$y_P = y_c - \frac{2f(1 - e^2)}{1 + e^2 - 2e\cos(\beta)}\tan(\theta_0/2)\sin(\phi_0) \tag{1A.13}$$

Equations (1A.12) and (1A.13) can also be written in the form

$$x_P = \pm\, r_c\cos(\phi_0) \tag{1A.14}$$

$$y_P = y_c \pm r_c\sin(\phi_0) \tag{1A.15}$$

where the plus and minus sign refers to the Cassegrainian and Gregorian systems respectively. This means that for the Cassegrainian system the angle ϕ_0 which defines the projection of a ray on the x_0, y_0-plane of the feed coordinate system, equals the azimuth angle ϕ_P of the image point on the main reflector aperture in the spherical coordinate system based on the x,y,x-system, while for the Gregorian system, $\phi_0 = \phi_P \pm \pi$. This difference between the Cassegrainian and Gregorian systems is, of course, attributable to the fact that in the Cassegrainian system, the rays do not pass through the focal point F_0, while for the Gregorian system, they do.

Turning now to the distribution of power in the paraboloid aperture, let $G(\theta_0,\phi_0)$ be the distribution of power on a unit sphere around the feed phase center and let $P(x,y)$ be the aperture power distribution. It can then be shown that

$$P(x,y) = \frac{1}{F^2} G(\theta_0, \phi_0) [1 + \cos(\theta_0)]^2 \qquad (1A.16)$$

where

$$F = \frac{2f(1 - e^2)}{1 + e^2 - 2e \cos(\beta)} \qquad (1A.17)$$

The feed pattern is thus transformed to the main reflector aperture distribution by multiplying by the factor

$$\frac{1}{F^2} [1 + \cos(\theta_0)]^2$$

which is independent of ϕ_0. Hence, if the feed pattern is rotationally symmetric, then so is the main reflector aperture power distribution.

It is important to note that the condition, Equation (1A.5) which guarantees that the images of circular cones of rays around the feed axis are concentric circles on the main reflector aperture and that a rotationally symmetric feed pattern produces a rotationally symmetric aperture power distribution, also guarantees that a feed with no cross polarization gives rise to an aperture field with no cross polarization [23,25]. Cross polarization here is defined as in the third definition of Ludwig [32]. For a transmitted field polarized in the x_0 direction at $\theta_0 = 0$, this definition implies that the θ_0 and ϕ_0 components of the transmitted electric field satisfy the equation

$$E_{\theta_0} \sin(\phi_0) = -E_{\phi_0} \cos(\phi_0)$$

while for a transmitted field polarization in the y_0-direction,

$$E_{\theta_0} \cos(\phi_0) = E_{\phi_0} \sin(\phi_0)$$

The field of a Huygens source—that is, a combination of crossed electric and magnetic dipoles of equal strength—satisfies these equations. If Equation (1A.5) is satisfied, an x_0-polarized transmitted field gives rise to a paraboloid aperture field with $E_y = 0$, and a y_0-polarized field to an aperture field with $E_x = 0$.

Appendix 1B
Computation of Corrugated Conical Horn Near-Zone Patterns

In this appendix we give details of the computational procedure used to calculate the near-zone pattern of a conical corrugated horn based on Equations (1.32) and (1.33) of Section 1.5.

We begin with the evaluation of $f_\nu^1(\theta)$ where

$$f_\nu^1(\theta) = \frac{P_\nu^1(\cos\theta)}{\sin\theta} + \frac{dP_\nu^1(\cos\theta)}{d\theta} \tag{1B.1}$$

where $P_\nu^1(\cos\theta)$ is the associated Legendre function of the first kind of degree ν and order 1. Equation (1B.1) can be written in the form

$$f_\nu^1(\theta) = \frac{P_\nu^1(\cos\theta)}{\sin\theta} - \sin\theta P_\nu^{1\prime}(\cos\theta) \tag{1B.2}$$

where we use $P_\nu^{1\prime}(\cos\theta)$ to mean $P_\nu^{1\prime}(x)|_{x=\cos\theta}$. Using the relation [41]:

$$(x^2 - 1)P_\nu^{1\prime}(x) = \nu x P_\nu^1(x) - (\nu + 1)P_{\nu-1}^1(x)$$

with $x = \cos\theta$ and substituting in (1B.2) yields

$$f_\nu^1(\theta) = (1 + \nu\cos\theta)\frac{P_\nu^1(\cos\theta)}{\sin\theta} - (\nu + 1)\frac{P_{\nu-1}^1(\cos\theta)}{\sin\theta}$$

The recursion relation [42]

$$\nu P_{\nu+1}^1(x) = (2\nu + 1)xP_\nu^1(x) - (\nu + 1)P_{\nu-1}^1(x)$$

or equivalently

$$\frac{\nu P^1_{\nu+1}(\cos\theta)}{\sin\theta} = (2\nu + 1)\cos\theta\,\frac{P^1_\nu(\cos\theta)}{\sin\theta} - (\nu + 1)\,\frac{P^1_{\nu-1}(\cos\theta)}{\sin\theta} \tag{1B.3}$$

can then be used to reduce the evaluation of $f^1_\nu(\theta)$ to the evaluation of $P^1_\nu(\cos\theta)/\sin\theta$ for $-1 \leqslant \nu \leq 1$.

The associated Legendre function of the first kind of degree ν and order 1, $P^1_\nu(x)$, is defined in terms of the (ordinary) Legendre function of the first kind of degree ν, $P_2(x)$, by [43]

$$P^1_\nu(x) = (1 - x^2)^{1/2}\,\frac{dP_\nu(x)}{dx} \tag{1B.4}$$

Combining (1B.4) with the relation [44]

$$(1 - x^2)\,\frac{dP_\nu(x)}{dx} = \nu P_{\nu-1}(x) - \nu x P_\nu(x)$$

and letting $x = \cos\theta$ gives

$$\frac{P^1_\nu(\cos\theta)}{\sin\theta} = \frac{\nu P_{\nu-1}(\cos\theta)}{\sin^2\theta} - \nu\cos\theta\,\frac{P_\nu(\cos\theta)}{\sin^2\theta}$$

The Legendre function $P_\nu(z)$ has the hypergeometric series representation [45]

$$P_\nu(z) = F\left(-\nu,\nu + 1;1;\frac{1 - z}{z}\right)$$

$$= \sum_{k=0}^{\infty} \frac{(-\nu)_k(\nu + 1)_k}{(k!)^2}\left(\frac{1 - z}{2}\right)^k$$

where $(\nu)_k$ is defined by

$$(\nu)_0 = 1$$

$$(\nu)_n = \nu(\nu + 1)\ldots(\nu + n - 1)$$

Letting $z = \cos\theta$ and using some elementary trigonometric relations then yields

$$\frac{P_\nu^1(\cos\theta)}{\sin\theta} = \frac{\nu}{2\cos^2\frac{\theta}{2}}\left\{1 + \frac{1}{2}\left[\sum_{k=1}^{\infty}\frac{[-(\nu-1)]_k(\nu)_k}{(k!)^2}\sin^{2(k-1)}\frac{\theta}{2}\right.\right.$$

$$\left.\left. - \cos\theta\sum_{k=1}^{\infty}\frac{(-\nu)_k(\nu+1)_k}{(k!)^2}\sin^{2(k-1)}\frac{\theta}{2}\right]\right\} \tag{1B.5}$$

For ν an integer we need treat only $\nu = 0$ and $\nu = 1$, since the recursion relation (1B.3) can be used to calculate $P_\nu^1(\cos\theta)/\sin\theta$ for ν any integer greater than 1 starting with $P_0^1(\cos)/\sin\theta$ and $P_1^1(\cos\theta)/\sin\theta$. From (1B.5) we obtain $P_0^1(\cos\theta)/\sin\theta = 0$ and $P_1^1(\cos\theta)/\sin\theta = 1$.

When ν is not an integer, we can restrict attention to the range $-1 < \nu < 1$. Let S_1 be the first sum in (1B.5) and let T_k be the kth term of S_1. Since

$$T_{k+1} = T_k\frac{(k+1-\nu)(\nu+k)}{(k+1)^2}\sin^2\frac{\theta}{2},\ k = 1, 2, \ldots$$

it follows that $|T_{k+1}/T_k| < \sin^2(\theta/2)$. Noting that all the terms of S_1 are positive or negative according as ν is positive or negative and letting $r = \sin^2\frac{\theta}{2}$

$$|S_1| < |T_1|\,(1 + r + r^2 + \ldots) < 2\,(1 + r + r^2 + \ldots)$$

so that the remainder after N terms of the series is smaller than $r^N/(1-r)$.

By choosing

$$N = \text{int}\left[\frac{\ln\frac{\varepsilon}{2} + \ln(1-r)}{\ln r}\right] + 1$$

we can ensure that the magnitude of the remainder is less than some small constant ε. The second sum in (1B.5) is similarly treated. The two sums converge very rapidly. For $\theta = 20°$ at most six terms are required if $\varepsilon = 10^{-9}$.

The integral in the expression for c_n in (1.33) can be evaluated in closed form [46]:

$$\int_0^{\theta_1} f_\nu^1(\theta) f_n^1(\theta) \sin\theta \, d\theta = \frac{\sin\theta_1}{\nu(\nu+1) - n(n+1)} \cdot [\nu(\nu+1) P_\nu^1(\cos\theta_1) f_n^1(\theta_1) - n(n+1) P_n^1(\cos\theta_1) f_\nu^1(\theta_1)] \quad (1B.6)$$

The spherical Hankel functions of the second kind, $h_n^{(2)}(x)$, which appear in (1.32) and (1.33), are defined by

$$h_n^{(2)}(x) = j_n(x) - iy_n(x)$$

where

$$j_n(x) = \left(\frac{\pi}{2x}\right)^{1/2} J_{n+1/2}(x)$$

and

$$y_n(x) = \left(\frac{\pi}{2x}\right)^{1/2} y_{n+1/2}(x)$$

are the spherical Bessel functions of the first and second kind respectively.

The method of calculation used [47] is the following: Let M be an integer somewhat larger than the maximum of n and x. Assume $F_M(x) = 0$, $F_{m-1}(x) = 1$ and using the recursion relation [48]

$$F_I(x) = \frac{2I+3}{x} F_{I+1}(x) - F_{I+2}(x) \quad (1B.7)$$

satisfied by all the spherical Bessel functions, calculate $F_I(x)$ for I from $M - 2$ to $-(n + 1)$. Provided that M was chosen sufficiently large the value of $F_I(x)$ will be proportional to $j_I(x)$ for I from $-(n + 1)$ to n. The constant of proportionality, P, is readily calculated since

$$P = j_0(x)/F_0(x)$$

and $j_0(x)$ is simply equal to sin (x)/x. Then

$$j_I(x) = PF_I(x),\ I = 1, 2, \ldots, n$$

and since

$$y_I(x) = (-1)^{I+1} j_{-I-1}(x)$$

we have

$$y_I(x) = (-1)^{I+1} PF_{-(I+1)}(x),\ I = 1, 2, \ldots, n$$

The proper choice of M in the above algorithm must be clarified a little. Clearly M must be larger than n, otherwise the backward recursion relations (1B.7) starting with $F_M(x) = 0$ and $F_{M-1}(x) = 1$ will not be able to develop a sufficient number of significant figures by the time $M = n$ is reached. In addition, if x is considerably larger than n and M is not chosen larger than x the algorithm will not yield correct values for $j_n(x)$ and $y_n(x)$. On the other hand, choosing M unnecessarily large adds to the computational cost and serves no useful purpose. In our calculations it was found that choosing M equal to the larger of n and x plus 20 gave excellent results.

It will be noted that the backward recursion method, when used to calculate $j_n(x)$ and $y_n(x)$, simultaneously yields values of $j_I(x)$ and $y_I(x)$ for all I from 0 to n. This fact can be put to good use in calculating the electric field of the horn from (1.32) since what is needed there are the values of $h_n^2(kR_0)$ and $h_n^2(kR_1)$ for n from 1 to n. Given N these values can all be generated in just two operations (one for the $h_n^2(kR_0)$ and one for the $h_n^2(kR_1)$) and then stored for use as needed in calculating the individual terms of the series.

Appendix 1C
List of Chapter 1 Computer Programs and Source Code Listings of Selected Programs

In this appendix the names of computer programs developed especially for Chapter 1 are listed. The source code for some of the simpler programs is listed. These tutorial programs are discussed in the text. All programs are written in Microsoft BASIC and FORTRAN for IBM PC and compatible computers. The source code for all programs with instructions for operating them is available on 5¼-inch diskettes in the optional software package with user's manual and example booklet.

The computer program names are given in the first few lines of the source code. The suffix .FOR indicates a FORTRAN program and .BAS a basic program. Also .DAT is for a data file and .COM for special executable programs.

SOFTWARE LIST, CHAPTER 1

PROGRAM NAME: ASSOCIATED PROGRAMS, FILES AND SUB-ROUTINES:

*PATT.FOR; PHAMP.DAT, ALAM.DAT, CJS.DAT
LINEPL.BAS; GRAPHICS.COM
CIRC.BAS; CIRC.DAT
HANSEN.FOR; BESSI0, BESSI1, BESSJ1, ZBRENT
DRANE.FOR; BESSI1
HORN1.FOR, F1N.FOR, ASLEGS.FOR, ZBRENT
HORN2.FOR, F1N.FOR, ASLEGS.FOR
HORN3.FOR; SPHBES.FOR
HORN4.FOR; F1N.FOR, ASLEGS.FOR, SPHEBES.FOR
HORN5.FOR; SVDFIT, SVDCMP, SVBKSB
CHRN.FOR; CORH.DAT, SDIM.DAT
CASS.BAS
*GREG.BAS
FFOC.BAS
*SPILL.BAS
*CROSS.BAS
STOCH.BAS

* Source code for these programs is printed in this appendix.

The programs BESSI0, BESSI1, BESSJ1, ZBRENT, SVDFIT, SVDCMP, and SVBKSB are Numerical Recipes subroutines in FORTRAN and these must be linked to the programs they follow. The programs PATT.FOR, LINEPL.BAS, HORN1-5.FOR, CASS.BAS, GREG.BAS, SPILL.BAS, and CROSS.BAS are "computational utilities" that are used frequently in later chapters.

```
C       Program computes radiation patterns from data
C       in PHAMP.DAT file.


        PROGRAM PATT
        COMMON AVAR
        REAL LAM
        INTEGER AVAR
        EXTERNAL POWER
        DATA PI/3.141592654/
        DATA PDB0,KEYSET/0.,0/

        OPEN(6,FILE='CJS.DAT',STATUS='NEW')
        OPEN(5,FILE='ALAM.DAT',STATUS='OLD',FORM='UNFORMATTED')
        REWIND 5
        READ(5) A,LAM
        READ(5) NRAD,NANG
        WRITE(*,101) A,LAM,NRAD,NANG
101     FORMAT(' A = ',F8.2,' LAMBDA = ',F8.2,' NRAD = ',I3,
     1 ' NANG = ',I3)
        OPEN(4,FILE='PHAMP.DAT',STATUS='OLD',FORM='UNFORMATTED',
     $ IOSTAT=M,ERR=220)
        REWIND 4

1       WRITE(*,102)
102     FORMAT(' ENTER THETAI,THETAF,DTHETA')
        READ(*,103)THETAI,THETAF,DTHETA
103     FORMAT(3F10.6)
        WRITE(*,104)
104     FORMAT(' ENTER PHI')
        READ(*,105)PHI
105     FORMAT(F10.4)
        NCOUNT=(THETAF-THETAI)/DTHETA+1
        WRITE(6,106)NCOUNT
106     FORMAT(I4)
        THETA=THETAI

        AVAR=3
        THETAR=THETA*PI/180
        PHIR=PHI*PI/180
        P=POWER(THETAR,PHIR,A,LAM,NRAD,NANG)
        PDB=10*ALOG10(P)
        PDB0=PDB
        WRITE(*,109)PDB0
109     FORMAT(F10.5)
        AVAR = 3
        GAIN = 4*PI*(A/LAM)**2*P/APINT(A,LAM,NRAD,NANG)
        GAIN = 10*ALOG10(GAIN)
        WRITE(*,110) GAIN
110     FORMAT(' GAIN(DB) = ',E12.4/)

        DO 200 KLS=1,NCOUNT
        AVAR=3
        THETAR=THETA*PI/180
        PHIR=PHI*PI/180
        P=POWER(THETAR,PHIR,A,LAM,NRAD,NANG)
        PDB=10*ALOG10(P)-PDB0
        WRITE(6,107)THETA
107     FORMAT(F9.4)
        WRITE(6,108)PDB
108     FORMAT(F10.4)
```

```
        THETA=THETA+DTHETA
200     CONTINUE

220     WRITE(*,221) M
221     FORMAT(' IOSTAT=',I5)
2       STOP
        END
C
C
C
        REAL FUNCTION POWER(THETA,PHI,A,LAM,NRAD,NANG)
C       POWERW.FOR IS A MODIFIED VERSION OF POWER.FOR DESIGNED TO
C       BE USED FOR ASYMMETRICAL APERTURE DISTRIBUTIONS

C       CALCULATES POWER FOR A GIVEN PAIR OF ANGULAR COORDINATES

        COMMON AVAR
        DIMENSION AMP(75),PHAS(75),RINTG(4),VALINT(4)
        REAL LAM
        INTEGER AVAR
        EXTERNAL SIMCOF

        DATA PI/3.141592654/

C       INITIALIZATION

        U=2*PI/LAM*A*SIN(THETA)

        DO 10 I=1,4
10        VALINT(I)=0

        DELRAD=1/FLOAT(NRAD-1)
        DELPHI=2*PI/FLOAT(NANG-1)

        PHIPRO=-PI

        DO 100 I=1,NANG

          PHIPR=PHIPRO+(I-1)*DELPHI

          DO 110 J=1,NRAD
110       READ(4,IOSTAT=M,ERR=220) AMP(J),PHAS(J)

          CALL INTEGR(AMP,PHAS,NRAD,DELRAD,U,PHI,PHIPR,RINTG)

C         FIND COEFFICIENT FOR COMPOUND SIMPSON'S RULE

          II=I
          C=SIMCOF(II,NANG)

          DO 120 J=1,4
120       VALINT(J)=VALINT(J)+C*RINTG(J)

100     CONTINUE

        DO 150 J=1,4
150       VALINT(J)=VALINT(J)*DELPHI/3

        POWER=(VALINT(1)-VALINT(2))**2+(VALINT(3)+VALINT(4))**2

        POWER=POWER/PI**2
        REWIND 4

        RETURN

220     WRITE(*,221) M,I,J
```

```
221     FORMAT(' IOSTAT=',I5,' IFP=',I5,' JFP=',I5)
        STOP

        END

        REAL FUNCTION SIMCOF(I,N)

C       CALCULATES COEFFICIENT FOR THE ITH TERM IN THE COMPOUND
C       SIMPSON'S INTEGRATION FORMULA.

        IF((I.EQ.1).OR.(I.EQ.N)) GOTO 2
        IF(MOD(I,2).EQ.0) GOTO 1

C       ODD POINTS

        SIMCOF=2
        RETURN

C       EVEN POINTS

1       SIMCOF=4
        RETURN

C       END POINTS

2       SIMCOF=1
        RETURN

        END

        SUBROUTINE INTEGR(AMP,PHAS,NRAD,DELRAD,U,PHI,PHIPR,RINTG)

        DIMENSION AMP(1),PHAS(1),RINTG(4),AMPCOS(75),AMPSIN(75)
        EXTERNAL FILON

        A=0
        B=1
        R0=0

        DO 10 I=1,NRAD

          R=R0+(I-1)*DELRAD
          AMPCOS(I)=AMP(I)*COS(PHAS(I))*R
10        AMPSIN(I)=AMP(I)*SIN(PHAS(I))*R

        UMIN=U*COS(PHI-PHIPR)

        RINTG(1)=FILON(AMPCOS,UMIN,A,B,NRAD,1)
        RINTG(2)=FILON(AMPSIN,UMIN,A,B,NRAD,2)
        RINTG(3)=FILON(AMPCOS,UMIN,A,B,NRAD,2)
        RINTG(4)=FILON(AMPSIN,UMIN,A,B,NRAD,1)

        RETURN
        END
        REAL FUNCTION FILON(F,T,A,B,M,KEY)
        DIMENSION F(M)
        N=M-1
        H=(B-A)/N
        TH=T*H
        IF(TH.LT.0.1) GOTO 1
        TH2=TH*TH
        TH3=TH2*TH
        STH=SIN(TH)
        CTH=COS(TH)
        STH2=STH*STH
        CTH2=CTH*CTH
```

```
      AL=(TH2+TH*STH*CTH-2*STH2)/TH3
      BE=2*(TH*(1+CTH2)-2*STH*CTH)/TH3
      GA=4*(STH-TH*CTH)/TH3
      GOTO 2
1     TH2=TH*TH
      TH3=TH2*TH
      AL=TH3*(2./45.-TH2*(2./315.-TH2*2./4725.))
      BE=2./3.+TH2*(2./15.+TH2*(4./105.+TH2*2./567.))
      GA=4./3.-TH2*(2./15.-TH2*(1./210.-TH2/11340))
2     SUM=0.0
      F1=F(1)
      F2=F(M)
      S1=SIN(A*T)
      S2=SIN(B*T)
      C1=COS(A*T)
      C2=COS(B*T)
      A1=A+H
      GOTO (10,20),KEY
10    SU=F2*S2-F1*S1
      SU1=-.5*(F2*C2-F1*C1)
      DO 30 I=2,N,2
      SUM=SUM+F(I)*COS(A1*T)
      A1=A1+H
      SU1=SU1+F(I+1)*COS(A1*T)
30    A1=A1+H
      GOTO 40
20    SU=F1*C1-F2*C2
      SU1=-.5*(F2*S2-F1*S1)
      DO 50 I=2,N,2
      SUM=SUM+F(I)*SIN(A1*T)
      A1=A1+H
      SU1=SU1+F(I+1)*SIN(A1*T)
50    A1=A1+H
40    FILON=H*(AL*SU+BE*SU1+GA*SUM)
      RETURN
      END
      REAL FUNCTION APINT(A,LAM,NRAD,NANG)
C----PERFORMS INTEGRAL OF APERTURE INTENSITY
      COMMON AVAR
      REAL LAM
      INTEGER AVAR
      EXTERNAL SIMCOF
      DATA PI/3.141592654/

      DELRAD = 1./FLOAT(NRAD-1)
      DELPHI = 2.*PI/FLOAT(NANG-1)
      APINT = 0.
      DO 100 I=1,NANG
        RADINT = 0.
        DO 110 J=1,NRAD
          JJ = J
          C = SIMCOF(JJ,NRAD)
          READ(4,IOSTAT=M,ERR=220) AMP,PHAS
          R = (J-1)*DELRAD
110       RADINT = RADINT + C*AMP*AMP*R

        II = I
        C = SIMCOF(II,NANG)
100     APINT = APINT + C*RADINT*DELRAD/3

      APINT = APINT*DELPHI/3
      APINT = APINT/(PI*PI)

      REWIND 4

      RETURN
```

```
220     WRITE(*,221) M,I,J
221     FORMAT('IOSTAT=',I5,' IPIN=',I5,' JPIN=',I5)
        STOP
        END
```

```
10 REM *CONFIGURES OFFSET ELLIPSOIDAL OR GREGORIAN ANTENNA*
20 LPRINT "GREG.BAS"
30 INPUT "FOR VERY WIDE-ANGLE FEED HORNS, ENTER W, OR N FOR NORMAL";W$
40 IF W$="W" THEN GOTO 900
50 LPRINT
60 LPRINT "INPUT CONFIGURATION CONSTANTS"
70 PI=3.141592654#
80 RANG=180/PI
90 I=1
100 E=.538 : Y3=62 : F=70 : C=35   : R=48
110 REM  INPUT "ENTER E,Y3,F,C,R";E,Y3,F,C,R
120 LPRINT  "E=";E;"YC=";Y3;"F=";F;"C=";C;"R=";R
130 LPRINT
140 T1=Y3/F  : A4=T1^2/4+1
150 B4=-T1/(2*E)*(1+E^2)
160 C4=(T1/(4*E)*(1+E^2))^2-T1^2/4
170 S3=(-B4-SQR(B4^2-4*A4*C4))/(2*A4)
180 A1=ATN(S3/SQR(1-S3^2))
190 S4=SIN(A1)*(1-E^2)/((1+E^2)*COS(A1)-2*E)
200 A2=ATN(S4)
210 D=C*(1-E^2)/(2*E^2)
220 B=2*F*(1-E^2)/(1+E^2-2*E*COS(A1))
230 S5=R/B : S6=ATN(S5) : A3=S6*2
240 AB4=A2-A1 : M1=B/(2*F) : A4=AB4
250 AA1=A1*RANG : AA2=A2*RANG : AA3=A3*RANG :AA4=AB4*RANG
260 LPRINT "FOR THE ELLIPSOID:" : LPRINT
270 LPRINT "BETA=";AA1
280 LPRINT "ALPHA=";AA2
290 LPRINT "HORN SEMIFLARE ANGLE=";AA3
300 LPRINT "ELLIPSOID CONSTANT, D=";D
310 LPRINT "MAGNIFICATION=";M1
320 LPRINT "ALPHA-BETA=";AA4
330 LPRINT "TRANSFORM CONSTANT, B=";B
335 LPRINT "ECCENTRICITY, E=";E
340 INPUT "To print all profiles enter C, to end E ";E$
350 IF E$="E" THEN GOTO 1120
370 REM INPUT "ENTER THETA AND PHI";AA6,AA7
380 LPRINT
390 LPRINT "CENTRAL PROFILE OF ELLIPSOID IN X1,Y1,Z1 COORDINATES"
400 LPRINT
410 ANGR=PI/180
420 FOR AA6=-AA3 TO AA3 STEP AA3/10
430 A7=90*ANGR
440      A6=AA6*ANGR : REM A7=AA7*ANGR
450 R1=E*D/(1-E*COS(A2)*COS(A6)+E*SIN(A2)*SIN(A6)*SIN(A7))
460 X0=R1*SIN(A6)*COS(A7)
470 Y0=R1*SIN(A6)*SIN(A7)
480 Z0=R1*COS(A6)
490 REM PRINT "X0=";X0;"Y0=";Y0;"Z0=";Z0
500 Y1=Y0*COS(A4)+Z0*SIN(A4)+C*SIN(A1)
510 Z1=-Y0*SIN(A4)+Z0*COS(A4)-C*COS(A1)
520 LPRINT "X1=";X1;"Y1=";-Y1;"Z1=";-Z1
530 IF I=0 GOTO 590
540 NEXT AA6
550 A6=0 : A7=0
```

```
560 LPRINT : LPRINT "CENTER OF SUBREFLECTOR IS AT:"
570 I=0
580 GOTO 450
590 LPRINT
600 LPRINT "PROFILE OF MAIN OFFSET PARABOLOID WITH CENTER AT Y1=";Y3
610 LPRINT
620 FOR Y=Y3-R TO Y3+R STEP R/10
630 Z1=F-Y^2/(4*F)
640 LPRINT "X1=";0;"Y1=";Y;"Z1=";Z1
650 NEXT Y
660 LPRINT : LPRINT "EDGE POINTS OF ELLIPSOID"
670 A6=A3 : A7=0
680 LPRINT
690 R1=E*D/(1-E*COS(A2)*COS(A6)+E*SIN(A2)*SIN(A6)*SIN(A7))
700 X0=R1*SIN(A6)*COS(A7)
710 Y0=R1*SIN(A6)*SIN(A7)
720 Z0=R1*COS(A6)
730 IF A7>360*ANGR THEN GOTO 810
740 LPRINT "PHI=";A7*RANG
750 A7=A7+30*ANGR
760 Y1=Y0*COS(A4)+Z0*SIN(A4)+C*SIN(A1)
770 Z1=-Y0*SIN(A4)+Z0*COS(A4)-C*COS(A1)
780 X1=X0
790 LPRINT "X1=";X1;"Y1=";-Y1;"Z1=";-Z1
800 GOTO 690
810 LPRINT
820 LPRINT "FOR COMPUTING Z0 WITH X0 & Y0 EQUALLY SPACED"
830 INPUT "ENTER X0,Y0";X0,Y0
840 A0=COS(A2)^2-1/E^2
850 B0=2*(D*COS(A2)-Y0*COS(A2)*SIN(A2))
860 C0=D^2-(X0^2+Y0^2)/E^2+Y0^2*SIN(A2)^2-2*D*Y0*SIN(A2)
870 Z0=(-B0-SQR(B0^2-4*A0*C0))/(2*A0)
880     LPRINT "X0=";X0;"Y0=";Y0;"Z0=";Z0
890 GOTO 830
900 LPRINT
910 LPRINT "INPUT CONFIGURATION CONSTANTS"
920 PI=3.141592654#
930 RANG=180/PI : ANGR=PI/180
940 I=1
950 BE=48.5037 : Y3=35 : F=30 : C=4.7 : R=30
960     INPUT "ENTER:  BETA,Y3,F,C,R";BE,Y3,F,C,R
970 LPRINT  "BETA=";BE;"YC=";Y3;"F=";F;"C=";C;"R=";R
980 A1=BE*ANGR
990 LPRINT
1000 T1=Y3/F/4
1010 B4=-(2*COS(A1)+SIN(A1)/T1)
1020 E=(-B4-SQR(B4^2-4))/2
1030 S4=SIN(A1)*(1-E^2)/((1+E^2)*COS(A1)-2*E)
1040 A2=ATN(S4)
1050 IF A2<=0 THEN A2=A2+PI
1060 D=C*(1-E^2)/(2*E^2)
1070 B=2*F*(1-E^2)/(1+E^2-2*E*COS(A1))
1080 S5=R/B : S6=ATN(S5) : A3=S6*2
1090 AB4=A2-A1 : M1=B/(2*F) : A4=AB4
1100 AA1=A1*RANG : AA2=A2*RANG : AA3=A3*RANG :AA4=AB4*RANG
1110 GOTO 260
1120 END
```

```
10 REM Computes spillover around edge of dish or subreflector
20 REM when horn or subreflector pattern is given by cos(A)^N.
30 REM Theta is the polar angle to the reflector edge.
40 PRINT "SPILL.BAS"
50 PI=3.14159265359# : CL=.4342944819#
60 ANGRAD=PI/180
70 PRINT
80      INPUT "N= ",N
90 INPUT "Range of Theta (degs. enter 2 Nos.): ",A2,A3
100 FOR A=A2 TO A3
110 A1=A*ANGRAD
120 P=COS(A1)^N
130 P1=10*CL*LOG(P)
140 PRINT "EDGE ANGLE= ";A
150 PRINT "dB TAPER (no path attenuation)= ";P1
160 P2=COS(A1)^(N+1)/(N+1)
170 F=(N+1)*P2
180 L=10*CL*LOG(1-F)
190 F2=100*F
200 PRINT "Psp= ";P2
210 PRINT "%= ";F2
220 PRINT "dB LOSS=";L
230 PRINT
240 NEXT A
250 END
```

```
10 REM *THIS IS A DEMONSTRATION PROGRAM TO SHOW*
20 REM *THE CROSS POLARIZATION CAUSED BY DISH OFFSET*
30 PRINT "CROSS.BAS"
40 PI=3.14159265359#
50 F=30 : EP=.00001 : LC=.4342944819#
60 INPUT "Tilt angle of feed. A0, DEGS.=",A0
70 A0=A0*PI/180
80 REM The Feed is a Y-polarized Huygen's Source.
90 PRINT : INPUT "Enter x,y: ",X,Y
100 X=X+EP
110 Z=F-(X*X+Y*Y)/4/F
120 R1=SQR(X*X+Y*Y+4*F*F)
130 L=X/R1 : M=Y/R1 : N=2*F/R1
140 REM Rotation of coordinates to tilt, Z1-axis
150 Y1=Y*COS(A0)-Z*SIN(A0)
160 Z1=Y*SIN(A0)+Z*COS(A0)
170 X1=X
180 PRINT "X1=";X1;"Y1=";Y1;"Z1=";Z1
190 REM PRINT "L,M,N=";L;M;N
200 R=SQR(X*X+Y*Y+Z*Z)
210 T=Z1/R
220 A=ATN(SQR(1-T*T)/T)
230 B=ATN(Y1/X1)
232 IF X1<=0 THEN B=B+PI
240 REM Huygen source polarization
250 L1=-(1-COS(A))*SIN(B)*COS(B)
260 M1=1-SIN(B)^2*(1-COS(A))
270 N1=-SIN(A)*SIN(B)
280 L2=L1
290 REM Rotation of polarization vectors back to Z-axis.
300 M2=M1*COS(A0)+N1*SIN(A0)
310 N2=N1*COS(A0)-M1*SIN(A0)
320 B2=2*(L*L2+M*M2+N*N2)
330 L3=B2*L-L2
340 M3=B2*M-M2
350 N3=B2*N-N2
360 REM PRINT "A,B=";A;B
370 REM PRINT "L1,M1,N1=";L1;M1;N1
380 REM PRINT "L2,M2,N2=";L2;M2;N2
390 PRINT "EAX,EAY,EAZ=";L3;M3;N3
400 P0=10*LC*LOG(M3*M3)
410 P1=10*LC*LOG(L3*L3+EP)
420 PRINT "Po=";P0;"Px=";P1
510 END
```

Chapter 2
PRIMARY FEEDS FOR REFLECTORS AND LENSES
Miguel Calvo

2.1 INTRODUCTION

This chapter mainly deals with the analysis of horn antennas as primary feeds for reflectors and lenses. Dipoles and slots as primary feeds are treated in Chapter 3, and some other types of feed will be considered at the end of this chapter.

The pyramidal and conical horns shown in Figure 2.1 can be considered as a taper section between the feeding rectangular or circular waveguide and an opening aperture with a cross-sectional dimension to accommodate a mode impedance which approximates that of free space. Such structures are easy to feed, simple to manufacture, and closely approximate the characteristics predicted by simple theories. These properties make horn antennas ideal, not only as feeds for reflectors and lenses, but also as reference gain standards and, when used by themselves, medium-gain antennas.

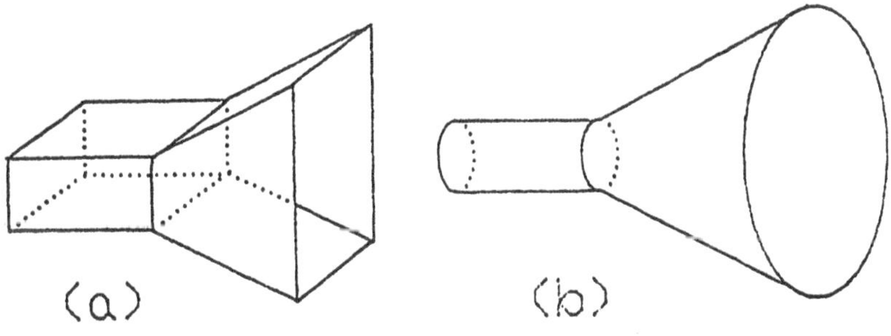

Figure 2.1 (a) Pyramidal and (b) Conical Horns

The long history of horn antennas can be found in the collection of papers edited by A.W. Love [1]. Recent textbooks on antennas [2,3] treat smooth-walled pyramidal horns and give an introduction to corrugated pyramidal horns and smooth-walled conical horns. A recent book by Clarricoats and Olver [4] contains a great amount of information on conical corrugated horns. The book by Milligan [5] contains a lot of practical data on smooth-walled pyramidal horns, conical horns, and corrugated conical horns.

In this chapter, the emphasis will be placed on the theoretical background used in the programs provided for the IBM PC, and on the use of those programs for the analysis and design of feed horns.

The analysis of horns is usually performed by considering the mouth of the horn as an open aperture in an infinite ground plane. Either calculated or estimated aperture fields are used to predict the radiated field by using the equivalence theorems, which will be established first.

2.2 UNIQUENESS AND EQUIVALENCE THEOREMS

Time harmonic electromagnetic fields will be assumed and complex phasor quantities, space dependent, used. In this way, time dependence $e^{j\omega t}$ can be eliminated from Maxwell equations, which appear as

$$\nabla \cdot \mathbf{D} = \rho \tag{2.1}$$

$$\nabla \cdot \mathbf{B} = 0 \tag{2.2}$$

$$\nabla \times \mathbf{E} = -j\omega\mathbf{B} \tag{2.3}$$

$$\nabla \times \mathbf{H} = \mathbf{J}^i + j\omega(\varepsilon + \sigma/j\omega)\mathbf{E} \tag{2.4}$$

where $\mathbf{J}^i$ are the currents imposed by the generators.

It is convenient to introduce a fictitious magnetic current density $\mathbf{M}$ and use, instead of (2.3), the generalized equation:

$$\nabla \times \mathbf{E} = -j\omega\mu\mathbf{H} - \mathbf{M} \tag{2.5}$$

Boundary conditions on the transition between two homogeneous media with parameters ε_1, μ_1, σ_1, and ε_2, μ_2, σ_2 are

$$\mathbf{n} \times (\mathbf{H}_2 - \mathbf{H}_1) = \mathbf{J}_s \tag{2.6}$$

$$(\mathbf{E}_2 - \mathbf{E}_1) \times \mathbf{n} = \mathbf{M}_s \tag{2.7}$$

where $\mathbf{J}_s$ and $\mathbf{M}_s$ are surface currents flowing in the boundary surface and $\mathbf{n}$ is the unit normal vector to that surface directed from medium 1 to medium 2.

Uniqueness theorems establish the information needed to obtain the field and the correspondence between the field and its sources. It can be easily shown [6] that "a field in a lossy region is uniquely specified by the sources within the region plus the tangential components of the electric field over the boundary, or the tangential components of the magnetic field over the boundary, or the former over part of the boundary and the latter over the rest of the boundary." A nondissipative medium should be considered as the limit case of a lossy medium when the dissipation tends to zero. If the region of interest is unbounded, the conditions over the surface of infinity are called radiation conditions and should be consistent with the fact that the power associated with the fields should propagate away from the sources and will either be constant for a nondissipative medium or decrease, transformed into losses, in a lossy medium.

Different source distributions outside a given region can produce the same field inside the region and are said to be equivalent within that region.

Consider a source distribution internal to surface S and free space external to S as in Figure 2.2(a). An equivalent problem in the external region, enforcing zero field internal to S and free space everywhere, is shown in Figure 2.2(b). There must exist surface currents, $\mathbf{J}_s$, $\mathbf{M}_s$ according to (2.6) and (2.7) with $\mathbf{E}_2$, $\mathbf{H}_2$ the original fields on S, $\mathbf{E}_1$, $\mathbf{H}_1$ the desired null field inside S, and $\mathbf{n}$ the unit normal vector to S. From the uniqueness theorem we know that the tangential electric *and* magnetic field components on S are the same as in the original problem, so the solution outside S will be the same if the radiation condition is also satisfied.

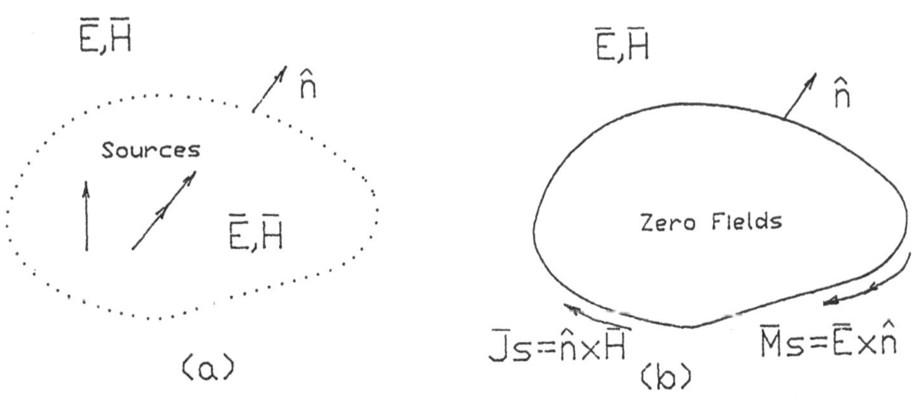

Figure 2.2 **(a) Fields Produced by Sources Inside Surface *S* and (b) Equivalent Problem External to *S* with Zero Fields Inside *S***

We have used tangential components of both **E** and **H** in the previous equivalent problem. From uniqueness we know that only the tangential components of **E** or **H** are needed to determine the field. The use of both has allowed us to put free space inside S. Let us now see how to construct equivalent problems with only **E** or **H**.

If we place a perfect electric conductor over S and on top of it a sheet of magnetic current $\mathbf{M}_s = \mathbf{E} \times \mathbf{n}$, the tangential components of **E** in this equivalent problem will be the same as in the original problem.

On the other hand, if we replace S with a perfect magnetic conductor and place over it an electric current sheet $\mathbf{J}_s = \mathbf{n} \times \mathbf{H}$, we will have the same tangential magnetic field components on S and so the same fields outside S as in the original problem.

In general, it is difficult to obtain the fields from the equivalent sources over an arbitrary surface S in the presence of electric or magnetic conductors. The problem is greatly simplified, if the surface S is planar, by using image theory which is summarized in Figure 2.3. With respect to the $z > 0$ half-space, an electric current sheet $\mathbf{J}_s$ in the presence of a $z = 0$ plane perfect magnetic conductor is equivalent to a $2\mathbf{J}_s$ current sheet radiating in free space as indicated in Figure 2.3(a). A magnetic current sheet $\mathbf{M}_s$ over a plane perfect electric conductor is equivalent, for $z > 0$, to a $2\mathbf{M}_s$ magnetic current sheet radiating in free space as indicated in Figure 2.3(b).

Horn feeds have a planar aperture ($z = 0$) on which we can know approximations of the electric and magnetic fields $\mathbf{E}_a$ or $\mathbf{H}_a$. We calculate the radiated fields in $z > 0$, assuming zero fields on the rest of the $z = 0$ plane (see Figure 2.4), and construct equivalent problems in the way previously indicated.

The radiated fields produced by an electric current sheet $\mathbf{J}_s = \mathbf{n} \times \mathbf{H}_a$ can be obtained from the magnetic vector potential **A**:

$$\mathbf{A} = \frac{e^{-j\beta r}}{4\pi r} \iint_S \mathbf{J}_s(r') \cdot e^{j\beta \mathbf{r} \cdot \mathbf{r}'} dS' = \frac{e^{-j\beta r}}{4\pi r}\, \mathbf{n} \times \mathbf{Q} \tag{2.8}$$

$$\mathbf{Q} = \iint_S \mathbf{H}_a(\mathbf{r}') \cdot e^{j\beta \mathbf{r}\cdot\mathbf{r}'} dS' \tag{2.9}$$

as the components of $\mathbf{E}_A$:

$$\mathbf{E}_A = -j\omega\mu\mathbf{A} \tag{2.10}$$

transverse to the unitary vector **r**.

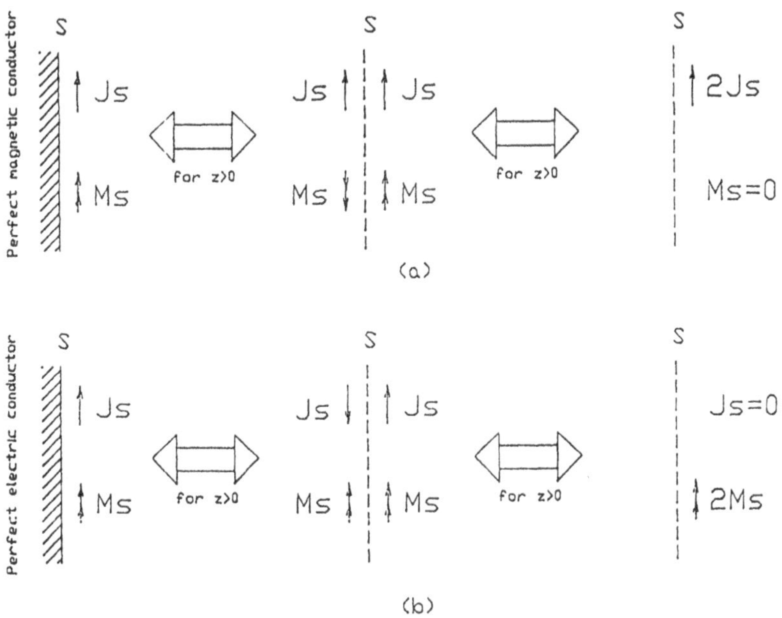

Figure 2.3 (a) Equivalent Problem of Sources over an Infinite Planar Perfect Magnetic Conductor and (b) Perfect Electric Conductor Using Images

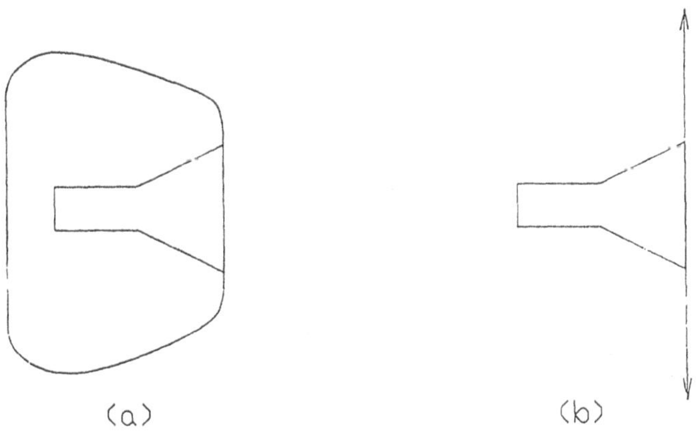

Figure 2.4 Idealized Infinite Planar Aperture (b) of a Finite Horn Model (a)

The radiated fields produced by a magnetic current sheet $\mathbf{M}_s = \mathbf{E}_a \times \mathbf{n}$ can be obtained through the electric vector potential $\mathbf{F}$:

$$\mathbf{F} = \frac{e^{-j\beta r}}{4\pi r} \iint_S \mathbf{M}_s(r') \cdot e^{j\beta \mathbf{r} \cdot \mathbf{r}'} dS' = \frac{e^{-j\beta r}}{4\pi r} \mathbf{P} \times \mathbf{n} \tag{2.11}$$

$$\mathbf{P} = \iint_S \mathbf{E}_a(\mathbf{r}') \cdot e^{j\beta \mathbf{r} \cdot \mathbf{r}'} dS' \tag{2.12}$$

as the components of $\mathbf{H}_F$:

$$\mathbf{H}_F = -j\omega\varepsilon\mathbf{F} \tag{2.13}$$

transverse to $\mathbf{r}$.

The electric field associated with $\mathbf{H}_F$ can be obtained from the TEM characteristics of the radiated fields as

$$\mathbf{E}_F = \eta\mathbf{H}_F \times \mathbf{r} \tag{2.14}$$

where η is the free-space characteristic impedance.

Integrals $\mathbf{P}$ and $\mathbf{Q}$ are two-dimensional Fourier transforms of the aperture fields $\mathbf{E}_a$ and $\mathbf{H}_a$ respectively. Since the aperture lies in the x-y plane, $\mathbf{P}$ and $\mathbf{Q}$ will have $\mathbf{x}$ and $\mathbf{y}$ components. Spherical $\boldsymbol{\theta}$ and $\boldsymbol{\phi}$ components of $\mathbf{A}$ and $\mathbf{F}$, and the same ones of the radiated $\mathbf{E}$ fields, will be obtained by expressing $\mathbf{x}$ and $\mathbf{y}$ in spherical coordinates and retaining the transverse to $\mathbf{r}$ ($\boldsymbol{\theta}$ and $\boldsymbol{\phi}$) components. With this procedure one obtains the following expressions for the radiated field components:

(a) Using $\mathbf{E}_a$ and $\mathbf{H}_a$ in the aperture:

$$E_\theta = j\beta \frac{e^{-j\beta r}}{4\pi r} [P_x \cos\phi + P_y \sin\phi + \eta\cos\theta\,(Q_y \cos\phi - Q_x \sin\phi)] \tag{2.15}$$

$$E_\phi = j\beta \frac{e^{-j\beta r}}{4\pi r} [\cos\theta(P_y \cos\phi - P_x \sin\phi) - \eta(Q_y \sin\phi + Q_x \cos\phi)] \tag{2.16}$$

When $\mathbf{E}_a$ and $\mathbf{H}_a$ are supposed to be related as a TEM wave in free space it is said that the aperture field is a Huygens source. In this case $Q_y = P_x/\eta$, $Q_x = -P_y/\eta$ and

$$E_\theta = j\beta \frac{e^{-j\beta r}}{4\pi r} (1 + \cos\theta) \cdot (P_x \cos\phi + P_y \sin\phi) \tag{2.17}$$

$$E_\phi = j\beta \frac{e^{-j\beta r}}{4\pi r} (1 + \cos\theta) \cdot (P_y \cos\phi - P_x \sin\phi) \tag{2.18}$$

(b) Using only the electric field in the aperture $\mathbf{E}_a$:

$$E_\theta = j\beta \frac{e^{-j\beta r}}{2\pi r} (P_x \cos \phi + P_y \sin \phi) \tag{2.19}$$

$$E_\phi = j\beta \frac{e^{-j\beta r}}{2\pi r} \cos \theta \, (P_y \cos \phi - P_x \sin \phi) \tag{2.20}$$

(c) Using only the magnetic field at the aperture $\mathbf{H}_a$:

$$E_\phi = j\beta\eta \frac{e^{-j\beta r}}{2\pi r} \cos \theta \, (Q_y \cos \phi - Q_x \sin \phi) \tag{2.21}$$

$$E_\phi = -j\beta\eta \frac{e^{-j\beta r}}{2\pi r} (Q_y \sin \phi + Q_x \cos \phi) \tag{2.22}$$

2.3 APERTURE *E* FIELD MODEL

The three formulations produce the same results if the exact fields over the infinite $z = 0$ plane are known. Only approximate fields are assumed on the aperture, however, and the three formulations provide different results. The differences are small for aperture sizes of some wavelengths, but for small aperture sizes the electric field formulation gives results which agree better with measurements [7], and formulation (b), known as the *E*-field model, is the most widely used.

The spherical field components are easily calculated using the previously described procedures, but they are not usually directly measured. We need to transform between E_θ, E_ϕ and the measured components (and *vice versa*) to allow comparisons between theory and measurements.

In general the far field radiated by an antenna is a local plane wave with an elliptical polarization, which means that the electric and magnetic field vectors in a phase front rotate with time, describing ellipses. It is well known that such a field can be described as the superposition of either two orthogonal linearly polarized fields or two circularly polarized fields with opposite sense of rotation. Those linear or circular orthogonal components of the radiated fields are the ones measured by using appropriate linear or circularly polarized probes.

Conversely the antennas are designed to produce linearly or circularly polarized radiated fields in space regions as broad as possible. To produce linearly polarized fields the horns are usually excited with the dominant mode of the corresponding input waveguide. For circular polarization the aperture field should be circularly polarized (unless an external polarizer screen is used) which implies two orthogonal (perpendicular) modes with equal amplitude and a 90° phase shift between them. The input waveguide must support those modes (square or circular waveguides are adequate), and a device (polarizer) is required to launch both modes and produce the adequate phase shift between them.

The definition of the desired (co-polar) and unwanted (cross-polar) components for circular polarization is unique: the one with the desired rotation sense and the one with the opposite sense. For linear polarization, however, the definition of co-polar and cross-polar components is not unique [8], although the so-called third Ludwig definition is the most widely used because it coincides with the most extended method of measuring linearly polarized antenna patterns as described in Silver [9] and Hollis *et al.* [10].

From the $\boldsymbol{\theta}$ and $\boldsymbol{\phi}$ radiated field components the right-handed (RH) and left-handed (LH) circular components are obtained as:

$$E_{RH} = (E_\theta - jE_\phi)e^{-j\phi}/\sqrt{2} \tag{2.23a}$$

$$E_{LH} = (E_\theta + jE_\phi)e^{-j\phi}/\sqrt{2} \tag{2.23b}$$

and the co-polar E_{co} and cross-polar E_{xp} components are

$$E_{co} = E_\theta \sin\phi + E_\phi \cos\phi \tag{2.24a}$$

$$E_{xp} = E_\theta \cos\phi - E_\phi \sin\phi \tag{2.24b}$$

for a (normally used) **y** directed aperture field, and

$$E_{co} = E_\theta \cos\phi - E_\phi \sin\phi \tag{2.25a}$$

$$E_{xp} = E_\theta \sin\phi + E_\phi \cos\phi \tag{2.25b}$$

for an **x** directed aperture field.

2.4 HORN APERTURE FIELD ESTIMATION

Approximations for the aperture fields in horn antennas can be obtained from the solution of Maxwell equations in infinite wedge or conical waveguides, depending on the horn cross section. The mode fields in those structures can be expressed in terms of Hankel, spherical Hankel, and Legendre functions, and so can the aperture fields. For moderate semiflare angle horns, however, the aperture field can be expressed in terms of the (more simple) modes of the cylindrical waveguide with the cross section of the horn aperture. A doubly curved cylindrical or a spherical phase variation in the aperture is included to account for the effect of the semiflare section. In general, only fundamental mode fields are considered in the aperture, although sometimes high order modes are deliberately produced to modify the horn radiation characteristics.

The far fields obtained using the E-field model are adequate in the main beam region (close to $\theta = 0$) and for moderate aperture sizes (greater than λ). For lower aperture sizes the radiation pattern is strongly influenced by the horn aperture flange structure. When we approach the $\theta = \pi/2$ zone the error in the predictions increases and obviously the model is no longer valid beyond $\theta = \pi/2$. To obtain the radiated field in those regions for which the aperture E-field model produces weak results (or no results at all), one can use GTD techniques [11,12,13]. Usually, this is not the case when horns are used as feeds, so we do not use these techniques here.

2.5 FUNDAMENTAL MODE SMOOTH-WALLED PYRAMIDAL HORNS

The geometry of this horn is shown in Figure 2.5, as is that of the two main longitudinal sections. The horn aperture is rectangular with dimensions $A \times B$, the semiflare angles are α_1 and α_2, and R_1 and R_2 are respectively the distances from the aperture to the two apex lines.

The field at the aperture is considered to be distributed in amplitude as the TE_{10} mode of the rectangular waveguide. The phase is approximated considering the field as the superposition of two cylindrical waves emanating from the apex lines. The field distribution is then

$$\mathbf{E}_{ap} = \mathbf{y}\, E_0 \cos \frac{\pi x}{A}\, e^{-j(\pi/\lambda)(x^2/R_1 + y^2/R_2)} \tag{2.26}$$

where λ is the wavelength. This aperture field is linearly polarized in $\mathbf{y}$ direction. Sometimes circular polarization is required and equal amplitude TE_{10} and TE_{01} modes with 90° phase shift are excited. The aperture field will be then

$$\mathbf{E}_{ap} = E_0 \left(\mathbf{y} \cos \frac{\pi x}{A} + j\, \mathbf{x} \cos \frac{\pi y}{B} \right) e^{-j(\pi/\lambda)(x^2/R_1 + y^2/R_2)} \tag{2.27}$$

where usually $A = B$ and $R_1 = R_2$ as square horns are most normally used.

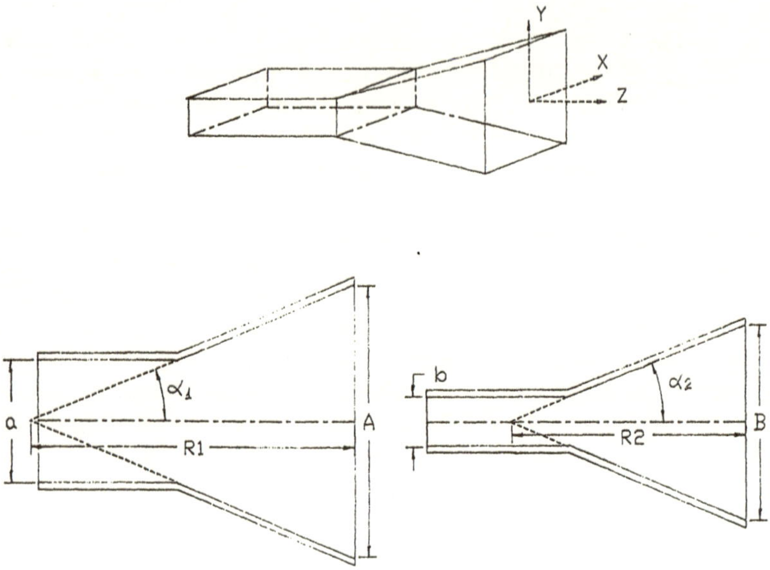

Figure 2.5 Smooth-Walled Pyramidal Horn Geometry

The total power radiated by the horn can be obtained from the flux of the Poynting vector through the aperture as

$$P_r = \int_{-A/2}^{A/2} \int_{-B/2}^{B/2} \frac{|E_{ap}|^2}{2\eta} \, dx \, dy = \frac{E_0^2}{2\eta} \cdot \frac{AB}{2} \tag{2.28}$$

for linear polarization and

$$P_r = \int_{-A/2}^{A/2} \int_{-B/2}^{B/2} \frac{(|E_{apx}|^2 + |E_{apy}|^2)}{2\eta} \, dx \, dy = \frac{E_0^2}{2\eta} \cdot AB \tag{2.29}$$

for circular polarization (η is the free-space characteristic impedance).

The radiated far field will be given by Equations (2.19) and (2.20) as follows.

(a) Linear polarization:

$$\mathbf{P}(\theta,\phi) = \mathbf{y} \, E_0 \int_{-A/2}^{A/2} \int_{-B/2}^{B/2} \cos\left(\frac{\pi x}{A}\right) e^{-j(\pi/\lambda)(x^2/R_1 + y^2/R_2)} \, e^{j\beta(xu+yv)} \, dx \, dy$$

$$= P_y \, \mathbf{y} = E_0 \, I_{1x} \cdot I_{2y} \, \mathbf{y} \tag{2.30}$$

$$I_{1x} = \int_{-A/2}^{A/2} \cos\left(\frac{\pi x}{A}\right) \mathrm{e}^{-\mathrm{j}[(\pi x^2/\lambda R_1) - (\beta u x)]}\, \mathrm{d}x \tag{2.31}$$

$$I_{2y} = \int_{-B/2}^{B/2} \mathrm{e}^{-\mathrm{j}[(\pi y^2/\lambda R_2) - (\beta v y)]}\, \mathrm{d}y \tag{2.32}$$

with $u = \sin\theta\cos\phi$, $v = \sin\theta\sin\phi$.

(b) Circular polarization:

$$\begin{aligned} P(\theta,\phi) &= \mathbf{y}\int_{-A/2}^{A/2}\int_{-B/2}^{B/2} E_{\mathrm{apy}} \cdot \mathrm{e}^{\mathrm{j}\beta(xu+yv)}\, \mathrm{d}x\, \mathrm{d}y \\ &\quad + \mathrm{j}\mathbf{x}\int_{-A/2}^{A/2}\int_{-B/2}^{B/2} E_{\mathrm{apx}} \cdot \mathrm{e}^{\mathrm{j}\beta(ux+vy)}\, \mathrm{d}x\, \mathrm{d}y \\ &= P_y\,\mathbf{y} + jP_x\,\mathbf{x} \end{aligned} \tag{2.33}$$

where P_y is the same as for linear polarization.

The above integrals can be expressed in terms of cosine and sine Fresnel integrals as follows:

$$\begin{aligned} P_y = E_0 \bigg\{ \frac{1}{2} \cdot \sqrt{\lambda R_1/2} \cdot \Big[&\mathrm{e}^{\mathrm{j}(R_1\pi/\lambda)(u+(\lambda/2A)^2} \cdot (C(s_2) \\ &- \mathrm{j}S(s_2) - C(s_1) + \mathrm{j}S(s_1)) + \mathrm{e}^{\mathrm{j}(R_1\pi/\lambda)(u-(\lambda/2A)^2)} \\ &\cdot (C(t_2) - \mathrm{j}S(t_2) - C(t_1) + \mathrm{j}S(t_1))\Big] \cdot \Big[\sqrt{\lambda R_2/2} \cdot \mathrm{e}^{\mathrm{j}(R_2\pi/\lambda)\cdot v^2} \\ &\cdot (C(p_2) - \mathrm{j}S(p_2) - C(p_1) + \mathrm{j}S(p_1))\Big]\bigg\} \end{aligned} \tag{2.34}$$

where

$$P_2 = \sqrt{2/(\lambda R_2)}\,(B/2 - R_2 v) \qquad \text{(a)}$$

$$s_2 = \sqrt{2/(\lambda R_1)}\left(A/2 - R_1 u - \frac{\lambda R_1}{2A}\right) \qquad \text{(b)}$$

$$t_2 = \sqrt{2/(\lambda R_1)}\left(A/2 - R_1 u + \frac{\lambda R_1}{2A}\right) \qquad \text{(c)}$$

$$p_1 = \sqrt{2/(\lambda R_2)}\,(-B/2 - R_2 v) \qquad \text{(d)}$$

$$s_1 = \sqrt{2/(\lambda R_1)}\left(-A/2 - R_1 u - \frac{\lambda R_1}{2A}\right) \quad \text{(e)}$$

$$t_1 = \sqrt{2/(\lambda R_1)}\left(-A/2 - R_1 u + \frac{\lambda R_1}{2A}\right) \quad \text{(f)} \tag{2.35}$$

and

$$C(x) = \int_0^x \cos\left(\frac{\pi}{2} t^2\right) \mathrm{d}t \tag{2.36}$$

$$S(x) = \int_0^x \sin\left(\frac{\pi}{2} t^2\right) \mathrm{d}t \tag{2.37}$$

can be evaluated numerically [2,14].

The result for P_x is

$$\begin{aligned} P_x = \mathrm{j}E_0 \cdot \Big\{ \frac{1}{2} \cdot \sqrt{\lambda R_2/2} \cdot \Big[&\mathrm{e}^{\mathrm{j}(R_2\pi/\lambda)(v+(\lambda/2B)^2)} \cdot (C(s_2') \\ &- \mathrm{j}S(s_2') - C(s_1') + S(s_1')) + \mathrm{e}^{\mathrm{j}(R_2\pi/\lambda)(v-(\lambda/2B)^2)} \cdot (C(t_2') \\ &- \mathrm{j}S(t_2') - C(t_1') + \mathrm{j}S(t_1'))\Big] \cdot \Big[\sqrt{\lambda R_1/2} \cdot \mathrm{e}^{\mathrm{j}(R_1\pi/\lambda)\cdot u^2} \\ &\cdot (C(p_2') - \mathrm{j}S(p_2') - C(p_1') + \mathrm{j}S(p_1'))\Big]\Big\} \end{aligned} \tag{2.38}$$

where

$$p_2' = \sqrt{2/\lambda R_1)}\,(A/2 - R_1 u) \quad \text{(a)}$$

$$s_2' = \sqrt{2/\lambda R_2)}\left(B/2 - R_2 v - \frac{\lambda R_2}{2B}\right) \quad \text{(b)}$$

$$t_2' = \sqrt{2/\lambda R_2)}\left(B/2 - R_2 v + \frac{\lambda R_2}{2B}\right) \quad \text{(c)}$$

$$p_1' = \sqrt{2/\lambda R_1}\,(-A/2 - R_1 u) \quad \text{(d)}$$

$$s_1' = \sqrt{2/\lambda R_1)}\left(-B/2 - R_2 v - \frac{\lambda R_2}{2B}\right) \quad \text{(e)}$$

$$t_1' = \sqrt{2/\lambda R_1)}\left(-B/2 - R_2 v + \frac{\lambda R_2}{2B}\right) \quad \text{(f)} \tag{2.39}$$

When α_1 or α_2 are 0 the horns are called E- and H-plane sectoral horns. Then R_1 or R_2 tends to ∞, and the integrals can be of the types:

$$I'_{1x} = \int_{-A/2}^{A/2} \cos\left(\frac{\pi x}{A}\right) e^{j\beta ux}\, dx = -\left(\frac{\pi A}{2}\right)\left[\frac{\cos(\beta uA/2)}{(\beta uA/2)^2 - (\pi/2)^2}\right] \tag{2.40}$$

$$I'_{2y} = \int_{-B/2}^{B/2} e^{j\beta vy}\, dy = B\,\frac{\sin(\beta vB/2)}{(\beta vB/2)} \tag{2.41}$$

which can be analytically evaluated.

The co-polar and cross-polar field components can be obtained by substituting E_θ and E_ϕ in Equations (2.24) or (2.23) for linear or circular polarization respectively.

2.6 CORRUGATED PYRAMIDAL HORNS

In the 1960s Kay [15] realized that making corrugations on the walls of a horn antenna would taper the aperture field distribution in both planes reducing diffractions from the edges, especially from those that are perpendicular to the E-field. The first paper on pyramidal corrugated horns is Lawrie and Peters [16] who introduced corrugations onto the E-plane walls to taper the aperture electric field and reduce the E-plane sidelobes. An extended background description of the developments in pyramidal corrugated horns can be found in [4].

The analysis of the propagation characteristics along the horn and the mode conversion from the rectangular input waveguide can be done using mode matching techniques [17,18]. The aim of the design is to produce a hybrid mode in the rectangular horn aperture with an electric field variation of cosine type in both principal planes. This aperture field creates a far field with low co-polar sidelobes, low cross-polar level, and an elliptical main beam contour. For a correct function several corrugations per wavelength are required with a depth of about $\lambda/4$. A mode converter section is required to launch the hybrid mode from the dominant mode of the rectangular input waveguide. This section consists of several slots with variable depths ranging from $\lambda/2$ at the input to $\lambda/4$ at the output.

The analysis of the radiated far field can be done using the E-field aperture model [19]. Considering the horn geometry shown in Figure 2.6 the aperture field can be expressed as

$$\mathbf{E}_{ap} = \mathbf{y}\, E_0 \cos\left(\frac{\pi x}{A}\right) \cos\left(\frac{\pi y}{B}\right) e^{-j(\pi/\lambda)(x^2/R_1 + y^2/R_2)} \tag{2.42}$$

for linear polarization, and

$$\mathbf{E}_{ap} = E_0 \cos\left(\frac{\pi x}{A}\right) \cos\left(\frac{\pi y}{B}\right) e^{-j(\pi/\lambda)(x^2/R_1 + y^2/R_2)} (K_{ry}\, \mathbf{y} + jK_{rx}\, \mathbf{x}) \tag{2.43}$$

for circular polarization. K_{rx} and K_{ry} account for the different propagation constant of the two hybrid-mode fields required for circular-polarization operation and are defined as

$$K_{rx} = [(2\pi/\lambda)^2 - (\pi/B)^2]^{1/2} \tag{2.44a}$$

$$K_{ry} = [(2\pi/\lambda)^2 - (\pi/A)^2]^{1/2} \tag{2.44b}$$

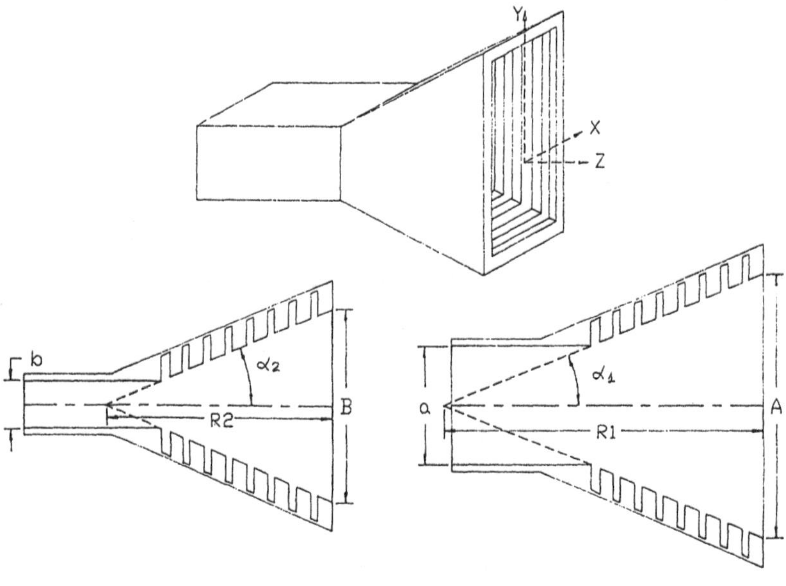

Figure 2.6 Corrugated Pyramidal Horn Geometry

The total radiated power can be easily obtained from the flux of the Poynting vector associated with the aperture field as

$$P_r = \int_{-A/2}^{A/2} \int_{-B/2}^{B/2} \frac{E_0^2 \cos^2(\pi x/A) \cos^2(\pi y/B)}{2\eta}\, dx\, dy = \frac{E_0^2}{2\eta} \cdot \frac{AB}{4} \tag{2.45}$$

for linear polarization, and

$$P_r = \int_{-A/2}^{A/2} \int_{-B/2}^{B/2} \frac{(|E_{apx}|^2 + |E_{apy}|^2)}{2\eta}\, dx\, dy = \frac{E_0^2}{2\eta} \cdot \frac{AB}{4}(K_{rx}^2 + K_{ry}^2) \tag{2.46}$$

for circular polarization.

As in the case of smooth-walled pyramidal horns, the integrals necessary to obtain the radiation field can be expressed in terms of Fresnel sine and cosine integrals. It can be easily found that for linear polarization:

$$\begin{aligned}
\mathbf{P} = P_y\mathbf{y} = y \cdot E_0 \int_{-A/2}^{A/2} \int_{-B/2}^{B/2} \cos(\pi x/A) \\
\cos(\pi y/B)\, e^{-j(\pi/\lambda)(x^2/R_1 + y^2/R_2)}\, e^{j\beta(xu+yv)}\, dx\, dy \\
= \mathbf{y}E_0 \left\{\frac{1}{2} \cdot \sqrt{\pi R_1/\beta} \cdot \left[e^{j(R_1/2\beta)(\beta u+\pi/A)^2} \cdot (C(s_2) - jS(s_2) - C(s_1)\right.\right. \\
+ jS(s_1)) + e^{j(R_1/2\beta)(\beta u-\pi/A)^2} \cdot (C(t_2) - jS(t_2) - C(t_1) \\
\left.\left. + jS(t_1))\right]\right\} \cdot \left\{\frac{1}{2}\sqrt{\pi R_2/\beta} \cdot \left[e^{j(R_2/2\beta)\cdot(\beta v-\pi/B)^2} \cdot (C(s_2') - jS(s_2')\right.\right. \\
- C(s_1') + jS(s_1')) + e^{j(R_2/2\beta)(\beta v + \pi/B)^2} \cdot (C(t_2') \\
\left.\left. - jS(t_2') - C(t_1') + jS(t_1'))\right]\right\}
\end{aligned} \tag{2.47}$$

where $C(x)$ and $S(x)$ are the cosine and sine Fresnel integrals and

$$s_2 = \frac{1}{\sqrt{\pi\beta R_1}}\left(\frac{A\beta}{2} - R_1\beta u - \frac{\pi R_1}{A}\right) \qquad \text{(a)}$$

$$s_2' = \frac{1}{\sqrt{\pi\beta R_2}}\left(\frac{B\beta}{2} - R_2\beta v - \frac{\pi R_2}{B}\right) \qquad \text{(b)}$$

$$t_2 = \frac{1}{\sqrt{\pi\beta R_1}}\left(\frac{A\beta}{2} - R_1\beta u + \frac{\pi R_1}{A}\right) \qquad \text{(c)}$$

$$t_2' = \frac{1}{\sqrt{\pi\beta R_2}}\left(\frac{B\beta}{2} - R_2\beta v + \frac{\pi R_2}{B}\right) \qquad \text{(d)}$$

$$s_1 = \frac{1}{\sqrt{\pi\beta R_1}}\left(-\frac{A\beta}{2} - R_1\beta u - \frac{\pi R_1}{A}\right) \quad \text{(e)}$$

$$s_1' = \frac{1}{\sqrt{\pi\beta R_2}}\left(-\frac{B\beta}{2} - R_2\beta v - \frac{\pi R_2}{B}\right) \quad \text{(f)}$$

$$t_1 = \frac{1}{\sqrt{\pi\beta R_1}}\left(-\frac{A\beta}{2} - R_1\beta u + \frac{\pi R_1}{A}\right) \quad \text{(g)}$$

$$t_1' = \frac{1}{\sqrt{\pi\beta R_2}}\left(-\frac{B\beta}{2} - R_2\beta v + \frac{\pi R_2}{B}\right) \quad \text{(h)} \qquad (2.48)$$

Substituting **P** in (2.19) and (2.20), the spherical far-field components for linear polarization will be

$$E_\theta(\theta,\phi) = \mathrm{j}\beta \frac{\mathrm{e}^{-\mathrm{j}\beta r}}{2\pi r} P_y \sin\phi \qquad (2.49)$$

$$E_\phi(\theta,\phi) = \mathrm{j}\beta \frac{\mathrm{e}^{-\mathrm{j}\beta r}}{2\pi r} P_y \cos\theta\cos\phi \qquad (2.50)$$

For circular polarization, the vector **P** will also have **x** component and can be expressed as

$$\mathbf{P}(\theta,\phi) = P_y\,(\mathrm{j}K_{rx}\mathbf{x} + K_{ry}\mathbf{y}) \qquad (2.51)$$

with P_y given in (2.47).

When α_1 or α_2 are 0, then R_1 or R_2 are ∞ and we find integrals similar to I_{1x} in (2.40) that can be analytically evaluated.

Co-polar and cross-polar field components are obtained from E_θ and E_ϕ substituting in (2.24) or (2.23) for linear or circular polarization.

2.7 FUNDAMENTAL MODE SMOOTH-WALLED CONICAL HORN

In the E-field model the aperture field of a fundamental mode conical horn, with the geometry shown in Figure 2.7, is approximated by the field amplitude distribution of the mode in a waveguide with the radius of the aperture corrected with a spherical phase distribution, as if the field emanates from the cone apex. The aperture field will then be

$$\mathbf{E}_{\mathrm{ap}} = E_{\mathrm{apx}}\mathbf{x} + E_{\mathrm{apy}}\mathbf{y} \qquad (2.52)$$

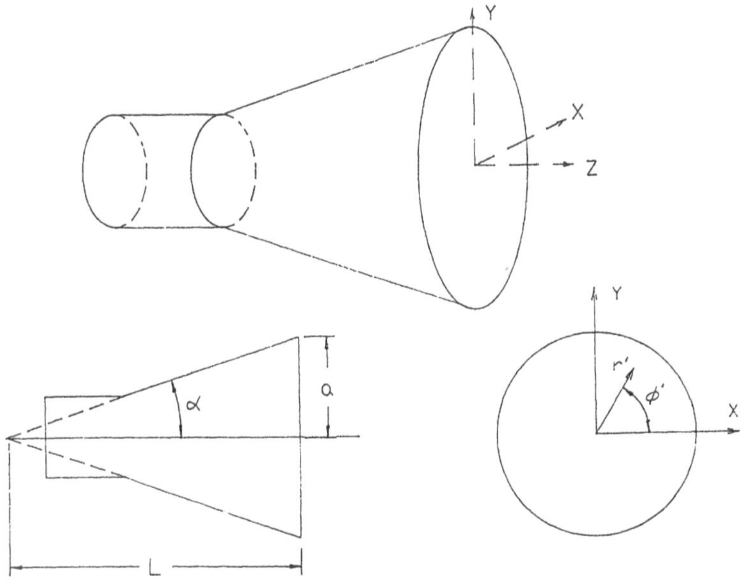

Figure 2.7 Geometry of a Smooth-Walled Conical Horn

$$E_{\text{apx}} = E_0 J_2(K_{11}r') \sin(2\phi')\, e^{-j(\pi/\lambda)(r'^2/L)} \tag{2.53a}$$

$$E_{\text{apy}} = E_0[J_0(K_{11}r') - J_2(K_{11}r') \cos(2\phi)]\, e^{-j(\pi/\lambda)(r'^2/L)} \tag{2.53b}$$

where E_0 is a constant, J_0 and J_2 are Bessel functions of the first kind and orders 0 and 2, respectively, $K_{11} = 1.8412/a$, a being the aperture radius, L is the distance from the cone apex to the aperture, and r' and ϕ' are polar coordinates over the aperture.

To obtain the radiated field we need to evaluate the integrals:

$$\mathbf{P} = \iint_S \mathbf{E}_{\text{ap}}(\mathbf{r}') \cdot e^{j\beta\mathbf{r}\cdot\mathbf{r}'}\, dS' = \int_{r'=0}^{a} \int_{\phi'=0}^{2\pi} \mathbf{E}_{\text{ap}}\, e^{j\beta r' \sin\theta \cos(\phi-\phi')}\, r'\, dr'\, d\phi' \tag{2.54}$$

which can be solved analytically [9] if $\alpha = 0$ ($L = \infty$), which is the case of an open-ended waveguide with no semiflare angle. In the general case the integrals must be evaluated numerically. A very efficient numerical integration method for this type of integral [20] is the Jacobi-Bessel series method [21].

The Jacobi-Bessel series method consists of the substitution of the subintegral function by its expansion in a double series of circular sine and cosine functions in the angular coordinate ϕ' and orthogonal modified Jacobi polynomials in the radial r' coordinate. In our case:

$$\mathbf{E}_{\text{ap}}(r',\phi') = \sum_{n=0}^{\infty} \sum_{m=0}^{\infty} (\mathbf{C}_{nm} \cos(n\phi') + \mathbf{D}_{nm} \sin(n\phi'))\, F_m^n(r') \tag{2.55}$$

where $\mathbf{C}_{nm}$ and $\mathbf{D}_{nm}$ are the expansion coefficients which can be obtained from the orthogonality conditions of circular functions:

$$\int_0^{2\pi} \cos(\mathrm{j}\phi') \cos(k\phi')\, \mathrm{d}\phi' = 2\pi \cdot \delta_{\mathrm{j}k}/\varepsilon_n,\ (n = \mathrm{j} = k) \tag{2.56a}$$

$$\int_0^{2\pi} \sin(\mathrm{j}\phi') \cos(k\phi')\, \mathrm{d}\phi' = 0 \tag{2.56b}$$

$$\int_0^{2\pi} \sin(\mathrm{j}\phi') \sin(k\phi')\, \mathrm{d}\phi' = 2\pi \cdot \delta_{\mathrm{j}k},\ (\mathrm{j},\ k \neq 0) \tag{2.56c}$$

where $\delta_{\mathrm{j}k}$ is the Kronecker delta (equal to 0 if $i \neq \mathrm{j}$ and equal to 1 otherwise) and ε_n is the Neumann number (equal to 1 if $n = 0$ and equal to 2, otherwise), and from the orthogonality properties of the modified Jacobi polynomials:

$$\int_0^1 F_j^n(s) \cdot F_k^n(s)\, s\, \mathrm{d}s = \delta_{\mathrm{j}k} \tag{2.57}$$

$$\left.\begin{matrix}\mathbf{C}_{nm}\\ \mathbf{D}_{nm}\end{matrix}\right\} = \frac{\varepsilon_n}{2\pi} \int_0^1 \int_0^{2\pi} \mathbf{E}_{\text{ap}}(as,\phi') \left\{\begin{matrix}\cos(n\phi')\\ \sin(n\phi')\end{matrix}\right\} F_m^n(s)\, s\, \mathrm{d}s\, \mathrm{d}\phi' \tag{2.58}$$

where the radius has been normalized by means of the variable change $r' = a \cdot s$.

Substituting the series representation of the aperture field in the integral, we obtain

$$\mathbf{P}(\theta,\phi) = a^2 \sum_{n=0}^{\infty} \sum_{m=0}^{\infty} \int_0^1 \int_0^{2\pi} (C_{nm} \cos(n\phi') + \mathbf{D}_{nm} \sin(n\phi'))\, F_m^n(s)\, \mathrm{e}^{\mathrm{j}\xi \cos(\phi-\phi')}\, s\, \mathrm{d}s\, \mathrm{d}\phi' \tag{2.59}$$

where $\xi = \beta as \cdot \sin(\theta)$. The integrations in ϕ' can be done taking into account the following results:

$$\int_0^{2\pi} \begin{Bmatrix} \cos(n\phi') \\ \sin(n\phi') \end{Bmatrix} e^{j\xi \cos(\phi-\phi')} \, d\phi' = 2\pi \, j^n \begin{Bmatrix} \cos(n\phi) \\ \sin(n\phi) \end{Bmatrix} J_n(\xi) \tag{2.60}$$

We then obtain

$$\mathbf{P}(\theta,\phi) = 2\pi a^2 \sum_{n=0}^{\infty} \sum_{m=0}^{\infty} j^n \int_0^1 [\mathbf{C}_{nm} \cos(n\phi) + \mathbf{D}_{nm} \sin(n\phi)] \cdot F_m^n(s) \cdot J_n[\beta a \sin(\theta) s] \cdot s \cdot ds \tag{2.61}$$

The remaining integrals can be obtained from the result:

$$\int_0^1 F_m^n(s') \, J_n[\beta a \sin(\theta) s'] \, s' \, ds' = \frac{\sqrt{2(n + 2m + 1)}}{\beta a \sin(\theta)} \cdot J_{n+2m+1}[\beta a \sin(\theta)] \tag{2.62}$$

Finally we obtain

$$\mathbf{P}(\theta,\phi) = 2\pi a^2 \sum_{n=0}^{\infty} \sum_{m=0}^{\infty} j^n [\mathbf{C}_{nm} \cos(n\phi) + \mathbf{D}_{nm} \sin(n\phi)] \frac{\sqrt{2(n + 2m + 1)}}{\beta a \sin(\theta)} J_{n+2m+1} [\beta a \sin(\theta)] \tag{2.63}$$

In practice the summations are truncated to N and M values. The Bessel functions needed can be prestored in a matrix for a set of equally spaced values of $u = \beta a \sin(\theta)$. The key problem remains in the calculation of the expansion coefficients from the integrals (2.58).

From the orthogonality of circular functions, however, it can be observed that

- $C_{nmx} = 0$ for all n different from 0 or 2;
- $C_{nmy} = 0$ always;
- $D_{nmx} = 0$ always;
- $D_{nmy} = 0$ except for $n = 2$ and $D_{2my} = C_{2mx}$.

Then only C_{0mx} and C_{2mx} need to be evaluated as

$$C_{0mx} = E_0 \int_0^1 J_0(1.8412\, s) \cdot F_m^0(s) \cdot e^{-j(\pi/\lambda L)(as)^2} \, s \, ds \tag{2.64a}$$

$$C_{2mx} = -E_0 \int_0^1 J_2(1.8412\ s) \cdot F_m^2(s) \cdot e^{-j(\pi/\lambda L)(as)^2}\ s\ ds \tag{2.64b}$$

The integrals can be obtained numerically by means of the Gauss-Legendre quadrature [22] as

$$\int_{s=a}^{b} f(s)\ ds \approx \frac{b-a}{2} \sum_{i=1}^{N} f(s_i)\ w_i \tag{2.65}$$

with

$$s_i = \frac{b-a}{2} x_i + \frac{b+a}{2}$$

and x_i, w_i are the abscissas and weights of the method.

We obtain

$$C_{nmx} = \frac{1}{2} \sum_{k=1}^{K} T_{nm}(k) \cdot e^{-jV(k)},\ (n = 0, 2) \tag{2.66}$$

where

$$T_{0m}(k) = E_0 J_0(1.8412\ s_k) \cdot F_m^0(s_k) \cdot s_k \cdot w_k$$

$$T_{2m}(k) = E_0 J_2(1.8412\ s_k) \cdot F_m^2(s_k) \cdot s_k \cdot w_k$$

$$V(k) = \pi(a\ s_k)^2/(\lambda\ L)$$

and K is the number of integration points used, w_k are the weights factors, and $s_k = (x_k + 1)/2$, with x_k the abscissas of the Gaussian quadrature. As can be seen, the elements $T_{nm}(k)$ are not dependent on the horn geometry and can be precalculated and stored for later use. The $V(k)$ are the only elements which depend on the geometry through a and L and can be easily calculated.

As a result, the evaluation of the expansion coefficients has been reduced to a straightforward summation of the precalculated T elements multiplied by the easily obtained geometrically dependent V factors. The far-field values in a set of predetermined u_i values ($\beta a \sin \theta_i$) for any ϕ cut are also straightforwardly obtained using (2.63). The values for intermediate θ values can be obtained by interpolation techniques.

The total radiated power can be obtained from the flux of the Poynting vector through the aperture as

$$P_r = \frac{E_0^2}{2\eta} \int_0^a \int_0^{2\pi} [J_0^2 (K_{11}r) + J_2^2(K_{11}r)$$

$$- 2J_0(K_{11}r) \cdot J_2(K_{11}r) \cos(2\phi)] \, r \, dr \, d\phi$$

$$= \frac{E_0^2}{2\eta} 2\pi a^2 \int_0^1 [J_0^2(1.8412 \, s) + J_2^2(1.8412 \, s)] \, s \, ds \tag{2.67}$$

As can be seen, the remaining integral in (2.67) is a constant independent of the horn geometry. It can be evaluated numerically and the resulting value is 0.47738 so $P_r = 0.47738 \, E_0^2 \, 2\pi a^2/2\eta$.

For circular polarization, we consider an aperture field consisting of two orthogonal TE_{11} modes with a 90° phase shift. The radiation field produced by the second mode is the one produced by the first in the plane $\phi + \pi/2$, phase shifted by $\pi/2$. The total power radiated will be twice that radiated by a single mode.

2.8 MULTIMODE CONICAL HORNS

High-order modes may appear at the horn aperture either being intentionally produced, as in the case of the Potter horn [23] and the trimode matched feed [24], or as a consequence of bad machining or mutual coupling effects as in cluster feeds [25]. The effects of those modes can then be either desirable or unwanted and need to be analyzed.

The electric field at the aperture can be expressed as the addition of the considered mode fields:

$$\mathbf{E}_{ap}(r',\phi') = \sum_m \sum_n (A_{mn}\mathbf{E}_{TEmn} + B_{mn}\mathbf{E}_{TMmn}) \tag{2.68}$$

where A_{mn} and B_{mn} are the complex amplitudes of the modes and $\mathbf{E}_{TEmn}$ and $\mathbf{E}_{TMmn}$ are the electric fields of the TE_{mn} and TM_{mn} modes of the cylindrical waveguide with the same radius of the horn aperture and corrected with a spherical phase variation. The Cartesian components of those fields are functions of $J_N(k_{mn}r')$, $J_N(k'_{mn}r')$, $\cos(N\phi')$, and $\sin(N\phi')$ with $N = m \pm 1$, $k_{mn} = p_{mn}/a$, $k'_{mn} = q_{mn}/a$, being p_{mn} and q_{mn} are nth root of $J_m(x) = 0$ and $J'_m(x) = 0$, respectively, and a the aperture radius. Those components can be written as

$$E_x = E_{mn}[J_{m+1}(k_{mn}r') \sin(m + 1)\phi'$$

$$+ J_{m-1}(k_{mn}r') \sin(m - 1)\phi'] \cdot e^{-j(\pi r'^2/\lambda L)} \tag{2.69}$$

$$E_y = E_{mn}[-J_{m+1}(k_{mn}r')\cos(m+1)\phi' + J_{m-1}(k_{mn}r')\cos(m-1)\phi'] \cdot e^{-j(\pi r'^2/\lambda L)} \tag{2.70}$$

for the TE_{mn} modes and

$$E_x = M_{mn}[J_{m+1}(k'_{mn}r')\cos(m+1)\phi' - J_{m-1}(k'_{mn}r')\cos(m-1)\phi'] \cdot e^{-j(\pi r'^2/\lambda L)} \tag{2.71}$$

$$E_y = M_{mn}[J_{m+1}(k'_{mn}r')\sin(m+1)\phi' + J_{m-1}(k'_{mn}r')\sin(m-1)\phi'] \cdot e^{-j(\pi r'^2/\lambda L)} \tag{2.72}$$

for the TM_{mn} modes. In these equations L is the distance from the horn apex to the aperture center and r' and ϕ' are polar coordinates in the aperture.

Using the Jacobi-Bessel series method, the aperture field is expanded in a Fourier series in ϕ' and a series of modified Jacobi polynomials in the variable $s = r'/a$ as

$$\mathbf{E}_{ap} = \sum_k \sum_i [\mathbf{C}_{ki}\cos(k\phi') + \mathbf{D}_{ki}\sin(k\phi')] \cdot F_k^i(s) \tag{2.73}$$

We can now obtain $\mathbf{P}$ as

$$\mathbf{P}(\theta,\phi) = \iint_{S_{ap}} \mathbf{E}_{ap}\, e^{j\beta\mathbf{r}\cdot\mathbf{r}'}\, dS = 2\pi a^2 \sum_k \sum_i j^k(\mathbf{C}_{ki}\cos k\phi + \mathbf{D}_{ki}\sin k\phi) \cdot \sqrt{2(k+2i+1)}\,\frac{J_{k+2i+1}(u)}{u} \tag{2.74}$$

with $u = \beta a \sin\theta$.

The $\mathbf{C}$ and $\mathbf{D}$ coefficients are obtained from the orthogonality conditions as

$$\mathbf{C}_{ki} = \sum_m \sum_n (A_{mn}\mathbf{C}_{TEki}^{mn} + B_{mn}\mathbf{C}_{TMki}^{mn}) \tag{2.75a}$$

$$\mathbf{D}_{ki} = \sum_m \sum_n (A_{mn}\mathbf{D}_{TEki}^{mn} + B_{mn}\mathbf{D}_{TMki}^{mn}) \tag{2.75b}$$

where

$$\mathbf{C}^{mn}_{\text{TE}ki} = \frac{\varepsilon_k}{2} \int_0^1 \int_0^{2\pi} \mathbf{E}_{\text{TE}mn} \cos k\phi' \, F^i_k(s) \, s \, \mathrm{d}s \, \mathrm{d}\phi' \tag{2.76a}$$

$$\mathbf{C}^{mn}_{\text{TM}ki} = \frac{\varepsilon_k}{2} \int_0^1 \int_0^{2\pi} \mathbf{E}_{\text{TM}mn} \cos k\phi' \, F^i_k(s) \, s \, \mathrm{d}s \, \mathrm{d}\phi' \tag{2.76b}$$

$$\mathbf{D}^{mn}_{\text{TE}ki} = \frac{\varepsilon_k}{2} \int_0^1 \int_0^{2\pi} \mathbf{E}_{\text{TE}mn} \sin k\phi' \, F^i_k(s) \, s \, \mathrm{d}s \, \mathrm{d}\phi' \tag{2.76c}$$

$$\mathbf{D}^{mn}_{\text{TM}ki} = \frac{\varepsilon_k}{2} \int_0^1 \int_0^{2\pi} \mathbf{E}_{\text{TM}mn} \sin k\phi' \, F^i_k(s) \, s \, \mathrm{d}s \, \mathrm{d}\phi' \tag{2.76d}$$

with ε_k the Neumann number.

The ϕ' variation of the Cartesian components of the fields reduces drastically the number of the nonzero coefficients. It can be seen that for TE_{mn} modes:

- $(C_{ki})_x = (D_{ki})_y = 0$ for every k.
- $(C_{ki})_y$ and $(D_{ki})_x$ are $\neq 0$ only for $k = m \pm 1$.
- $(C_{(m+1)i})_y = -(D_{(m+1)i})_x$.
- $(C_{(m-1)i})_y = (D_{(m-1)i})_x$.

Similarly for TM_{mn} modes:

- $(C_{ki})_y = (D_{ki})_x = 0$ for every k.
- $(C_{ki})_x$ and $(D_{ki})_y$ are $\neq 0$ only for $k = m \pm 1$.
- $(C_{(m+1)i})_x = -(D_{(m+1)i})_y$.
- $(C_{(m-1)i})_x = (D_{(m-1)i})_y$.

The resulting integrations in ϕ' are analytical. The integrals on s can be performed numerically using a Gaussian quadrature integration method. A general coefficient can be expressed in the form:

$$C_{k1} = \sum_i T_{k1}(i) \cdot \mathrm{e}^{-\mathrm{j}V(i)} \tag{2.77}$$

in which

$$T_{k1}(i) = g_{mn} J_k(h_{mn} s_i) \cdot F^1_k(s_i) \cdot s_i \cdot w_i \tag{2.78a}$$

$$V(i) = \pi(a \cdot s_i)^2 / \lambda L \tag{2.78b}$$

g_{mn} being a mode-dependent constant, $h_{mn} = p_{mn}$ or q_{mn} for TE or TM modes, and s_i and w_i the abscissas and weights of the integration method. As can be seen, $T_{k1}(i)$ depends only on the mode and not on the geometry, so they can be precalculated and stored to be used in any horn calculation. On the other hand, $V(i)$ is independent of the mode and depends only on the horn geometry. With those considerations, the calculation of the coefficients is straightforward and can be performed very quickly on a microcomputer.

2.9 CORRUGATED CONICAL HORNS

The aperture field is supposed to be the hybrid HE_{11} of the corrugated waveguide with the aperture radius corrected in phase assuming a spherical front with center in the cone apex. We use the field (given by equation (38) in [26]), which has the following Cartesian components:

$$E_{apx} = E_0 \frac{\exp[-j\beta(\sqrt{r'^2 + L^2} - L)]}{\sqrt{r'^2 + L^2}} f_s(\cos\theta')(\cos^2\phi' \cos\theta' + \sin^2\phi') \tag{2.79a}$$

$$E_{apy} = E_0 \frac{\exp[-j\beta(\sqrt{r'^2 + L^2} - L)]}{\sqrt{r'^2 + L^2}} f_s(\cos\theta')(\cos\theta' - 1)\sin\phi' \cos\phi' \tag{2.79b}$$

where $\cos\theta' = L/\sqrt{r'^2 + L^2}$, L is the distance from the cone apex to the aperture, r' and ϕ' are polar coordinates at the aperture, $\theta' \leqslant \alpha$ is the angle shown in Figure 2.8, and

$$f_s(\cos\theta') = \frac{P_s^1(\cos\theta')}{\sin\theta'} + \frac{\partial P_s^1(\cos\theta')}{\partial\theta'} \tag{2.80}$$

with $P_s^1(\cos\theta')$ the associated Legendre function of first kind, order s and degree 1, being s the first root of

$$\left[\frac{P_s^1(\cos\theta')}{\sin\theta'} + \frac{\partial P_s^1(\cos\theta')}{\partial\theta'}\right]_{\theta'=\alpha} = 0 \tag{2.81}$$

Applying the Jacobi-Bessel series method, the aperture field is expanded in a Fourier series in ϕ' and in a series of modified Jacobi polynomials in r'. Then the integral $\mathbf{P}(\theta,\phi)$ can be obtained as

$$\mathbf{P}(\theta,\phi) = 2\pi a^2 \sum_{n=0}^{\infty} \sum_{m=0}^{\infty} \mathrm{j}^n \, [\mathbf{C}_{nm} \cos n\phi$$

$$+ \mathbf{D}_{nm} \sin n\phi] \sqrt{2(n + 2m + 1)} \, \frac{J_{n+2m+1}(\beta a \sin \theta)}{\beta a \sin \theta} \qquad (2.82)$$

The **C** and **D** coefficients are obtained applying the orthogonality conditions. From the ϕ' variation of the aperture field it can be seen that

- $C_{nmx} = 0$ for every n;
- $C_{nmy} = 0$ for $n \neq 0$ or $n \neq 2$;
- $D_{nmy} = 0$ for every n;
- $D_{nmx} = 0$ for $n \neq 2$;
- $D_{2mx} = -C_{2my}$.

The ϕ' integrals are obtained analytically and the remaining r' integrals can be obtained numerically using the Gauss-Legendre quadrature.

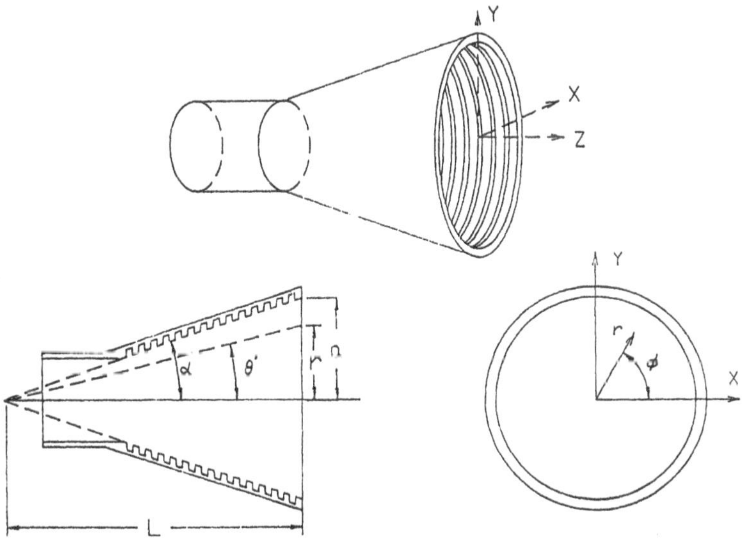

Figure 2.8 **Geometry of a Corrugated Conical Horn**

In this case the expansion coefficients cannot be separated into factors depending only on the geometry and factors independent of the geometry which can be precalculated. In these circumstances the use of the Zernike-Bessel series method instead of the Jacobi-Bessel method [30] is advantageous. In this method the subintegral function is expanded in a double series of complex exponentials in the azimuth ϕ' coordinate and in circular Zernike polynomials in the radial r' coordinate.

Then if we have an integral of the type:

$$\mathbf{P}(\theta,\phi) = \int_0^1 \int_0^{2\pi} \mathbf{g}(s,\phi')\, e^{jU(\theta)\cdot s\cdot\cos(\phi-\phi')}\, s \cdot ds \cdot d\phi' \tag{2.83}$$

we expand

$$\mathbf{g}(s,\phi') = \sum_m \sum_n \mathbf{B}_m^n \cdot R_m^n(s) \cdot e^{jn\phi'} \tag{2.84}$$

where B_m^n are the expansion coefficients and $R_m^n(s)$ are the Zernike polynomials defined as

$$R_m^n(s) = \sum_{k=0}^{(m-n)/2} \frac{(-1)^k \cdot (m-k)!}{k! \cdot \left[\frac{m+n}{2} - k\right]! \cdot \left[\frac{m-n}{2} - k\right]!} \cdot s^{m-2k} \tag{2.85}$$

with

$$m = 0, 1, \ldots, \infty$$

$$n = \begin{cases} \pm 1, \pm 3, \pm 5, \ldots, \pm m \text{ for } m \text{ odd} \\ 0, \pm 2, \pm 4, \ldots, \pm n \text{ for } n \text{ even} \end{cases}$$

Substituting (2.84) in (2.83) we obtain

$$\mathbf{P}(\theta,\phi) = \sum_m 2(m+1) \cdot j^m \cdot \frac{J_{m+1}(U(\theta))}{U(\theta)} \cdot \sum_n \mathbf{b}_m^n\, e^{jn\phi} \tag{2.86}$$

with

$$\mathbf{b}_m^n = \int_0^1 \int_0^{2\pi} \mathbf{g}(s,\phi') \cdot R_m^n(s) \cdot e^{-jn\phi'} \cdot s \cdot ds \cdot d\phi' \tag{2.87}$$

In the case of a pure hybrid HE_{11} mode in a conical corrugated horn, we use

$$\mathbf{g}(s,\phi') = E_{apx}\mathbf{x} + E_{apy}\mathbf{y} \tag{2.88}$$

with E_{apx} and E_{apy} as given in (2.79a) and (2.79b) using the variable change $r' = a \cdot s$ (a is the aperture radius).

The b^n_{mx} coefficients will be

$$\begin{aligned} b^n_{mx} &= \int_0^1 \int_0^{2\pi} E_{apx} R^n_m(s)\, e^{-jn\phi'} \cdot s \cdot ds \cdot d\phi' \\ &= \int_0^1 \int_0^{2\pi} E_0 \cdot \frac{\exp[-j\beta(\sqrt{r'^2 + L^2} - L)]}{\sqrt{r'^2 + L^2}} \\ &\quad \cdot f_s(\cos\theta') \cdot (\cos^2\phi' \cdot \cos\theta' + \sin^2\phi') \\ &\quad \cdot R^n_m(s) \cdot e^{-jn\phi'} \cdot s \cdot ds \cdot d\phi' \end{aligned} \tag{2.89}$$

which can be written as

$$\begin{aligned} b^n_{mx} &= \int_0^1 E_0 \frac{e(-j\beta(\sqrt{r'^2 + L^2} - L))}{\sqrt{r'^2 + L^2}} \\ &\quad \cdot f_s(\cos\theta') \cdot R^n_m(s) \cdot I_1 \cdot s \cdot ds \end{aligned} \tag{2.90}$$

where I_1 is the integral on ϕ'

$$I_1 = \int_0^{2\pi} (\cos\phi' \cos\theta' + \sin\phi') \cdot e^{-jn\phi'} \cdot d\phi' \tag{2.91}$$

As

$$\cos^2\phi' = \frac{1 + \cos 2\phi'}{2}$$

and

$$\sin^2\phi' = \frac{1 - \cos 2\phi'}{2}$$

I_1 can be written as

$$I_1 = \int_0^{2\pi} \left[\frac{1}{2}(\cos\theta' + 1) + \frac{1}{2}(\cos\theta' - 1)\cdot\cos 2\phi'\right]$$

$$\cdot (\cos n\phi' - \mathrm{j}\sin n\phi')\cdot \mathrm{d}\phi'$$

$$= \int_0^{2\pi} \left[\frac{1}{2}(\cos\phi' + 1) + \frac{1}{2}(\cos\phi' - 1)\cdot\cos 2\phi'\right]$$

$$\cdot \cos n\phi' \cdot \mathrm{d}\phi' - \mathrm{j}\int_0^{2\pi}\left[\frac{1}{2}(\cos\theta' + 1\right.$$

$$\left. + \frac{1}{2}(\cos\theta' - 1)\cdot\cos 2\phi'\right]\cdot \sin n\phi' \cdot \mathrm{d}\phi' \tag{2.92}$$

The second term of I_1 is zero for all n so we have

$$I_1 = \frac{1}{2}(\cos\theta' + 1)\int_0^{2\pi} \cos n\phi' \cdot \mathrm{d}\phi'$$

$$+ \frac{1}{2}(\cos\theta' - 1)\int_0^{2\pi} \cos 2\phi' \cdot \cos n\phi' \cdot \mathrm{d}\phi' \tag{2.93}$$

The first term of the last equation is zero for all $n \neq 0$. For $n = 0$ its value is 2π. The second term is zero for $n \neq \pm 2$. The integral value is π for $n = \pm 2$. The only nonzero coefficients are then b^0_{mx}, b^2_{mx} and b^{-2}_{mx} and are given by

$$b^0_{mx} = \pi E_0 \int_0^1 \frac{\exp[-\mathrm{j}\beta(\sqrt{r'^2 + L^2} - L)]}{\sqrt{r'^2 + L^2}}$$

$$\cdot f_s(\cos\theta')\cdot R^0_m(s)\cdot(1 + \cos\theta')\cdot s\cdot \mathrm{d}s \tag{2.94}$$

$$b^2_{my} = b^{-2}_{my} = \frac{\pi E_0}{2}\int_0^1 \frac{\exp[-\mathrm{j}\beta(\sqrt{r'^2 + L^2} - L)]}{\sqrt{r'^2 + L^2}}$$

$$\cdot f_s(\cos\theta')\cdot(\cos\theta' - 1)\cdot R^2_m(s)\cdot s\cdot \mathrm{d}s \tag{2.95}$$

The b^n_{my} coefficients will be

$$b_{my}^{n} = \int_0^1 \int_0^{2\pi} E_0 \frac{\exp[-j\beta(\sqrt{r'^2 + L^2} - L)]}{\sqrt{r'^2 + L^2}} \cdot f_s(\cos\theta') \cdot (\cos\theta' - 1)$$

$$\cdot \sin\phi' \cdot \cos\phi' \cdot R_m^n(s) \cdot e^{-jn\phi'} \cdot s \cdot ds \cdot d\phi' \qquad (2.96)$$

which can be written as

$$b_{my}^{n} = \int_0^1 E_0 \frac{\exp[-j\beta(\sqrt{r'^2 + L^2} - L)]}{\sqrt{r'^2 + L^2}}$$

$$\cdot f_s(\cos\theta') \cdot (\cos\theta' - 1) \cdot R_m^n(s) \cdot I_2 \cdot s \cdot ds \qquad (2.97)$$

where I_2 is the ϕ' integral:

$$I_2 = \int_0^{2\pi} \sin\phi' \cdot \cos\phi' \cdot (\cos n\phi' - j\sin n\phi') \cdot d\phi'$$

$$= \int_0^{2\pi} \frac{\sin 2\phi'}{2} \cdot \cos n\phi'$$

$$\cdot d\phi' - j\int_0^{2\pi} \frac{\sin 2\phi'}{2} \cdot \sin n\phi' \cdot d\phi' \qquad (2.98)$$

The first term is always zero and the second term is zero for $n \neq \pm 2$. The integral value is then $-j\pi/2$ for $n = 2$ and $j\pi/2$ for $n = -2$. Then the only nonzero b_{my}^{n} coefficients are b_{my}^{2} and b_{my}^{-2} and are given by

$$b_{my}^{2} = -j\frac{\pi E_0}{2}\int_0^1 \int_0^{2\pi} \frac{\exp[-j\beta(\sqrt{r'^2 + L^2} - L)]}{\sqrt{r'^2 + L^2}}$$

$$\cdot f_s(\cos\theta') \cdot (\cos\theta' - 1) \cdot R_m^2(s) \cdot s \cdot ds \qquad (2.99)$$

$$b_{my}^{-2} = -b_{my}^{2} \qquad (2.100)$$

As can be seen b_{my}^{2} and b_{my}^{-2} only differ from b_{mx}^{2} by a constant. It is only necessary to calculate b_{mx}^{0} and b^2m_x and the rest of the nonzero coefficients will be:

$$b_{my}^{2} = jb_{mx}^{2} \qquad (2.101a)$$

$$b_{my}^{-2} = -jb_{mx}^{2} \qquad (2.101b)$$

$$b_{mx}^{-2} = b_{mx}^{2} \tag{2.101c}$$

The total radiated power is obtained from the flux of the Poynting vector through the horn aperture as

$$P_r = \frac{1}{2\eta} \iint_{S_{ap}} |\mathbf{E}_{ap}|^2 \, dS' = \frac{E_0^2}{2\eta} \int_0^a \int_0^{2\pi} \left[\frac{F_s^1(\cos\theta')}{\sqrt{r'^2 + L^2}} \right]^2$$

$$\cdot \left[(\cos^2\phi' \cos\theta' + \sin^2\phi')^2 + (\cos\theta' - 1)^2 \sin^2\phi' \cdot \cos^2\phi' \right]$$

$$\cdot \, r' \, dr' \, d\phi'$$

$$= \frac{E_0^2}{2\eta} \pi \int_0^a \left[\frac{F_s^1(\cos\theta')}{\sqrt{r'^2 + L^2}} \right]^2 (\cos^2\theta + 1) \, r' \, dr' \tag{2.102}$$

which must be evaluated numerically.

2.10 HORN ANALYSIS PROGRAMMING

The analysis of the horn antennas described in the previous paragraphs has been implemented in FORTRAN in the program HORN.

The main program simply asks the operator for the type of horn to be analyzed: multimode smooth-walled conical horn (including the TE_{11} fundamental mode), corrugated conical horn, smooth-walled pyramidal horn (including sectoral horns), and corrugated pyramidal horn. Then it gives control to the corresponding subroutine that performs the computations: SWPH for the analysis of smooth-walled pyramidal horns, CPH for corrugated pyramidal horns, MSWCH for multimode (up to five modes) smooth-walled conical horns, and CCH for conical corrugated horns.

Each of those subroutines can be easily transformed into an independent program if desired.

2.10.1 Subroutine SWPH

This subroutine performs the analysis of smooth-walled pyramidal horns, including sectoral horns. The subroutine asks for the input and output data file names, reads wavelength, and gives control to subroutines ESPH and RADSPH.

Subroutine ESPH is in charge of reading the geometrical and polarization data of the horn from the input data file. It also performs the calculation of the total power radiated by the horn.

Subroutine RADSPH reads the data that control the pattern calculations and the output for drawing those patterns. Then it calculates the horn directivity and the constant ϕ radiation patterns specified in the input data file. The obtained results are written to the output file specified and to the drawing files. The far-field E_θ and E_ϕ components are obtained from subroutine RFSPH. A first call to RFSPH for $\theta = 0$, $\phi = 0$ allows the calculation of the radiation intensity at broadside and the horn directivity. The patterns are obtained from successive calls to RFSPH. E_θ and E_ϕ are converted to co-polar and cross-polar components in RADSPH.

Subroutine RFSPH calculates E_θ and E_ϕ for any given direction θ, ϕ. The expressions obtained in previous paragraphs are used and the Fresnel sine and cosine functions are obtained from subroutine SCFRES.

Subroutine SCFRES provides the sine and cosine Fresnel functions for any x argument. The method used has been suggested in Abramowitz and Stegun [29].

2.10.2 Subroutine CPH

This subroutine performs the analysis of corrugated pyramidal horns. It asks for the input and output data file names, reads wavelength, and gives control to subroutines ECPH and RADCPH.

Subroutine ECPH reads the geometrical and polarization data of the horn from the input data file. It also calculates some constants for later use including the total power radiated by the horn.

Subroutine RADCPH controls the ϕ constant patterns calculations. It reads the pattern data as well as the drawing information from the input file, then calculates the directivity from the E_θ and E_ϕ far field components in the direction $\theta = 0$, $\phi = 0$. The far field components are obtained by the subroutine RFCPH. Successive calls to RFCPH allow the patterns calculation and their output to the output file and drawing files if any.

The subroutine RFCPH obtains the E_θ and E_ϕ far field components using the equations shown in the theory. The cosine and sine Fresnel functions are calculated by the SCFRES subroutine.

2.10.3 Subroutine CCH

This subroutine performs the analysis of conical corrugated horns with a pure HE_{11} hybrid-mode excitation. The subroutine asks for the input and output data file names, reads the wavelength, and gives control to subroutines ECCH and RADCCH.

Subroutine ECCH initializes the horn calculations by obtaining the coefficients of the Zernike-Bessel expansion of the aperture field. To perform this it calls to subroutines CALCS and CALCFS. The required values of the Zernike polynomials have been precalculated by a different program and are obtained here from the BLOCK DATA subroutine. ECCH also obtains the total power radiated by the horn.

Subroutine CALCS obtains the value of the order, s, of the associated Legendre function in the aperture field expression.

Subroutine CALCFS obtains the values of the function f_s (cos θ').

Subroutine RADCCH obtains the directivity and radiation patterns of the horn by means of successive calls to RFCCH. As the field values are obtained in a set of predetermined U values the co-polar and cross-polar components for the desired θ values are obtained by interpolation using subroutine INTERP.

Subroutine RFCCH obtains the E_θ and E_ϕ far-field components. The Bessel functions needed for the summation of the far-field calculation have been prestored for a set of U values. Then the far field can be obtained for any ϕ but only for the U values of the selected set. This is not a severe limitation because the full pattern cut can be recovered by interpolation in the RADCCH subroutine.

2.10.4 Subroutine MSCH

This subroutine performs the analysis of smooth-walled conical horns with a multimode aperture field. After some data initialization, the subroutine reads the input and output data files from the console. From the input data file, it then reads the wavelength, the aperture radius, and the horn semiflare angle.

The first five lower-order modes can be combined at the aperture. The corresponding mode existence is controlled by the logical variables TE_{01}, TE_{11}, TE_{21}, TM_{01}, and TM_{11} from the input data file. Relative amplitude and phases of the modes must also be provided in the input file.

Linear or circular polarization can be obtained in a set of NFI constant ϕ plane cuts (from an initial FI0 value in steps DFI) at NTH θ values, starting at TH0 with DTH increments.

The subroutine obtains the Jacobi-Bessel coefficients from subroutines CTE01, CTE11, CTE21, CTM01, and CTM11. The far field is calculated in a set of $k \cdot a \cdot \sin\theta$ values from which the values at the desired θ values are interpolated on each ϕ plane cut.

2.11 NEAR FIELD OF CONICAL CORRUGATED HORNS

Although the Jacobi-Bessel series method can also be used to obtain the near field from apertures [27], we have found that for horns the series convergence is very slow and the method is not suitable. On the other hand, a direct numerical integration method can be used for the potential integral with reasonable response time on a microcomputer.

We use the geometry shown in Figure 2.8 in which the horn semiflare angle is α, the aperture radius is a, and L is the distance from the horn apex to the aperture plane, being $\tan(\alpha) = a/L$.

The electric aperture field on the aperture plane S_{ap} is supposed to be known as

$$\mathbf{E}_{ap} = E_{apx}(r',\phi')\mathbf{x} + E_{apy}(r',\phi')\mathbf{y} \tag{2.103}$$

where E_{apx} and E_{apy} are complex values which include the spherical phase front assumed for the field at the aperture.

Using images, the equivalent source for $z > 0$ will be the equivalent surface magnetic current $\mathbf{M}_{ap} = 2\mathbf{E}_{ap} \times \mathbf{z}$. The electric potential at any point P with a vector position $\mathbf{r}$ will be

$$\mathbf{F}(\mathbf{r}) = \frac{\varepsilon}{4\pi}\int_{S_{ap}} \frac{\mathbf{M}_{ap}e^{-j\beta|\mathbf{r}-\mathbf{r}'|}}{|\mathbf{r}-\mathbf{r}'|}\,dS' = \frac{\varepsilon}{4\pi}\mathbf{P} \tag{2.104}$$

And the electric field at point P will be

$$\mathbf{E}(\mathbf{r}) = \eta\mathbf{H}(\mathbf{r}) \times \mathbf{r} = j\omega\eta(\mathbf{r} \times \mathbf{F}) = \frac{j\beta}{4\pi}(\mathbf{r} \times \mathbf{P}) \tag{2.105}$$

In the case of a corrugated conical horn with a pure hybrid HE_{11} mode at the aperture the electric aperture field $\mathbf{E}_{ap}$ is given by Equations (2.79a) and (2.79b).

Usually, the near fields are expressed taking the origin of coordinates in the horn apex instead of in the aperture center. In this case:

$$\mathbf{r} = r(\sin\theta\cos\phi\mathbf{x} + \sin\theta\sin\phi\mathbf{y} + \cos\theta\mathbf{z})$$

$$\mathbf{r}' = r'(\cos\phi'\mathbf{x} + \sin\phi'\mathbf{y}) + L\mathbf{z}$$

$$|\mathbf{r} - \mathbf{r}'| = [(r\sin\theta\cos\phi - r'\cos\phi')^2 + (r\sin\theta\sin\phi - r'\sin\phi')^2 + (r\cos\theta - L)^2]^{1/2}$$

and the integrals in (2.104) will be

$$P_x = \int_0^a \int_0^{2\pi} \frac{2E_{apy}e^{-j\beta|\mathbf{r}-\mathbf{r}'|}}{|\mathbf{r}-\mathbf{r}'|}\,r'\,dr'\,d\phi' \tag{2.106}$$

$$P_y = \int_0^a \int_0^{2\pi} \frac{-2E_{apx}e^{-j\beta|\mathbf{r}-\mathbf{r}'|}}{|\mathbf{r} - \mathbf{r}'|} r' \, dr' \, d\phi' \tag{2.107}$$

The evaluation of the integrals can be done using the Gauss-Legendre quadrature for two-dimensional integrals as

$$\int_{a_x}^{b_x} \int_{a_y}^{b_y} f(x,y) \, dx \, dy = \frac{(b_x - a_x)}{2} \frac{(b_y - a_y)}{2} \sum_{i=1}^{n} \sum_{j=1}^{n} f(x_i, y_j) \, w_i W_j \tag{2.108}$$

where n is the number of integration points, x_i and y_j are

$$x_i = \frac{b_x - a_x}{2} z_i + \frac{b_x + a_x}{2} \tag{2.109a}$$

$$y_j = \frac{b_y - a_y}{2} z_j + \frac{b_y - a_y}{2} \tag{2.109b}$$

being z_i the ith zero of the Legendre polynomial $P_n(z)$ and the weights w_i are

$$w_i = \frac{2}{(1 - z_i^2)} [P_n'(z_i)]^2$$

which can be found in tables [22].

Our integrals have the following general form:

$$\int_0^a \int_0^{2\pi} f(r',\phi') \frac{\exp[-j\beta g(r,\theta,\phi,r',\phi')]}{g(r,\theta,\phi,r',\phi')} r' \, dr' \, d\phi'$$

$$\approx \frac{2\pi a}{4} \sum_{i=1}^{n} \sum_{j=1}^{n} f(r_i',\phi_j') \frac{\exp[-j\beta g(r,\theta,\phi,r_i',\phi_j')]}{g(r,\theta,\phi,r_i',\phi_j')} r_i' \, w_i \, w_j \tag{2.110}$$

in which the factors $f(r_i',\phi_j') \cdot r_i' \cdot w_i \cdot w_j$ are independent of the P coordinates (r, θ, ϕ) and can be precalculated and stored for the n^2 combinations r_i', ϕ_j'. For each point P in which we desire the field it will be necessary to evaluate the function $g(r,\theta,\phi,r_i',\phi_j')$ for the n^2 integration points.

The total radiated power is obtained from the flux of the Poynting vector through the horn aperture as was done for far-field calculations.

2.11.1 Program NEARF.FOR

This method of computing the near field from apertures has been implemented for a conical corrugated horn in a program named NEARF.FOR. The main program initializes constants, reads the names of the input and output data files from the console, and reads the wavelength from the input data file. It then gives control to subroutines EBOC and DIRAD described below.

The EBOC subroutine initializes abscissa and weight data for a 32-point Gauss-Legendre integration method. It then reads the horn geometry and the polarization from the input file. The aperture field is then obtained at the integration points and the total radiated power is calculated as the flux of the Poynting vector over the aperture.

The DIRAD subroutine reads the data for drive pattern calculations and for the generation of drawing files from the input file. It controls the pattern calculations and the outputting of the obtained results. The required field values are obtained from subroutine CAMRAD. This subroutine obtains the E_θ and E_ϕ field components at any given distance R in any direction (θ,ϕ).

2.12 PHASE-CENTER CALCULATION

In general, the fields obtained have been referred to a coordinate system centered at the horn aperture or located in the horn apex. The phase of the obtained fields is not constant with θ and ϕ, and this produces phase error losses in reflectors and lenses if the horn aperture center is positioned at the system focus. A point must be found such that when the fields are referred to it the new phase is nearly constant or has the minimum possible variation. This point is named the horn-phase center (usually referred to a plane wave) and the system focus should be positioned in this point to reduce phase-error losses.

The ϕ = constant theoretical radiation pattern cuts of horns are symmetrical with respect to the horn axis and the phase center, for each cut, will be located on the axis. In general, each pattern not only has a different phase center, but the phase center of each cut will also depend on the minimization criterion used and on the angular extent of the pattern considered.

Following Rusch and Potter [28] the far field can be expressed in general as

$$\mathbf{E}(\theta,\phi) = \mathbf{e}\,\frac{e^{-j\beta R}}{R}\,|\mathbf{E}(\theta,\phi)|\,e^{j\Phi(\theta,\phi)} \tag{2.111}$$

where **e** is a unit polarization vector, $|\mathbf{E}(\theta,\phi)|$ is the field amplitude, and $\Phi(\theta,\phi)$ is the phase variation over a sphere of radius R centered in the origin O as shown in Figure 2.9.

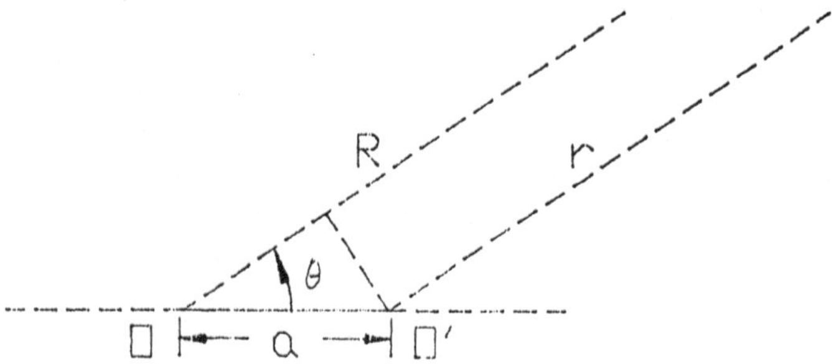

Figure 2.9 Relevant Geometry for the Phase-Center Calculation of a Far-Field, Plane-Cut Pattern

Ideally the phase center will be the point O' such that when the field is expressed from O' it has a constant phase:

$$\mathbf{E}(\theta,\phi) = \mathbf{e}\,\frac{e^{-j\beta r}}{r}\,|\mathbf{E}(\theta,\phi)|\,e^{jC} \tag{2.112}$$

In reality the phase variation will not be constant and the field variation will be

$$\mathbf{E}(\theta,\phi) = \mathbf{e}\,\frac{e^{-j\beta r}}{r}\,|\mathbf{E}(\theta,\phi)|\,e^{j(C+\Delta(\theta,\phi))} \tag{2.113}$$

where $\Delta(\theta,\phi) = -\beta R + \Phi(\theta,\phi) + \beta r - C = \Phi(\theta,\phi) - \beta a \cos\theta - C$, and a is the distance OO' as shown in Figure 2.9. In practice, we will know Φ in a set of N points corresponding to θ_i values and the problem is how to determine a and C such that Δ_i will be minimal in a least squares sense. This is equivalent to minimizing the sum of the squares of Δ_i, in general weighted with weights w_i if some control over different parts of the pattern is desired. Usually the weighting factors are selected proportional to the field amplitude on each θ_i but different criteria can be used. So we will minimize

$$\sigma^2 = \sum_{i=1}^{N} w_i \Delta_i^2 = \sum_{i=1}^{N} w_i[\Phi_i - \beta a \cos\theta_i - C]^2 \tag{2.114}$$

where $\Phi_i = \Phi(\theta_i, \phi)$. The solution is obtained by solving the system of equations:

$$\frac{\partial \sigma^2}{\partial C} = 0 \tag{2.115a}$$

$$\frac{\partial \sigma^2}{\partial a} = 0 \tag{2.115b}$$

as

$$a = \frac{1}{\beta} \frac{\Sigma\, w_i \cdot \Sigma\, w_i \Phi_i \cos\theta_i - \Sigma\, w_i \cos\theta_i \cdot \Sigma\, w_i \Phi_i}{\Sigma\, w_i \cdot \Sigma\, w_i (\cos^2\theta_i) - [\Sigma\, w_i \cos\theta_i]^2} \tag{2.116}$$

From the obtained value of a, the phase of the far field referred to the phase center O' will be

$$C_i = \Phi_i - \beta a \cos\theta_i \tag{2.117}$$

In the near-field region we will follow the procedure suggested in [4]. Given two points P_1 and P_2 at angles θ_1 and θ_2 and distance R from 0, as in Figure 2.10, we define $X = \Delta\Phi/\beta R$, with $\Delta\Phi$ the phase difference of the field at both points, i.e., $\Delta\Phi = \Phi_1 - \Phi_2$. The distance a from O to the phase center O' will be:

$$a = \frac{R\{-\tfrac{1}{2}(\cos\theta_2 + \cos\theta_1) \pm [\cos\theta_2 \cos\theta_1 - 1 + X^2/4 + (\cos\theta_2 - \cos\theta_1)^2/X^2]\}}{[(\cos\theta_2 - \cos\theta_1)^2/X^2 - 1]} \tag{2.118}$$

where the positive sign is for $\Delta\Phi$ positive and the negative sign for $\Delta\Phi$ negative.

Again we will obtain a different value of a for each two points considered. To obtain an average phase-center position from a set of measured or calculated field values one may use one of the following criteria:

1. Get the average value of the phase-center positions obtained considering a reference field value, for instance at broadside, and each of the other values.
2. The pairs of points are taken as consecutive values of the set.
3. All possible combinations of field values are considered.

2.12.1 Programming PHACEN.FOR

PHACEN.FOR is the program designed to obtain the phase center of a given plane cut pattern or refer that pattern to the obtained phase center. The program has the following options:

A. FAR FIELD
B. TRANSFORMATION OF FAR-FIELD FILES
C. NEAR FIELD (Method 1)
D. NEAR FIELD (Method 2)
E. NEAR FIELD (Method 3)

Option A permits calculation of the phase center of a far-field radiation pattern plane cut. Subroutine CFLEJ implements Equations 2.117 and 2.116 to obtain the phase center.

Option B gives control to subroutine TR which transforms the field values referring them to a new origin displaced along the z-axis.

Options C, D, and E give control to subroutine CFCER(I), $I = 1, 2,$ or 3, to obtain the near-field phase center estimation in accordance with the three criteria previously described.

2.13 SMOOTH-WALLED PYRAMIDAL HORN DESIGN

Horns can be used not only as feeds for reflectors and lenses but also as gain standards, as probes in field measurement setups, and as medium-gain antennas. Here we describe a design procedure for horns as feeds.

The main objective of the design is to determine the geometrical parameters of the horn that will produce the desired field levels (relative to the maximum pattern level) at given angular values (usually corresponding to the edges of the reflector or lens) in the two main planes: the one containing the electric field in the aperture (called E-plane) and the one orthogonal to this (called H-plane). Usually the electric field in the aperture is directed along the $\mathbf{y}$ axis, so the H-plane corresponds to the $\phi = 0$ plane cut and the E-plane to the $\phi = 90°$ plane cut, and the design objective is to produce a field value ET at angles θ_1 in the H-plane and θ_2 in the E-plane.

The design is based on an interpolation of the results obtained from many analyzed horns. Those results are better stored in the form of the so-called "universal E and H plane patterns" [31] as a function of the argument:

$$u = (W/\lambda) \cdot \sin\theta \tag{2.119}$$

(any other dependencies with θ,ϕ form what is called the *obliquity factor* $\mathrm{OF} = (1 + \cos\theta)/2$ not included in the universal patterns). In (2.119) W is A or B for the H-plane and E-plane respectively. The phase error s is taken as parameter in the curves. This phase error is defined as the extra distance along the horn sides with respect to the center of the aperture in wavelengths (the difference between semiflare length and axial length), as depicted in Figure 2.11. It will then be:

$$s = D/\lambda = \frac{W/2}{\lambda} \cdot \tan\alpha/2 \tag{2.120}$$

$$D = L - R = L - \sqrt{L^2 - (W/2)^2} = L\left[1 - \left(1 - \frac{(W/2)^2}{R^2}\right)^{1/2}\right]$$

$$= L\,[1 - (1 - \sin^2\alpha)^{1/2}] = (W/2)\,\frac{1 - \cos\alpha}{\sin\alpha} = (W/2) \cdot \tan\alpha/2 \tag{2.121}$$

Figures 2.12 and 2.13 show those universal curves for s values of 0.001, 0.025, 0.05, 0.075, 0.1, 0.125, 0.15, 0.175, 0.2, 0.3, 0.4, 0.5, 0.6, 0.7, 0.8, 0.9, and 1.0, and for the E and H plane respectively. The curves have been drawn with 41 points in the interval $0 \leq u \leq 4.0$ which are the values stored in the BLOCK DATA subroutine of the design program DSPH.

For the desired edge-field value ET each s = constant pattern will be cut at a certain u value as indicated in Figure 2.14. Those points form a constant edge-taper curve in the u,s plane as does the one shown in Figure 2.15. We can obtain two curves, one for the E-plane and the other for the H-plane. Each point of the curves represents a set of geometrical dimensions (W,α) in the corresponding E or H plane that can be obtained from (u,s) by using Equations (2.119) and (2.120).

Due to the suppression of the obliquity factor, the universal curves are broader than the real ones. To compensate for this effect we do the design for a broader angle θ', instead of the specified θ. This new angle is given by

$$\theta' = \frac{\sqrt{\text{ET(dB)} + \text{OF(dB)}}}{\sqrt{\text{ET(dB)}}}\,\theta \qquad (2.122)$$

as suggested by Kelleher [32].

Not all of the combinations of E and H plane dimensions are possible simultaneously, however, due to an additional restriction: the realizability condition. This condition is a consequence of the fact that the distance from the aperture center to the intersection plane with the input waveguide must be equal in both planes.

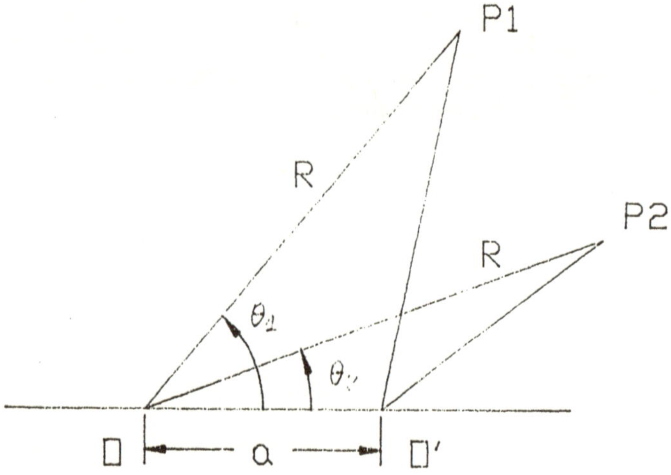

Figure 2.10 Relevant Geometry for the Phase-Center Calculation of a Near-Field, Plane-Cut Pattern

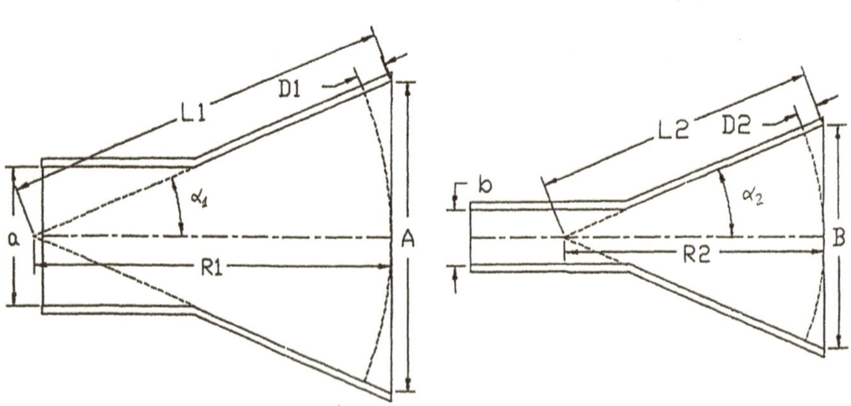

Figure 2.11 Phase-Error Definitions in Pyramidal Horns

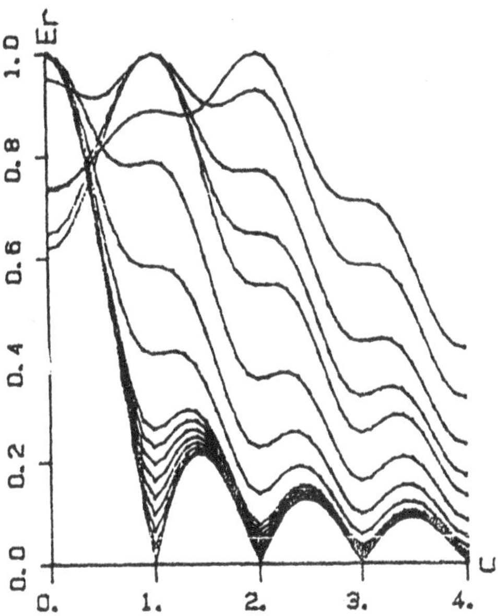

Figure 2.12 *E*-Plane Universal Patterns of a Pyramidal Horn for the *s* Values of Text

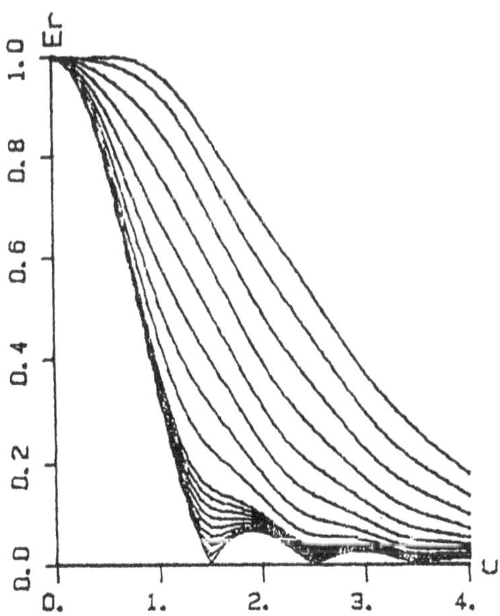

Figure 2.13 *H*-Plane Universal Patterns of a Pyramidal Horn for the *s* Values of Text

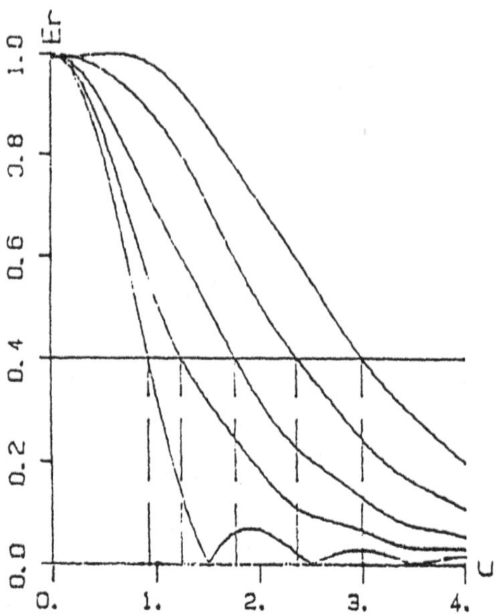

Figure 2.14 Cut Points of the s = Constant Universal Patterns at a Constant *ET* Field Value

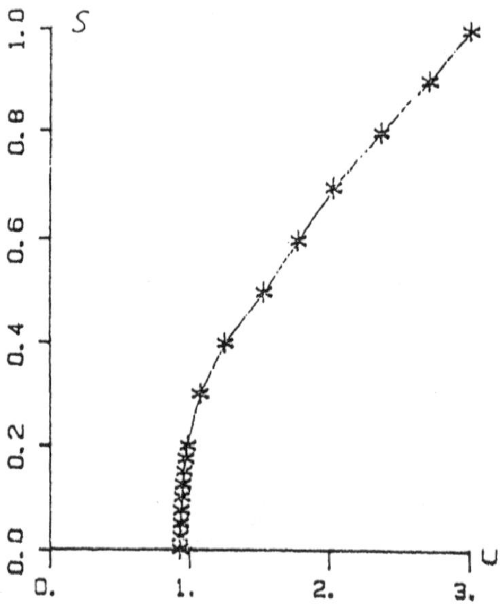

Figure 2.15 Constant-Edge Taper Curve in the *s*-*u* Plane

The realizability condition can be expressed as an additional relation between u_2 and s_2 of the E-plane or between u_1 and s_1 of the H-plane. It implies, as can be easily seen, that

$$\frac{A - a}{\tan(\alpha_1)} = \frac{B - b}{\tan(\alpha_2)} = C \tag{2.123}$$

where C is a constant and a and b are the input waveguide dimensions (known *a priori* in the design). Taking into account the trigonometric relation:

$$\tan \alpha = \frac{2 \tan \alpha/2}{1 - \tan^2 \alpha/2} \tag{2.124}$$

and using Equations (2.123), (2.120), and (2.119) we obtain the relation:

$$s_i^2 + \frac{2\,(u_i/\sin\theta_i)}{C[(2\lambda u_i/\sin\theta_i) - w]}\, s_i - (u_i/\sin\theta_i)^2 = 0 \tag{2.125}$$

where $i = 1$ for the H-plane and $i = 2$ for the E-plane and w is a or b, respectively.

We proceed then as follows. We select a point u_i, s_i in one of the planes (E or H) and get the corresponding dimensions by using (2.119) and (2.120). Then we are able to use (2.125) for the other plane, and we draw this parabolic type of relation in the u,s plane. We will or will not have a realizable horn if this curve does or does not intersect with the constant edge taper curve. From the intersection point values (u,s) we can derive the geometrical dimensions from Equations (2.119) and (2.120). In practice it is easier to obtain realizable horns if we start with the design of the narrower beam plane which we can call the "principal plane."

To obtain the intersection of the realizability condition curve and the constant edge taper level curve a simple search procedure can be used. We start in a point P_0 in the constant edge taper curve corresponding to a low s_0 value. We obtain the corresponding u_0 and, with Equation (2.125), the point P_0' (that has the same u_0) in the realizability condition curve. This last point has an $s = s_1$ value which corresponds to point P_1 in the constant edge taper curve as shown in Figure 2.16. We can repeat the procedure starting with P_1 to obtain a new point P_2, and so on. The procedure is stopped when the newly obtained point is close enough to the previous point.

For intermediate values of the phase error s the field value is obtained by means of an interpolation subroutine. In our case we have

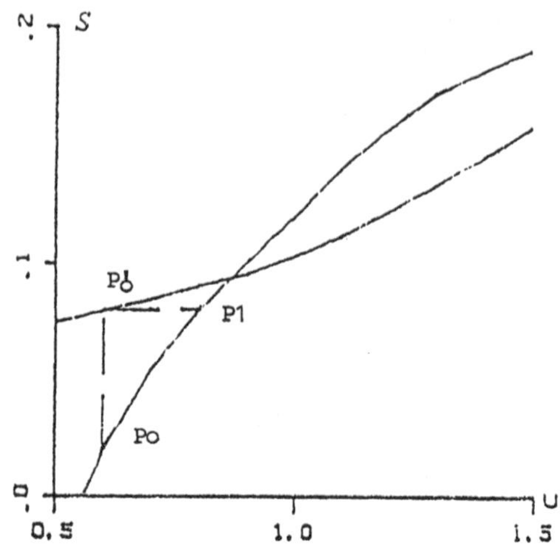

Figure 2.16 Intersection of the Edge-Taper Curve with the Realizability Condition Curve

successfully used the three-point Lagrange method. Given three values of the function Y corresponding to three values of the abscissa X, (X_0, Y_0), (X_1, Y_1) and (X_2, Y_2), the value Y corresponding to X is

$$Y = Y_0 \frac{(X - X_1)(X - X_2)}{(X_0 - X_1)(X_0 - X_2)} + Y_1 \frac{(X - X_0)(X - X_2)}{(X_1 - X_0)(X_1 - X_2)} + Y_2 \frac{(X - X_0)(X - X_1)}{(X_2 - X_0)(X_2 - X_1)} \tag{2.126}$$

Given a set of Y_i values of a function Y for the set X_i of abscissas, if we desire the value of Y corresponding to $X_i < X < X_{i+1}$ we interpolate with X_{i-1}, X_i and X_{i+1} and with X_i, X_{i+1} and X_{i+2} and use the average of both values. If X is in one of the extreme intervals of the set of values we use only the nearest three points.

The program DSPH that implements this design procedure provides the dimensions of ten horns with phase errors in the principal plane ranging from 0.1, 0.2, . . . , 1.0. Designs for intermediate s values can be also obtained.

When the horn dimensions have been obtained it is convenient to get an estimation of the horn parameters such as gain and phase centers on E and H planes.

Following Milligan [33] the directivity can be obtained from

$$D = 10 \log(4\pi AB/\lambda^2) - \mathrm{ATL}_h - \mathrm{ATL}_e - \mathrm{PEL}_h - \mathrm{PEL}_e \quad (2.127)$$

where the ATLs are the amplitude taper losses in the H and E planes (0.91 dB for the cosine aperture distribution in the H-plane and 0 dB for the uniform amplitude distribution in the E-plane) and the PELs are the corresponding phase-error losses (which have been tabulated in Table 2-21 of [33] for both quadratic phase uniform amplitude and quadratic phase cosine amplitude distributions). The BLOCK DATA file of the program contains all those values. For intermediate s values interpolation can be used to get the losses.

The phase-center axial displacement behind the aperture (normalized to the corresponding plane slant length) for each E and H plane can be obtained as a function of the phase error s (from Table 7-3 in [33]). Those values are also included in the program BLOCK DATA and for intermediate s values interpolation is used.

2.14 PYRAMIDAL HORN EXAMPLE

The purpose of this example is to compare measured radiation patterns with the theoretical predictions obtained with the analysis program. We describe here the design and measurements of a gain standard of high polarization purity (in the broadside direction) for the frequency band around 12.5 GHz.

Usually, this type of horn is manufactured by soldering plane metal pieces and also soldering their junction to the input waveguide. Any loss in orthogonality will produce high-order modes at the horn aperture that can reduce the on-axis polarization purity to −30 or −35 dB levels. To obtain higher polarization purity the horns must be manufactured by a more expensive, electro-forming procedure. In our case to allow for a more precise mould machining the two semiflare angles of the horn were made equal ($\alpha_1 = \alpha 2$).

The lower-aperture dimension B was selected of three wavelengths at the center frequency, so $B = 3 \times 24 = 72$ mm.

If $\alpha_1 = \alpha_2$, then Equation (2.123) gives the other aperture dimension $A = B + a - b$, a and b being the input waveguide dimensions. At 12.5 GHz the normalized WR90 waveguide with internal dimensions of 22.86×10.16 mm was used. Then we found $A = 84.70$ mm.

The length between the aperture plane and the intersection plane between the input waveguide and the horn was selected of 8λ (192 mm). The corresponding semiflare angles are 9.15°. Measured and predicted patterns have been superimposed in Figure 2.17.

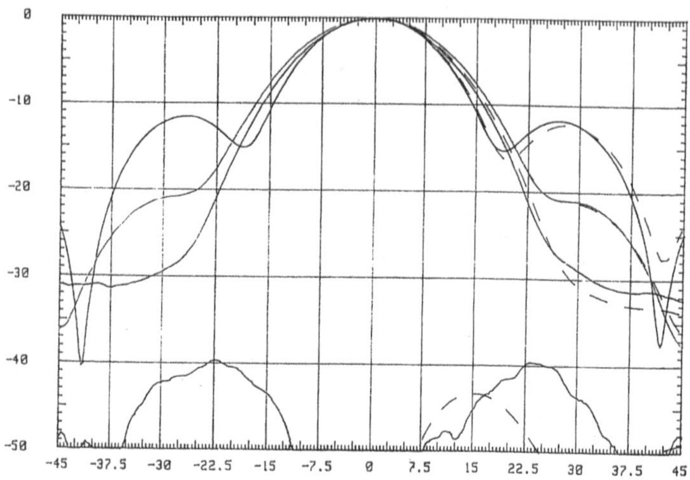

Figure 2.17 **Linear Co-polar ϕ = 0, 45° and 90° and Cross-polar ϕ = 45° Plane Cuts Measured (*solid lines*) and Calculated (*dashed lines*) of a Smooth-Walled Pyramidal Horn. The horn dimensions are 84.7 × 72 mm with α_1 = α_2 = 9.15°. Frequency was 12.5 GHz**

2.15 CORRUGATED PYRAMIDAL HORN DESIGN

The design procedure for corrugated pyramidal horns, implemented in program DCPH, closely follows the previously discussed procedure for smooth-walled pyramidal horns.

In the CPH case the universal patterns are the same for the E and H planes, since both have the same cosine amplitude type distribution as the H-plane of a smooth-walled pyramidal horn.

For the desired edge field value ET at θ_e and θ_h the constant s curves are cut at a set of u values from which we obtain a curve in the u,s plane similar to the one in Figure 2.15. The realizability condition is also expressed by Equations (2.123) and (2.125). Again we select first the narrower beam plane as the principal one and use Equations (2.119) and (2.120) to obtain the geometry in this plane. Also to account for the obliquity factor we use θ' values related to the θ values as in Equation (2.122). Then we obtain the intersection point, if any, of the realizability curve and the constant ET curve in the other plane. With the (u,s) values of this intersection point we obtain the geometry of this plane, again by using (2.119) and (2.120).

The directivity of the horn can be obtained from

$$D = 10 \log(4\pi AB/\lambda^2) - \text{ATL}_e - \text{ATL}_h - \text{PEL}_e - \text{PEL}_h \qquad (2.128)$$

where the amplitude taper losses in the E and H planes are now 0.91 dB, corresponding to cosine type distributions, and PEL_e and PEL_h are the tabulated values (in Table 2-21 of [33]), corresponding to quadratic phase cosine amplitude distribution.

Also, there are the axial displacements behind the aperture of the phase centers in the E and H planes, normalized to their corresponding slant lengths, as a function of s (in Table 7-3 of [33]), corresponding to the H-plane.

All the tabulated data and the universal patterns have been included in the BLOCK DATA subroutine of the program DCPH, which implements the design procedure.

2.16 CORRUGATED PYRAMIDAL HORN EXAMPLES

To show the accuracy of CPH pattern predictions two examples are included here. The first example is a corrugated pyramidal horn with aperture dimensions $A = B = 2.96\lambda$ and semiflare angles $\alpha_1 = \alpha_2 = 25°$ designed and measured at MBB in Germany [34]. The measured $\phi = 0$ plane cut co-polar pattern has been digitized and superimposed on the theoretical pattern prediction in Figure 2.18.

The second example is also a square corrugated horn that was constructed and measured at ERA in England and that has been analyzed with the mode-matching technique by Khün [17]. The aperture dimensions are $A = B = 2.91\lambda$ and the semiflare angles $\alpha_1 = \alpha_2 = 12°$. The measured co-polar and cross-polar $\phi = 45$ plane cut patterns have been superimposed on our theoretical predictions in Figure 2.19.

2.17 SMOOTH-WALLED CONICAL HORN DESIGN

Again the radiation patterns can be stored as universal curves taking the phase error s as a parameter. Now in Equation (2.120) W is the aperture diameter or $W/2$ is the aperture radius a.

For the TE_{11} fundamental-mode horn the pattern at any intermediate ϕ plane can be obtained from the E and H patterns. In particular at $\phi = 45°$ $E(45) = [E^2(0) + E^2(90)]^{1/2}/\sqrt{2}$.

The E-plane pattern is narrower than the H-plane pattern and the 45° plane pattern lies in between both. We have only two geometrical parameters to determine the horn (a,α), however, and we cannot select the

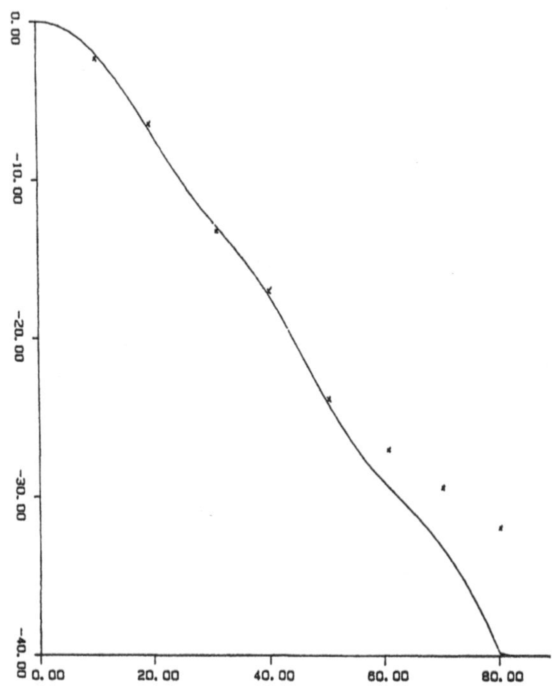

Figure 2.18 Linear Co-polar $\phi = 0°$ Plane Cut Measured (*crosses*) and Calculated (*solid line*) of a Corrugated Pyramidal Horn of $2.96\lambda \times 2.96\lambda$ Aperture Dimensions and $\alpha_1 = \alpha_2 = 25°$ Semiflare Angles

beamwidths at both planes independently. It is usual to perform the design in the 45° plane.

If we desire a field value ET at a given angle θ we force this condition in the 45° plane. As a result the *E*-plane pattern will be narrower and the *H*-plane pattern broader. The obliquity factor can be accounted for by considering a modified θ' angle as given by (2.122).

Cutting the universal $\phi = 45$ patterns with the constant *ET* desired level we obtain a set of (u,s) values and corresponding solutions. The geometrical parameters can be derived from (2.119) and (2.120).

Following Milligan [33], the gain of the conical horn can be obtained from

$$D = 20 \cdot \log(2\pi a/\lambda) - \text{ATL} - \text{PEL} \tag{2.129}$$

where ATL and PEL are the amplitude-taper loss and phase-error loss,

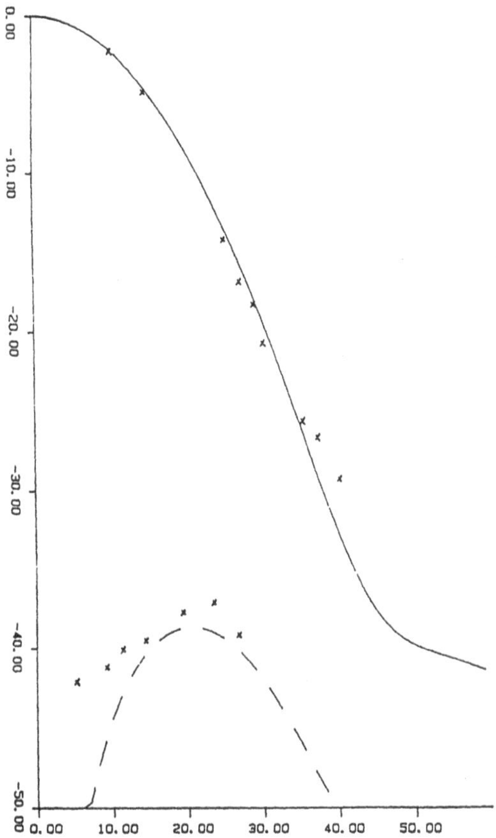

Figure 2.19 Linear Co-polar and Cross-polar ϕ = 45° Plane Cut Measured (*crosses*) and Calculated (*solid and dashed lines*) of a Corrugated Pyramidal Horn of 2.91λ × 2.91λ Aperture Dimensions and $\alpha_1 = \alpha_2 = 12°$

respectively. The combination of both is called the "gain factor," GF = ATL + PEL, and is listed in [33] (Table 7-4) for s values ranging from 0 to 0.72.

The E and H plane phase centers are also different. Their displacements behind the aperture (R_{pe} and R_{ph}), normalized to the horn slant length L, are tabulated (in Table 7-5 of [33]) for s values ranging from 0 to 0.48. With those data we have obtained a fourth-order interpolation polynomial for the H-plane and two third-order interpolation polynomials for the E-plane, subdividing its range of variation as follows

$$R_{ph}/L = a_i s^i, \quad 0 \leqslant s \leqslant 0.5$$

$$R_{pe}/L = b_i s^i, \quad 0 \leqslant s \leqslant 0.32$$

$$= c_i s^i, \quad 0.32 < s \leqslant 0.5$$

with the a, b, and c coefficients as given in Table 2.1.

Table 2.1

i	*0*	*1*	*2*	*3*	*4*
a	2.47 *E*-4	−5.96 *E*-2	4.109	−8.718	26.804
b	6.48 *E*-4	−0.0573	8.0832	−0.8162	—
c	−0.875	12.854	−33.562	29.937	—

2.18 SMOOTH-WALLED CONICAL HORN EXAMPLES

The accuracy in the radiation pattern prediction using the *E*-field model depends on the electrical aperture size [35]. Two examples of theoretical and measured patterns of horns with large electrical aperture size are shown in Figures 2.20 and 2.21. The horn geometries and frequencies are included in the figures. Excellent agreement between theory and measurements can be seen.

For small electrical aperture sizes, however, the flange dimensions strongly influence the radiation pattern. As an example, Figure 2.22 shows the measured (*solid lines*) and predicted (*dashed lines*) patterns, of a conical horn of 3.05 cm radius and 18.35° semiflare angle measured at 12.35 GHz. Good agreement in the *H*-plane cut can be seen, but the measured *E*-plane pattern is narrower than the predicted pattern. This disagreement can also be noticed in the cross-polar pattern in $\phi = 45°$. The same effect can be observed at lower frequencies and Figure 2.23 shows the comparison at 11.5 GHz.

2.19 CORRUGATED CONICAL HORN DESIGN

The design of a corrugated conical horn is done by the program DCCH. For a pure HE_{11} hybrid mode at the aperture the radiation pattern

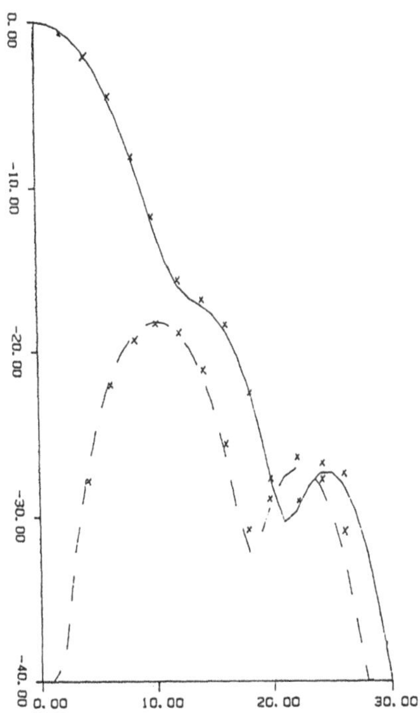

Figure 2.20 Linear Co-polar and Cross-polar $\phi = 45°$ Plane Cut Measured (*crosses*) and Calculated (*solid and dashed lines*) of a Smooth Conical Horn of 4.7-cm Aperture Radius and $\alpha = 7.9°$ Semiflare Angle at 20 GHz

is circularly symmetric and only one set of universal patterns is needed. For a desired ET value at a given angular direction θ, the intersection of the constant ET line with the universal patterns gives a curve in the u,s plane representing all the possible horns. Their corresponding geometrical radius and semiflare angles can be obtained from

$$a = (u\lambda)/\sin \theta' \tag{2.130}$$

$$\alpha = \tan^{-1}(2\lambda s/a) \tag{2.131}$$

where θ' is the modified edge angle to account for the obliquity factor as given by (2.122).

To obtain the horn directivity we use

$$D = 20 \log(2\pi a/\lambda) - \mathrm{GF} \tag{2.132}$$

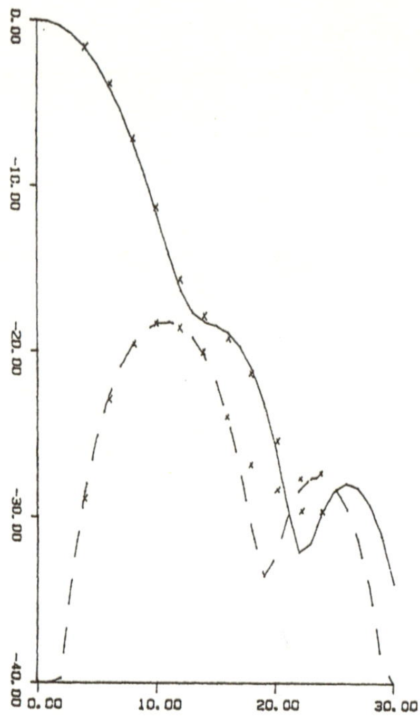

Figure 2.21 Linear Co-polar and Cross-polar $\phi = 45°$ Plane Cut Measured (*crosses*) and Calculated (*solid and dashed lines*) of a Smooth Conical Horn of 3.3-cm Aperture Radius and $\alpha = 7.1°$ Semiflare Angle at 30 GHz

where the gain factor GF is the summation of the amplitude taper loss ATL and the phase error loss PEL. GF can be obtained (from Table 7.7 in [33]) for s values ranging from 0 to 1.0.

The phase center location is now the same for all the planes. It starts at the aperture for $s = 0$ and moves toward the horn apex as s increases. The distance behind the aperture, normalized to the horn semiflare length, has been tabulated (in Table 7-8 of [33]) for s values up to 0.68. We use the following interpolation polynomial:

$$R_p/L = d_i s^i \tag{2.133}$$

with $d_0 = 6.8 \times 10^{-5}$, $d_1 = -6.98 \times 10^{-2}$, $d_2 = 4.821$, $d_3 = -11.951$, $d_4 = 28.368$, $d_5 = 8.734 \times 10^{-3}$, $d_6 = -91.595$, and $d_7 = 79.971$.

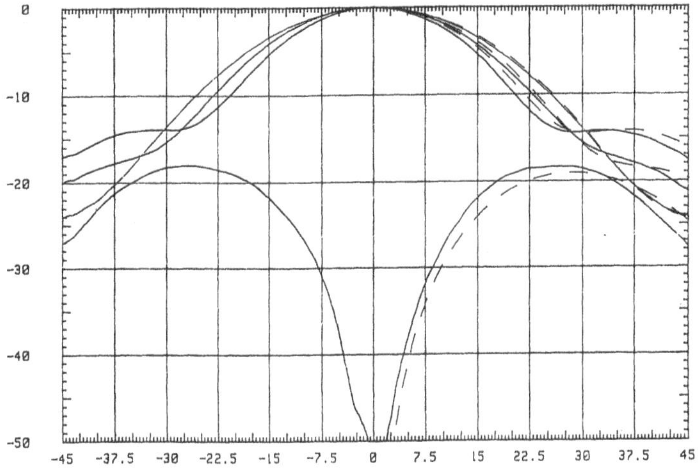

Figure 2.22 Linear Co-polar $\phi = 0°$, 45°, and 90° and $\phi = 45°$ Cross-polar Plane Cuts Measured (*solid lines*) and Calculated (*dashed lines*) of a Smooth Conical Horn of 3.05-cm Aperture Radius and $\alpha = 18.35°$ Semiflare Angle at 12.35 GHz

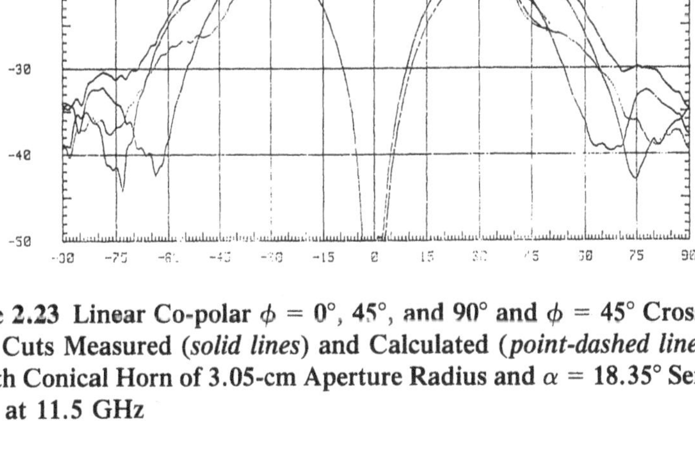

Figure 2.23 Linear Co-polar $\phi = 0°$, 45°, and 90° and $\phi = 45°$ Cross-polar Plane Cuts Measured (*solid lines*) and Calculated (*point-dashed lines*) of a Smooth Conical Horn of 3.05-cm Aperture Radius and $\alpha = 18.35°$ Semiflare Angle at 11.5 GHz

2.20 CORRUGATED CONICAL HORN EXAMPLE

As an example, Figure 2.24 shows a comparison between measured (*crosses*) and theoretical (*solid lines*) prediction of a corrugated horn at 12.5 GHz. The co-polar pattern prediction agrees well with the measured pattern. The cross-polar prediction was below −40 dB and a measured cross-polar peak of −28 dB was observed. This discrepancy is due to the generation of higher-order hybrid modes at the horn throat. A careful design of this mode-converter section is required so as to avoid this problem [36].

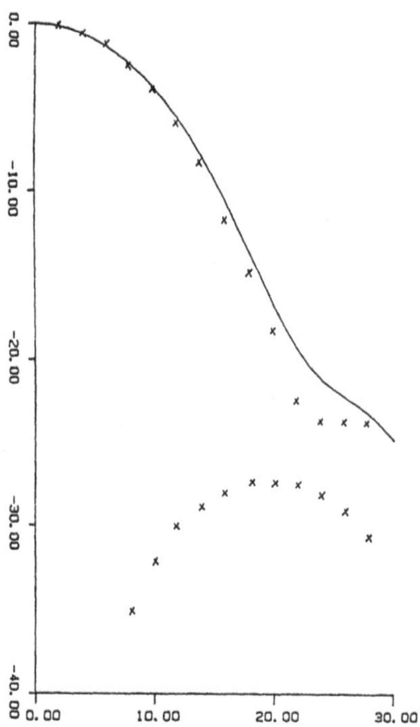

Figure 2.24 **Linear Co-polar and Cross-polar $\phi = 45°$ Plane Cut Measured (*crosses*) and Calculated (*solid line* Co-polar; the Cross-polar Level was below the −40-dB level) of a Corrugated Conical Horn of 5.3-cm Aperture Radius and $\alpha = 10.71°$ Semiflare Angle at 12.5 GHz**

Acknowledgments

The work contained in this chapter was done at the Radiation Group of the E.T.S.I.T. of the Polytechnic University of Madrid. The author wishes to recognize the general support of all the Radiation Group members. In particular thanks are due to J.L. Besada and P. Caballero for the antenna measurements and to E. Vega, D. Ferreras, J. Molina, R. Mompó, and R. Caballero for the programs that generated the source codes in this chapter.

References

[1] Love, A.W., *Electromagnetic Horn Antennas,* IEEE Press, 1976.
[2] Stutzman, W.L., and G.A. Thiele, *Antenna Theory and Design,* New York: John Wiley and Sons, 1981.
[3] Balanis, C.A., *Antenna Theory. Analysis and Design,* New York: Harper & Row, 1982.
[4] Clarricoats, P.J.B., and D.A. Olver, *Corrugated Horns for Microwave Antennas,* IEE Electromagnetic Waves Series 18, London: Peter Peregrinus, 1984.
[5] Milligan, T., *Modern Antenna Design,* New York: McGraw-Hill, 1985.
[6] Harrington, R.F., *Time Harmonic Electromagnetic Fields,* New York: McGraw-Hill, 1961.
[7] Love, A.W., A.W. Rudge, and A.D. Olver, "Primary Feed Antennas" in *The Handbook of Antenna Design,* Vol. I, Chap. 4, London: Peter Peregrinus, 1982.
[8] Ludwig, A.C., "The Definition of Cross Polarisation," IEEE Trans. Antennas and Propagation, Vol. AP-21, pp. 116–119, Jan. 1973.
[9] Silver, S., *Microwave Antenna Theory and Design,* New York: McGraw Hill, 1949.
[10] Hollis, J.S., T.J. Lyon, and L. Clayton, "Microwave Antenna Measurements," Scientific Atlanta, Inc., July 1970.
[11] Andersen, J. Bach, "Low and Medium Gain Microwave Antennas," in *The Handbook of Antenna Design,* Vol. I, Chap. 7, London: Peter Peregrinus, 1982.
[12] Russo, P.M., R.C. Rudduck, and L. Peters, "A Method for Computing E-plane Patterns of Horn Antennas," IEEE Trans., Vol. AP-13, 1965, pp. 219–224.
[13] Narasimhan, M.S., and M.S. Sheshadri, "GTD Analysis of the Radiation Patterns of Conical Horns," IEEE Trans., Vol. AP-26, 1978, pp. 774–778.
[14] Abramowitz, M., and I. Stegun, *Handbook of Mathematical Functions,* Chap. 9, New York: Dover, 1972.

[15] Kay, A.F., "The Scalar Feed," US AFCRL Rep. 62-347 AD60169, March 1964.

[16] Lawrie, R.E., and L. Peters, "Modifications of Horn Antennas for Low Sidelobe Levels," Trans. IEEE Antennas and Propagation, Vol. AP-14, 1966, p. 605.

[17] Khün, E., and B.K. Watson, "Rectangular Corrugated Horns: Analysis, Design and Evaluation," Proc. 14th European Microwave Conference, Sept. 1984, pp. 221–227.

[18] Khün, E., and B.K. Watson, "Computer Aided Design and Performance Evaluation of Fully Corrugated Rectangular Horns," Proc. IEEE Int. Symp. on Antennas and Propagation, June 1986, pp. 795–798.

[19] Calvo, M., D. Ferreras, and E. Vega, "Corrugated Pyramidal Horn Feed Model for Single-Offset Reflector Analysis," Proc. IEE Fourth Int. Conf. on Antennas and Propagation, Coventry (UK), April 1985, pp. 469–473.

[20] Calvo, M., J. Molina, and R.M. Lois, "Fast Analysis of Multimode Conical Horns by Bessel-Jacobi Series Method," Proc. IEEE Int. Symp. on Antennas and Propagation, June 1986, pp. 783–786.

[21] Galindo-Israel, V., and R. Mittra, "A New Series Representation for the Radiation Integral with Application to Reflector Antennas," IEEE Trans. Antennas and Propagation, Vol. AP-25, Sept. 1977, pp. 631–641.

[22] Abramowitz, M., and I. Stegun, *Handbook of Mathematical Functions,* Chap. 25, New York: Dover, 1972.

[23] Potter, P.D., "A New Horn Antenna with Suppressed Sidelobes and Equal Beamwidths," Microwaves Journal, June 1963, pp. 71–76.

[24] Rudge, A.W., and N.A. Adatia, "Matched Feeds for Offset Parabolic Reflector Antennas," Proc. Sixth European Microwave Conf., Rome, Italy, Sept. 1976, pp. 1–5.

[25] Clarricoats, P.J.B., S.M. Tun, and C.G. Parini, "Effects of Mutual Coupling in Conical Horn Arrays," IEE Proc., Vol. 131, Pt. H, No. 3, June 1984, pp. 165–171.

[26] Clarricoats, P.J.B., and P.K. Saha, "Propagation and Radiation Behaviour of Corrugated Feeds. Part II—Corrugated Conical Horn Feed," Proc. IEE, Vol. 118, Sept. 1971, pp. 1177–1186.

[27] Galindo-Israel, V., and Y. Rahmat-Samii, "A New Look at Fresnel Field Computation Using the Jacobi-Bessel Series," IEEE Trans. Antennas and Propagation, Vol. AP-29, No. 6, Nov. 1981, pp. 885–898.

[28] Rusch, W.V.T., and P.D. Potter, *Analysis of Reflector Antennas,* Sec. 4.25, New York: Academic Press, 1970.

[29] Abramowitz, M., and I. Stegun, *Handbook of Mathematical Functions,* Chap. 5, New York: Dover, 1972.
[30] Jamnejad, V., "A New Look at the Circle Polynomial Series Representation for the Radiation Integral of Reflector Antennas," Proc. 1984 IEEE/URSI Int. Antennas and Propagation Symp., Vol. I, pp. 309–313, May 1984.
[31] Stutzman, W.L., and G.A. Thiele, *Antenna Theory and Design,* Sect. 8.4, New York, John Wiley and Sons, 1981.
[32] Kelleger, K.S., *Antenna Engineering Handbook,* p. 12.5, New York: McGraw-Hill, 1961.
[33] Milligan, T., *Modern Antenna Design,* Chaps. 6 and 7, New York: McGraw-Hill, 1985.
[34] Fasold, D., and P. Pecher, "Gain of Rectangular Horns," Microwave Journal, March 1979, pp. 76–79.
[35] Hombach, V., "Radiation from Flanged Waveguides," Proc. URSI Int. Symp., Santiago de Compostela, Aug. 23–26, 1983, pp. 77–80.
[36] James, G.L., "Analysis and Design of TE_{11} to HE_{11} Corrugated Cylindrical Waveguide Mode Converters," IEEE Trans. Microwave Theory and Techniques, Vol. MTT-29, Oct. 1981, pp. 1059–1066.

Appendix 2A
List of Chapter 2 Computer Programs and Source Code Listings of Selected Programs

This appendix lists the names of computer programs developed especially for Chapter 2. Source code on 5¼-inch diskettes is available for all programs in the optional software package with user's manual (giving operating instructions) and example booklet. All programs are written in Microsoft (DOS) FORTRAN-77 and BASIC for IBM-PC type computers.

Various suffixes are used to designate the nature of the computer program or file. For example, .FOR is a FORTRAN source program and .BAS is a BASIC source program. Also .DAT is a data file, .DOC and .INF are information files, and .RES is a data or results file. Input and output data files can be given names interactively. These can include descriptive suffixes, such as, for example, .DRC for a drawing or graphics file for co-polarized power data and .DRX for a graphics file for cross-polarized power data. In the example provided in our optional software package, .DRF indicates a drawing file for phase data and .FED indicates the file generated for the feed data used in computing aperture antenna radiation patterns.

Types of horn feed are identified in the design program names as follows: CCH = Corrugated Conical Horn, CPH = Corrugated Pyramidal Horn, SCH = Smooth-Walled Corrugated Horn, and SPH = Smooth-Walled Pyramidal Horn. Similar systems of identifying letters are employed in other programs and files.

Program names with first letter D are for the preliminary design programs. First letters BD indicate a block-data program, SYNT***.RES indicates synthesis in the sense of choosing feed configuration, and ANAL***.RES means analytical evaluation of the SYNT***.RES horn configuration selected.

SOFTWARE LIST, CHAPTER 2

HORN DESIGN PROGRAMS:

BDSPH.FOR ; DSPH.FOR ANALSPH.RES, SYNTSPH.RES
BDCPH.FOR ; DCPH.FOR ANALCPH.RES, SYNTCPH.RES
BDSCH.FOR ; DSCH.FOR ANALSCH.RES, SYNTSCH.RES
(The BD files must be linked before the D files.)

HORN ANALYSIS PROGRAMS:

BDATA.FOR ; HORN.FOR ; SWPH.FOR ; CPH.FOR ; CCH.FOR ; MSCH.FOR

HORN.FOR (This is the driver program for the above horn analysis programs. Typical data and graphics files for an example horn corresponding to menu options A,B,C, and D are listed below.)

A: SWCOHO.DAT,name1.RES,name2.DRC,name3.DRX,name4.DRF, name5.FED
B: COCOHO.DAT,name1.RES,name2.DRC,name3.DRX,name4.DRF, name5.FED
C: PIHORN.DAT,name1.RES,name2.DRC,name3.DRX,name4.DRF, name5.FED
D: PIHORN.DAT,name1.RES,name2.DRC,name3.DRX,name4.DRF, name5.FED

NEARF.FOR (This program is for near-field analysis. Typical data and graphics files are named in the following example.)

NEARF0.DAT, NEARF0.RES, RES0.DRA

PHACEN.FOR (This is the program for phase center determinations. Names of data and graphics files used in an example are listed below.)

PHACEN0.DAT, RES0.DRA, RES1.DRA
CPHEX00.PHC, CPHEX00.DRF, CPHEX00.DRN

DRAWING PROGRAMS:

DRAWCGA.BAS ; DRAWEX.DAT
DRAWHGC.BAS ; DRAWEX.DAT

Graphics programs that operate with both the Hercules Graphics Cards and with the Color Graphics Adapter are provided.

Chapter 3
REFLECTOR ANTENNAS AND SURFACE SHAPING TECHNIQUES
C. J. Sletten

3.1 INTRODUCTION

This tutorial chapter is written primarily to provide the designers of single-beam reflector antennas with helpful computer techniques. Reflectors scanned by feed motions or fed by extended sources are treated in Chapter 7. Each computer program with numerical examples is fashioned to give insight and numerical data on some critical choice or parameter influencing the antenna's gain, near-in and far-out sidelobes, and polarization. The paraboloid mirror is the reference antenna, and its performance when prime-focus fed by simple feeds is compared with results obtained with offset and subreflector feeding. Methods for shaping subreflectors and the paraboloid sections are developed as the main tools for approaching the ideal gain *versus* sidelobe goals.

Is an offset-fed antenna significantly better than the axially symmetric dish? Does Cassegrain subreflector feeding improve performance of the symmetric dish antenna? Why is offset dual-reflector feeding better? Should the subreflector choice be a Cassegrain or Gregorian configuration? What will reflector shaping achieve? What are the shaping techniques and how can they be programmed? This chapter explores these questions using computer software with simple examples to try to give the answers.

In the computer illustrations of reflector antenna design principles, the antenna feed is usually assumed to be a conical corrugated horn having a radiation pattern that can be well approximated by a $\cos^N\theta$ function. Slot and dipole feeds are used to study effects of asymmetry and cross polarization. More complex feeds can be substituted by representing their radiation patterns by power series.

The combining of all the design factors for paraboloid antennas into an approximate CAD model is done in Chapter 4. In this chapter, each design parameter is considered separately and a number of configuration options are presented.

Two comprehensive review articles [1,2] are recommended for experienced antenna engineers interested in contemporary paraboloidal antenna design up to 1978. Most of the offset dual-reflector shaping work has been done since then, and several references to this later work are included. Of more importance are the references to "classic" papers that explain the design fundamentals and give derivations of basic expressions such as the radiation fields of slots and dipoles.

A number of tutorial computer programs are listed in Appendix 3A at the end of the chapter. Novel numerical computer techniques are introduced in this chapter that are best understood by reading the computer codes and noting the sequence of operations with the basic design equations given in the text. To make the programs easier to read the same symbols for variables are often used in the text and in the software. Computer codes do not allow subscripts, so you will often find the variables X1, Y1, Z1 in the text rather than X_1, Y_1, Z_1.

3.2 PRIME-FOCUS-FED PARABOLOIDS

The ubiquitous "dish" with its circular aperture is included in this prime-focus-fed category of antenna, usually with a focal length, f, to diameter, D, ratio of about $f/D = 1/3$. Although pattern gain and sidelobe levels deteriorate somewhat due to the blocking of feed and feed supports the symmetric paraboloid is a low-cost, compact antenna. Pattern performance depends on skillfully distributing the feed energy over the reflector aperture. The f/D ratio and feed pattern are the main design variables available for controlling gain and the sidelobe and cross-polarization levels.

In 1954, E.M.T. Jones [3] showed that when the symmetrical paraboloid is illuminated on-axis by a Huygens source the aperture cross-polarized fields disappear. This property can be demonstrated using the program CROSS.BAS (Chapter 1) with the feed-tilt angle set to zero. For the y-polarized Huygens source in this program it is seen that the EAX and EAZ aperture fields components are near zero. Because the conical corrugated horn has nearly Huygens source polarization, and, because by varying the semiflare angle of the horn the nearly symmetrical radiation pattern of the horn can be controlled, these horns are well suited for feeding paraboloids with circular apertures. The wide-angle corrugated horn or scalar feed is very wideband and has a rather well defined center of phase. Lower cost feeds such as slots and dipoles and many of the feeds described in Chapter 2 can also be used to feed the dish.

3.2.1 Path Length Attenuation Effects

The path length or $1/R^2$ losses from the feed phase center to points on the reflector surface influence how the feed pattern is transformed into the aperture distribution. For the paraboloid of focal length f shown in Figure 3.1, the path length attenuation function is listed in Table 3.1. The geometry [4] of the paraboloid determines the relation between the feed angle, θ, and the radial distance, r, across the circular aperture. The power density on the aperture is also easy to express as a function of θ:

$$r = 2f \tan(\theta/2), \text{ where } r = \sqrt{(x^2 + y^2)} \tag{3.1}$$

$$dP/dA \approx G(\theta,\phi) \cos^4(\theta/2)/f^2 = G/[1 + (r/(2f))^2]^2 \tag{3.2}$$

where dP/dA is power per unit area, $G(\theta,\phi)$ is the feed pattern, and the factor $\cos^4(\theta/2)$ is due to path length attenuation for the symmetrical paraboloidal section.

Table 3.1 Circular Reflector Aperture Power Distributions (Path Length Attenuation Dependence on Feed Angle, θ)

Reflector Type	*Radial Dependence on θ*	*Power Distribution Due to Path Attenuation*
Paraboloid	$r = 2f \tan(\theta/2)$	$dP/dA \approx \cos^4(\theta/2)$
Symmetric Cassegrain	$r = \dfrac{2f(e+1)}{(e-1)\tan(\theta/2)}$	$dP/dA \approx \cos^4(\theta/2)$
Offset Subreflector Feeding Paraboloids with Axis Tilt β	$r = \dfrac{2f(1-e^2)\tan(\theta/2)}{(1+e^2-2e\cos\beta)}$	$dP/dA \approx \cos^4(\theta/2)$

Note: Here f is focal length of paraboloid and e is eccentricity of conic section subreflector; dP/dA is power density on the circular paraboloid aperture.

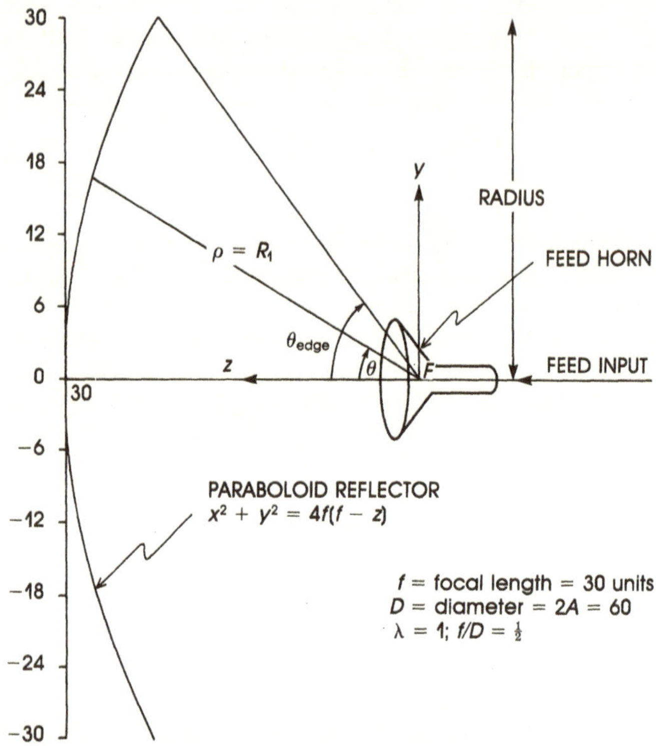

Figure 3.1 Prime Focus-Fed Axially Symmetric Paraboloidal Antenna

3.2.2 Design Approximations for the Symmetric Paraboloid

A first approximate dish design calculation can be done on a hand calculator or by using the program APRX.BAS. These simple formulas are given for use in the preliminary selection of the reflector and its feed. First, the antenna half-power beamwidth is approximately

$$\text{BW}(\tfrac{1}{2}\text{power}) \approx 1.1\lambda/D \text{ in radians} \tag{3.3}$$

where λ is the wavelength and D is the dish diameter.

The maximum power gain for the circular aperture is given by

$$G_0 = (2\pi A/\lambda)^2 \tag{3.4}$$

where $A = D/2$ is the radius of the dish. Realizable gains are generally about 0.2 dB to 3 dB less than G_0.

General information about the influence on gain and sidelobe levels of aperture taper, aperture distribution functions, and phase errors is given in Chapter 1 and in References [5, 6, and 7].

The power and phase (near-field) radiation patterns of corrugated horns can be computed quite accurately using HORN1,2,3,4 and the programs of Chapter 2. The conical horn power patterns can be well approximated with power series by using HORN5 or by using simple $\cos^N(\theta)$ or Gauss error functions. For preliminary design and for getting a feel for the trade-offs in gain, spillover, and edge illumination (edge taper), P_e, on the dish, the $\cos^N(\theta)$ approximation of the feed pattern is adequate and is used frequently in this chapter.

Choice of f/D gives the cone angle subtended by the edge of the dish at the focal point F, $\theta_e = \tan^{-1}(y/z)_e$. To find the coefficient N for a given edge taper in dB write,

$$\begin{aligned} P_e &= \cos^N(\theta_e)\cos^4(\theta_e/2);\ \text{dB} = 10\log P_e \\ N &= [\text{dB}/10 - 4\log\cos(\theta_e/2)]/\log\cos(\theta_e) \end{aligned} \tag{3.5}$$

Most conical corrugated horns have about −17-dB level radiation at the semiflare angle.

Using the dish configuration of Figure 3.1, we see that with $f = 30$ and $A = 30$, then $\theta_e = 53.13°$ and $N = 5.8877$ for a −15-dB dish edge taper.

The spillover loss is sensitive to edge taper. The program SPILL.BAS discussed in Chapter 1 gives an approximation of the gain loss due to energy not illuminating the paraboloid. The program also assumes a $\cos^N(\theta)$ horn pattern, and by entering the angles close to θ_e we can print out the spillover losses for our choice of N. The far-out sidelobes will be at about the level of the horn radiation at the edge when referred to the isotropic level.

For the example antenna of Figure 3.1 the spillover loss from SPILL.BAS is −0.13 dB.

We can improve the estimations of gain and −3-dB beamwidth obtained by APRX.BAS for the dish antenna by using SYMPF.FOR which carries out an integration of the scalar aperture distribution to compute the radiation pattern. The patterns will also give the first sidelobe positions and levels relative to the main-beam gain. The program SYMPF.FOR listed in Appendix 3A accepts as inputs the dish radius, A, and the horn pattern $\cos^N\theta$ and computes the antenna radiation pattern for a wavelength expressed in the units of A. This data is entered into data file DISH.DAT. SYMPF.FOR generates the dish aperture distribution in r,ϕ coordinates which it enters into the file PHAMP.DAT which is integrated by PATT.FOR to compute the antenna pattern. The gain is printed by PATT and the file CJS.DAT contains pattern profile data in dB as a function of θ for any ϕ-cut chosen. The radiation pattern for the $\phi = 0$ cut (because of the symmetry all the scalar patterns are the same for any ϕ-cut) is plotted in Figure 3.4(a). The gain is seen by running the program PATT to be 44.59 dBi (for $N = 5.8877$) which is about 1 dB less than the maximum gain calculated from Equation (3.4).

The program SYMPF.FOR is a simple prototype of many other programs in this book for computing radiation patterns. The code is very simple to read and the results are almost trival as the aperture phase is set to zero and all radial cuts have the same distribution. In the program SYMPF2.FOR a more interesting feed consisting of a pair of half-wave slots is introduced into the subroutine PHSAMP using the feed variables θ = (TH) and ϕ = (PHI) related to the aperture points, X_0, Y_0.

We have neglected the effects of blocking due to the feed and its supports and cables. Blocking geometry varies greatly from dish to dish and is best evaluated [8,9,9A] from the actual structural configuration. In Chapter 4 the scattering of three or four cylindrical struts and circular plates simulating feeds is done with some accuracy and the scattered fields are vectorially combined with the aperture radiation field. To estimate blockage loss, subtract the power scattered by the cross sections of the blockage structure. This loss must be subtracted from the gain of the antenna computed by PATT.FOR. Feed blockage caused by a circular feed or feed holder of radius R_f generates near-in sidelobes approximately given by

$$E_s = J_1(u)/u$$

where $u = 2\pi R_f \sin(\theta)/\lambda$ and λ is the wavelength. The $J_1(u)$ values are Bessel functions. These sidelobes have half-power beamwidths in radians of about $\lambda/(4R_f)$.

Blockage sidelobes tend to modulate the aperture sidelobes computed by PATT.FOR and of course make them generally higher. The circular struts produce broad conical lobes about the axis of each strut member. These effects generally cause about 0.3 dB gain loss and a 2–5 dB rise in sidelobe levels on well designed dishes.

There are other effects that may be important for some antenna applications. With the symmetric dish configuration, some of the energy reflected from the zone near the vertex of the reflector is intercepted by the feed. This reflected energy causes increases in the feed VSWR. A vertex plate with proper diameter and position can be installed to reduce the reflection back to the feed. Silver discusses this technique in ([5], pp.

443–447). The vertex plate solution to the input impedance problem causes a rise in sidelobe level.

3.3 DIPOLES AND SLOTS AS FEEDS FOR PARABOLOIDS

A resonant slot in a rectangular waveguide makes an acceptable, simple feed for a dish. The narrow slot on a metal ground plane has a broad (half-space omnidirectional) pattern in the *E*-plane. Wider slots are slightly more directive. We can use a pair of slots arrayed in the *E*-plane to get the directivity needed for efficient illumination of the symmetric paraboloid with a circular aperture. These elementary radiators are often used in arrays, and it is important to understand their characteristics and limitations as polarized sources for dishes.

Electric dipoles have (according to Babinet's principle) characteristics similar to half-wave slots but reflecting or directing elements are needed near the dipoles to produce unidirectional patterns appropriate for illuminating reflector surfaces. The dipole disk (reflector) feed and the waveguide double dipole feeds shown in ([5], p. 254) were some of the earliest dish illuminators.

Expressions for the radiated (far-fields) patterns of slots and dipoles in ([10], pp. 44, 560–563) are programmed in DIPOL.BAS AND SLOTS.BAS to illustrate the aperture (vector) fields they produce when used to feed the paraboloid. These programs are listed in Appendix 3A.

For the narrow resonant slot shown in Figure 3.2(a) we can write for radiated fields programmed in SLOTS.BAS,

$$E_\theta \approx \cos\phi F_0 = F_1 \tag{3.6}$$

$$E_\phi \approx -\sin\phi\cos\theta F_0 = F_2 \tag{3.7}$$

where

$$F_0 = \cos(\pi/2\sin\phi\sin\theta)/(1 - \sin^2\phi\sin^2\theta)$$

Because the radiation pattern of a single narrow slot is nearly omnidirectional in the $y = 0$ half-plane we can reduce spillover in this plane by using a pair of slots, as shown in Figure 3.2(b). A two-element in-phase array with half-wave spacing has a directional field pattern

$$F_7 = \cos(\pi/2\sin\theta\cos\phi)\cos^2(\theta/2)$$

where the path length attenuation term $\cos^2(\theta/2)$ has been included.

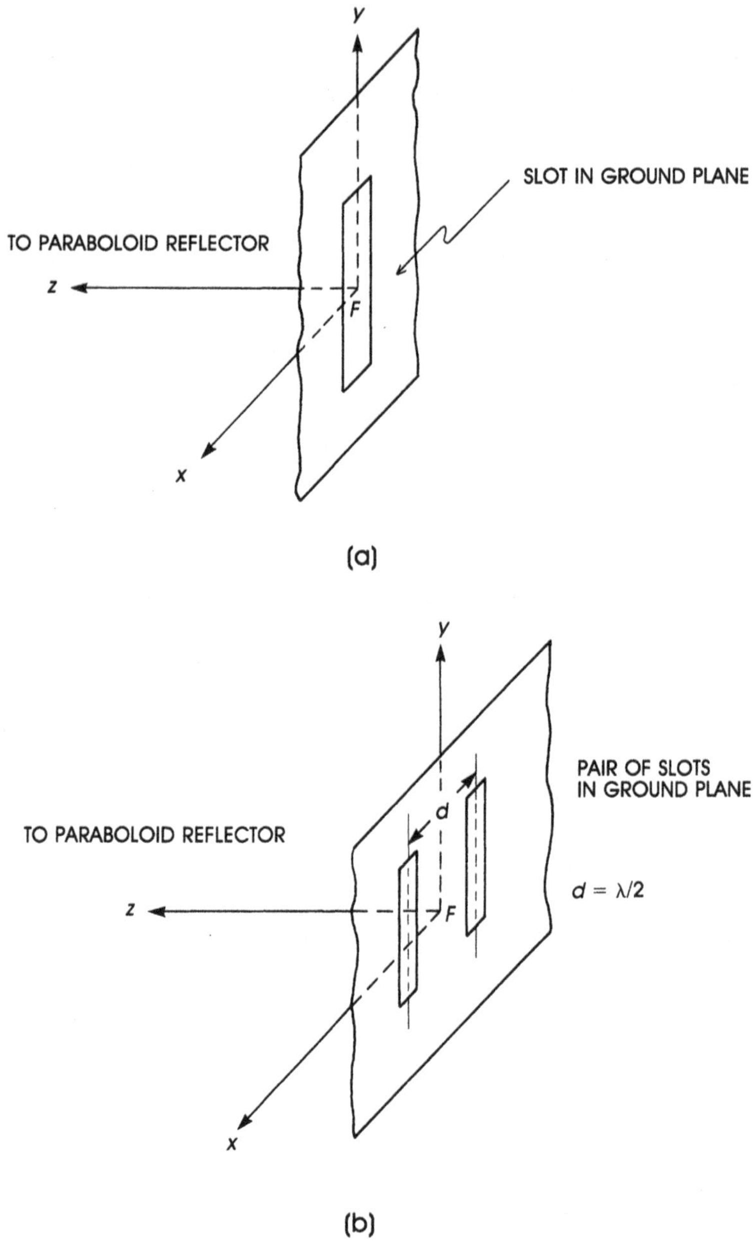

Figure 3.2 Resonant Half-Wave Slots Feeding Symmetric Paraboloid

Referring to SLOTS.BAS the functions F_0, F_1, F_2, and F_7 can be identified in lines 150–200. These field components are converted, using (3.12), to rectangular coordinates and the aperture fields EXA, EYA, and EZA found using Snell's law for *E*-vectors reflected from metal surfaces,

$$\mathbf{E}_A = 2(\mathbf{n} \cdot \mathbf{E}_P)\mathbf{n} - \mathbf{E}_P \tag{3.8}$$

where $\mathbf{n}$ is the unit vector normal to the paraboloid surface and $\mathbf{E}_P$ are the radiated field components of the slot pair. Lines 210–350 of SLOTS.BAS carry out these steps and print out the *X*-polarized fields and cross-polarized fields on the aperture in dBs referred to the central power density for any aperture point, x,y. The variations in edge taper can be seen by entering x,y at the edge points. By writing $F_7 = 1$ in a new line 175 of SLOTS.BAS, the wide *E*-plane pattern of the single slot can be studied as a feed for the symmetric dish.

The radiated fields of the slot feed are printed out as EXP, EYP,EZP, and the aperture fields in the paraboloid aperture of focal length, *F*, are displayed as EXA,EYA,EZA.

When an electric dipole or pair of dipoles are substituted for the slots as feeds the paraboloid aperture fields are displayed using the program DIPOL.BAS. The radiation pattern of the half-wave dipole is simply written in the coordinates of Figure 3.3(a) as

$$\mathbf{E} \approx \cos[(\pi/2)\cos\theta_1]/(\sin\theta_1)\boldsymbol{\theta}_1 \tag{3.9}$$

where $\boldsymbol{\theta}_1$ is a unit vector. To write the *E*-field in the coordinates of Figure 3.3(b) a 90° rotation is needed. This leads [11] to the simple relations between X, Y, Z and X_1, Y_1, Z_1

$$\begin{aligned} X_1 &= X \\ Y_1 &= Y\cos A_0 - Z\sin A_0 \\ Z_1 &= Y\sin A_0 + Z\cos A_0 \end{aligned} \tag{3.10}$$

where $A_0 = 90°$. The equations $\mathbf{x}_1 = \mathbf{x}$ and $\mathbf{y}_1 = -\mathbf{z}$ and $\mathbf{z}_1 = \mathbf{y}$ are the unit vector relations. Using the spherical coordinate relations in the two coordinate frames, we can relate θ_1 and ϕ_1 to θ and ϕ by these expressions

$$
\begin{aligned}
\cos\theta_1 &= \sin\theta\sin\phi \\
\sin\theta_1 &= \sqrt{(1 - \sin^2\theta\sin^2\phi)} \\
\tan\phi_1 &= -\cot\theta\sec\phi
\end{aligned}
\tag{3.11}
$$

The unit vectors $\boldsymbol{\theta}_1$ and $\boldsymbol{\phi}_1$ are related to $\mathbf{x}_1$, $\mathbf{y}_1$, $\mathbf{z}_1$ by

$$
\begin{aligned}
\boldsymbol{\theta}_1 &= \cos\theta_1\cos\phi_1\mathbf{x}_1 + \cos\theta_1\sin\phi_1\mathbf{y}_1 - \sin\theta_1\mathbf{z}_1 \\
\boldsymbol{\phi}_1 &= -\sin\phi_1\mathbf{x}_1 + \cos\phi_1\mathbf{y}_1
\end{aligned}
\tag{3.12}
$$

The expressions (3.8) through (3.12), after some algebra, are programmed into DIPOL.BAS for a pair of *Y*-polarized dipoles spaced $\lambda/2$ apart. By running SLOTS.BAS and DIPOLE.BAS it is seen that the pair of slots and dipoles produce identical aperture fields when feeding an axially symmetrical paraboloid, but with orthogonal polarizations.

Exercising programs SLOTS.BAS and DIPOL.BAS demonstrates that these elemental feeds do not give the constant edge tapers and low aperture cross polarization that are obtained by feeding with conical corrugated horns. The dipole pair feed is bidirectional so that 3 dB is lost unless passive elements are used to reflect the dipole pattern(s) forward toward the paraboloid vertex.

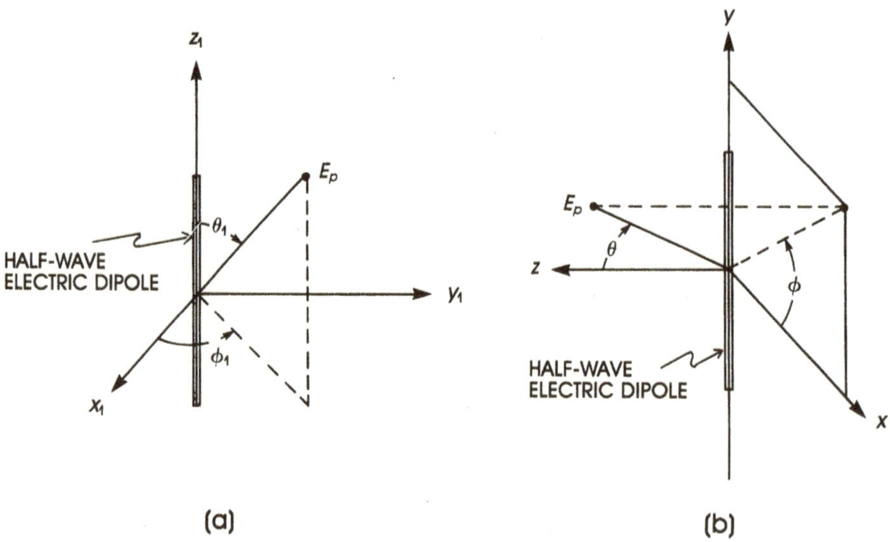

Figure 3.3 Electric Dipoles with z-Axis Parallel and Normal to Radiators

3.4 THE OFFSET PRIME-FOCUS-FED PARABOLOID

Offset or off-axis (Spencer's term) feeding of a paraboloid section with circular aperture eliminates nearly all of the strut and feed blocking but introduces more aperture cross polarization than for the symmetrical fed dish. (It is the paraboloid section that is offset; the feed is on the axis but tilted toward the center of the reflector.) The program CROSS.BAS demonstrates that even for the Huygens source feed polarization, well approximated by a conical corrugated horn pattern, there are aperture cross polarization components when the tilt angle A_0 is not zero. (Correctly tilted offset subreflector feeding again restores the theoretical zero cross polarization when the subreflector is illuminated by a Huygens source.) Another disadvantage of the offset geometry is that the path length attenuation and feed angles to different edge points on the offset paraboloid are generally different, meaning that the "edge taper" and spillover varies on the antenna aperture.

These effects can be studied with the programs ASYMM.BAS and OFFSET.FOR. First the aperture fields for an offset paraboloid as shown in Figure 3.4 are displayed by ASYMM.BAS. Here again (as for the symmetric dish) the radiated fields E2X, E2Y, E2Z for the feed are calculated first. The feed is a Huygens polarization source with power pattern given by $\cos^{MM}$, where θ is measured from the horn axis. The horn is tilted by an angle A_0 toward the midpoint of the paraboloid section at Y_3. Here we enter the aperture points in cylindrical coordinates RA,PHI (in degrees). The scalar aperture power distribution, P_{oo}, in dB, is shown with the copolar power P_o. The cross-polar component levels P_x are generally quite low so that the scalar aperture computation for the offset antenna done in OFFSET.FOR is fairly accurate. Unlike for the symmetric paraboloid, the scalar patterns are not the same for all ϕ-cuts in pattern space. Very small changes in patterns and gain result from small changes in A_0. Therefore the midpoint, Y_3, Z_3 ($X_3 = 0$), is a good point to direct the horn axis.

We see, at least for the example presented, that the offset prime-focus-fed paraboloid gives excellent patterns and gains. Data files are computed to compare the performance in aperture gain and near-in sidelobes for a symmetric dish with radius of 30 inches and focal length of 30 inches with an offset paraboloid with the same radius and focal length, using the programs SYMPF.FOR and OFFSET.FOR. Both reflectors are illuminated by horn patterns that give edge tapers of about -15 dB. Of course the offset antenna has variable edge tapers ranging from -13.2 dB to -19.6 dB. The offset antenna has a gain of 44.55 dB (at 11.81 GHz) and its $\phi = 0$ cut pattern computed by PATT.FOR is shown in Figure 3.5. The patterns for the symmetric dish are shown in this figure and the gain is

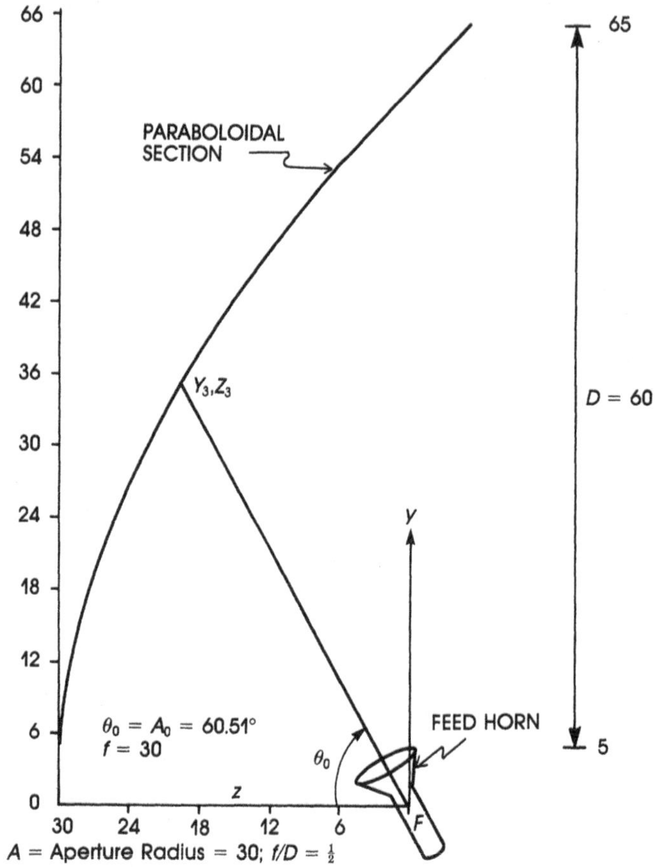

Figure 3.4 The Prime Focus Offset-Fed Paraboloidal Antenna

44.59 dB at the same frequency. It is apparent for this example that the symmetric and offset antennas give comparable performance. Only when the blocking effects and polarization components are included in the diffraction pattern computations made with the programs of Chapter 4 can a precise comparison be made between the two designs.

The cross polarization in the aperture fields of the offset antenna as calculated by ASYMM.BAS causes beam squint with changes in frequency. The cross polarization on circularly polarized offset antenna is essentially zero as x and y components are affected equally. In Chapter 8 it is seen that by using Gaussian modes the cross-polarized fields induced by an offset paraboloid reflector are circularly polarized in the far field. The beam squint effects, however, which can be studied with the vector

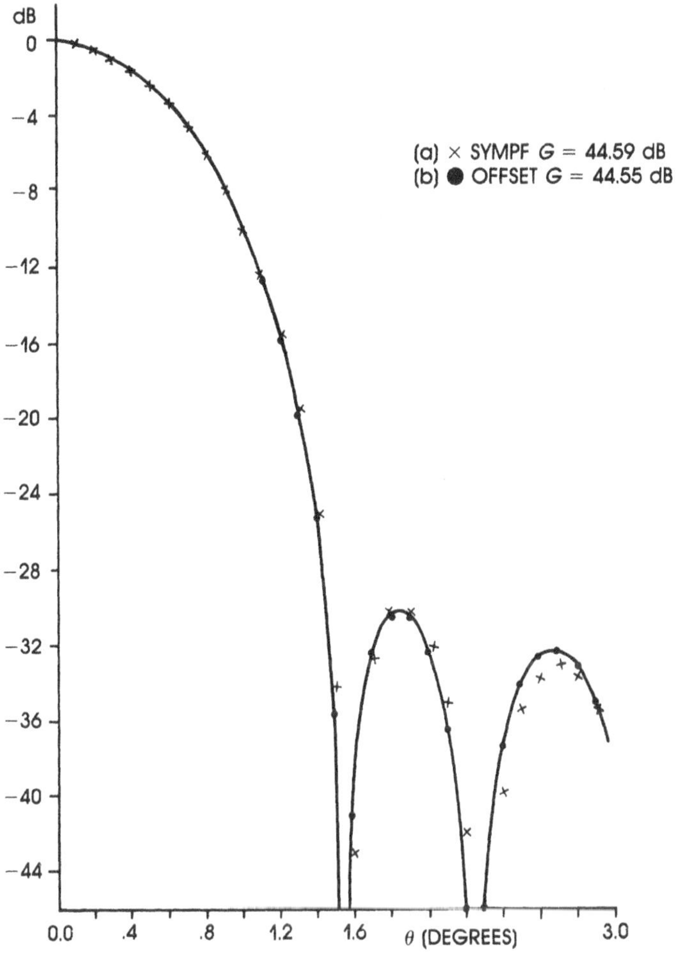

Figure 3.5 Radiation Patterns, Symmetric Paraboloid Compared with Offset Paraboloid (for Circular Apertures)

diffraction methods of Chapter 4, still cause problems on some communication links. Adatia and Rudge [2,12] have shown that the change in beam angle Φ for circularly polarized antennas is given approximately by

$$\sin \Phi = \lambda \sin(A_0)/(4\pi f) \tag{3.13}$$

where f = focal length, λ = wavelength, A_0 = tilt angle of the offset feed.

3.5 SYMMETRIC SUBREFLECTOR FEEDING, CASSEGRAIN ANTENNAS

Cassegrain feeding of the symmetric paraboloid was perhaps the first subreflector-fed antenna of importance. Although the Cassegrain feed blocking is generally greater than for prime-focus feeding, the same low cross-polarization characteristics for horn or Huygens source feeding are preserved. The Cassegrain geometry, Figure 3.6, allows the transmission line or beam waveguide connecting the power source to be located near the paraboloid vertex. Another advantage of the axially symmetric Cassegrain antenna configuration is that when the subreflector and main reflector are shaped for controlling the aperture distribution, these surfaces are still surfaces of revolution about the central axis.

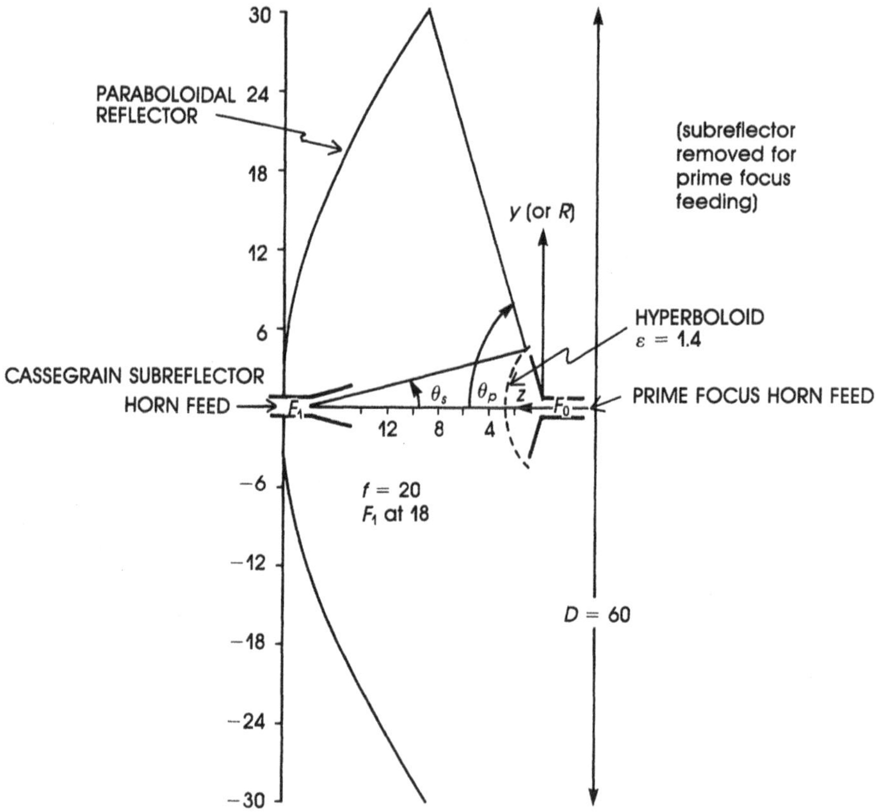

Figure 3.6 Prime Focus versus Cassegrain Feeding of Symmetric Dish

Program PFCAS.BAS illustrates that the aperture distribution for a hyperboloidal subreflector-fed symmetric dish can be quite similar to that of the prime-focus paraboloid of circular aperture, using the configurations of Figure 3.6 for the example input data. From Table 3.1 it is clear that ray pathlength attenuation has the same dependence of feed angle, θ. However the horn semiflare angles to obtain -10-dB edge tapers are quite different for Cassegrain compared to prime-focus feeding. The prime-focus feed must illuminate for the example shown in Figure 3.6 at an angle of 73.74° while the subreflector horn semiflare angle is only 14.25° for a hyperboloid of eccentricity 1.4. These parameters were calculated using Equation (3.5) where θ_e is now the Cassegrain subreflector edge angle.

The program PFCAS.BAS computes the magnification M:

$$M = (E + 1)/(E - 1)$$

where E is the eccentricity. The subreflector transform constant, B, used in the program is given by

$$B = 2fM$$

where $f = 20$ is the focal length of the same symmetric dish illuminated by the prime focus horn and the Cassegrain feed. These expressions follow from the symmetric Cassegrain equation given in ([4], p. 53),

$$R = 2f[(E + 1)/(E - 1)] \sin \theta/(1 + \cos \theta) = 2fM \tan(\theta/2) = B \tan(\theta/2)$$

Both horn patterns are approximated by $\cos^N\theta$ or $\cos^{MM}\theta$ patterns, where $N = 1.1076$ (prime-focus horn) and $MM = 72.6849$ (Cassegrain horn). There is a -10-dB taper on the main reflector for both prime-focus and Cassegrain feeding.

Plotted in Figure 3.7 are the aperture field profiles for the dish of radius $R = 30$. It is evident that, for this example, adding a Cassegrain subreflector need not greatly modify the antenna aperture distribution. The popularity of Cassegrain (and in rare cases Gregorian, or ellipsoidal subreflector) feeds for the axially symmetric dish is due primarily to shaping to increase antenna aperture efficiency and the other advantages mentioned at the beginning of this discussion.

Using SPILL.BAS the spillover loss for the prime-focus feeding is found to be -0.31 dB for the example data. Spillover is less about the edges of the Cassegrain subreflector but there is a second spillover loss near the rim of the main reflector. More precise accounts of feed blockage

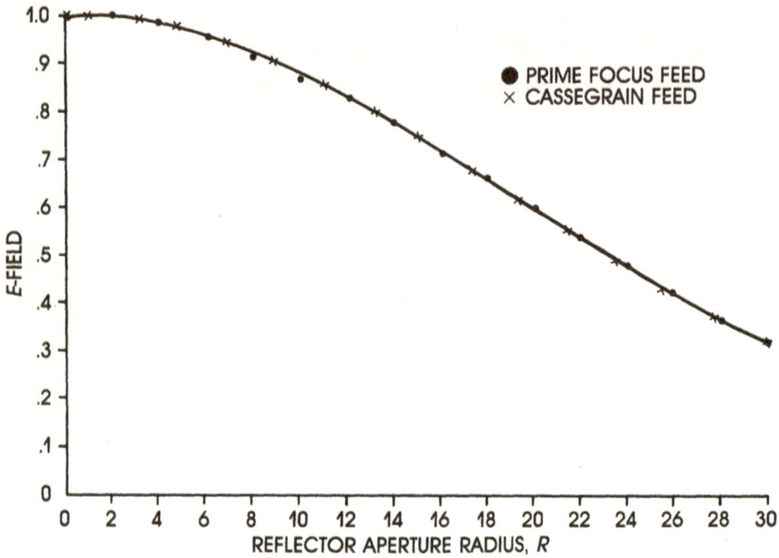

Figure 3.7 Comparison of Radial Aperture Distributions, Prime Focus versus Cassegrain Feeds

and edge diffraction can be found using the GTD methods of Chapter 4. Cross-polarization losses are very low for both symmetrically fed antennas.

3.6 OFFSET DUAL-REFLECTOR ANTENNAS

Offset feeding with subreflectors largely eliminates feed and strut blocking. When the feed horns and offset subreflectors are correctly oriented as explained in Chapter 1, low cross polarization and beam symmetry, without beam squint with frequency change, can be achieved. These are advantages in comparison with offset prime-focus feeding of a circular paraboloidal section. Shaping can also be done to control aperture distributions but now the shaped subreflectors and main reflector are not axially symmetric. The performance improvements with offset subreflectors are important. The remainder of this chapter is devoted primarily to their design. The offset Gregorian or ellipsoidal subreflector can be fed by wide semiflare angle conical corrugated horns (the "scalar" horn) with wide frequency bandwidth and center of phase near the apex of the horn. The offset Cassegrain antenna, generally fed by horns of smaller semiflare angles, is developed for wide-angle scanning in Chapter 7.

The unshaped offset ellipsoidal and hyperboloidal subreflectors can be configured using the programs GREG.BAS and CASS.BAS. The main reflector, an offset paraboloidal section, is first selected to fix the parameters R, Y_c, and F. R is the circular aperture radius, Y_c is the midpoint height above the axis, and F (or f) is the focal length of the paraboloid. The subreflectors have the parameters: E = eccentricity, C = distance between subreflector foci, and angle θ_e = the semiflare angle, is the θ angle subtended by the subreflector edge. Once the main reflector is selected, then the eccentricity of the subreflector is varied to obtain the semiflare angle desired for the corrugated feed horn. The distance C is varied to adjust the size or area of the subreflectors; since changing C does not change the angular relations, it can be varied last to assure that the subreflector diameter is 15 wavelengths or greater. Parameter variations are more restricted for the offset Cassegrain to avoid aperture blocking by the top of the subreflector.

Radiation patterns for offset dual-reflector antennas can be computed using the program SUBPAT.FOR. Two example designs are presented for two offset subreflector feeds. The "constants" programs GREG.BAS and CASS.BAS allow some flexibility in antenna configuration and always assure the correct tilt angles β (tilt of the subreflector axis with respect to the paraboloid z-axis) and α (tilt of the horn axis with respect to the subreflector axis) for beam symmetry and low cross polarization. The printouts for the offset Gregorian and Cassegrain are listed by the programs. For the example ellipsoid shown in Figure 3.8(a) the input configuration constants are:

$$E = 0.425,\ Y_c = 40.8333,\ F = 36.6667,\ C = 17,\ R = 30$$

For the example hyperboloid in Figure 3.8(b) the input configuration constants are:

$$E = 1.7,\ Y_c = 60,\ F = 55,\ C = 45,\ R = 30$$

These two antennas are fed with horns producing $\cos^{NN}\theta$ patterns assumed without phase errors. The coefficient $NN = 73.05$ is for the ellipsoidal subreflector and $NN = 420.7267$ for the hyperboloid. These are calculated from Equation (3.5) (where $\theta_e = \theta_h$ is now the horn angle subtended by the subreflector) to give a -20-dB edge taper on the main reflectors which both have radii of $R = 30$. To compute the radiation patterns for these antennas the program SUBPAT.FOR accepts the values of NN and B for either antenna. The gain is found to be 44.02 dB for both. The antenna profiles are shown in Figures 3.8(a) and 3.8(b). Their nearly identical radiation patterns are plotted in Figure 3.9 as computed by SUBPAT.FOR and PATT.FOR. Parameter in DISH2.DAT files was used by SUBPAT.FOR.

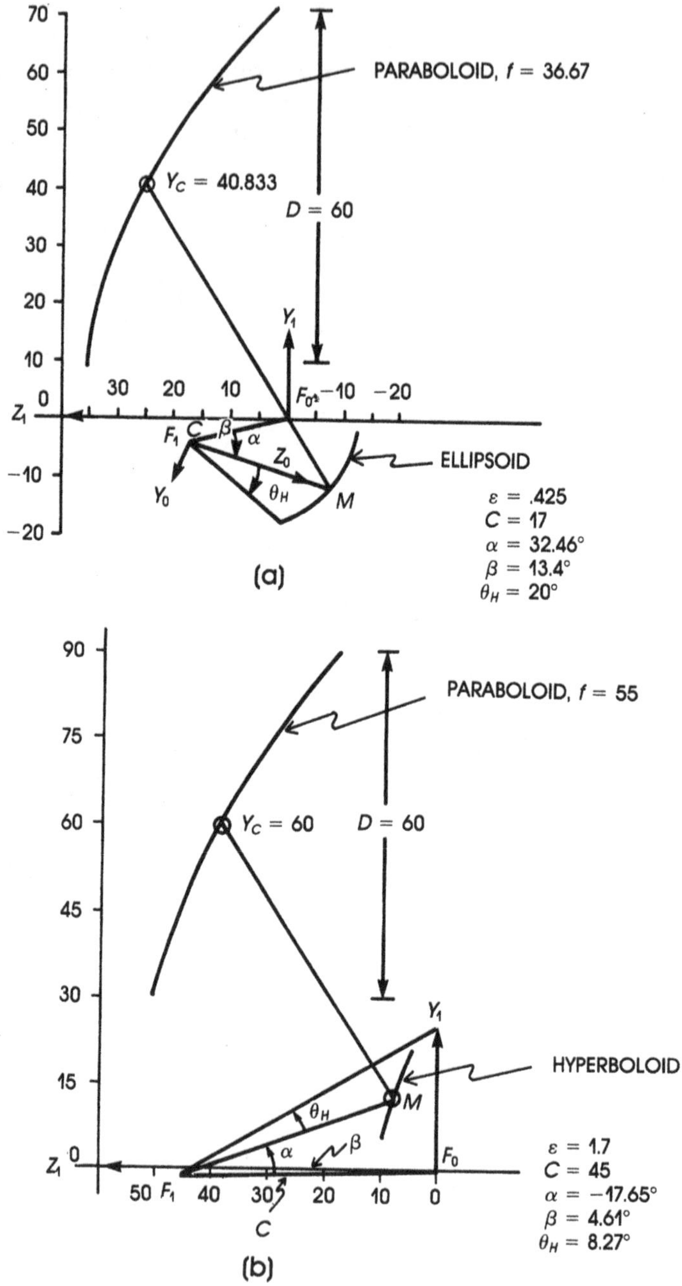

Figure 3.8 Offset Gregorian (a) and Cassegrain (b) Antenna Profiles

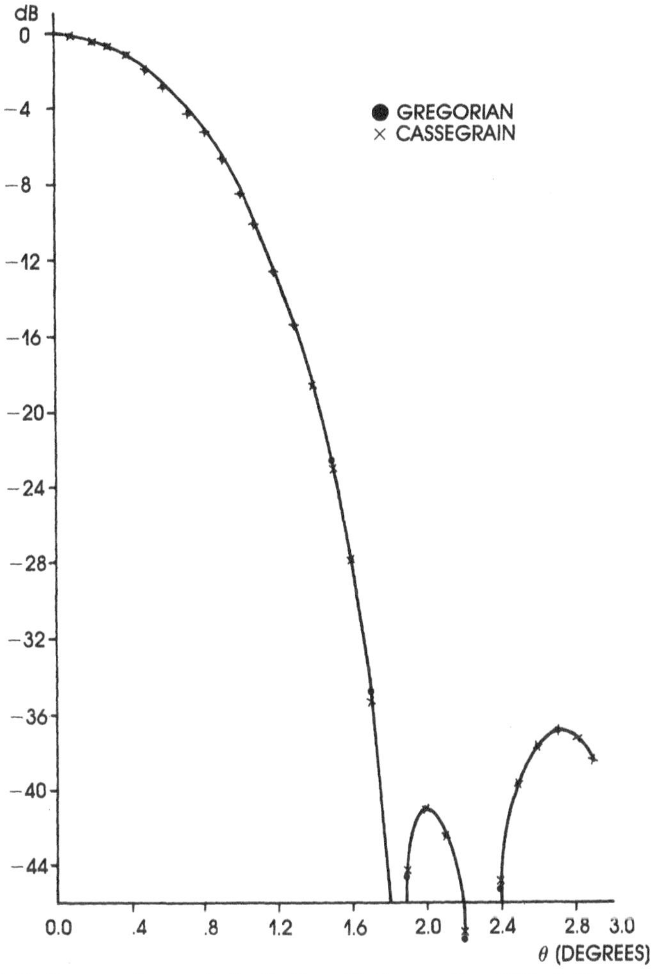

Figure 3.9 Comparison of Radiation Patterns for Offset Cassegrain and Gregorian Antennas

Because of the beam and pattern symmetry all the ϕ-cuts on the patterns are the same as for the symmetric paraboloid. No aperture phase errors are included, although it is very easy to enter the horn phase errors and bring them to the aperture using the feed θ angle. The near-field horn patterns computed by HORN1–4 can be closely represented by power series for both their amplitude and phase patterns using HORN5. We always assume that the feed polarization is approximately that of a linear polarized Huygens source. The inaccuracies in gain and sidelobes due to cross polarization for the optimally tilted feeds with conical corrugated horns illuminating large subreflectors are very small, as can be checked with the vector equations programmed in Chapter 4.

3.7 A REFOCUSING TECHNIQUE FOR HORN PHASE ERRORS

Horn phase errors can be partially corrected by repositioning the horn with respect to the (second) subreflector focus, F_1 shown in Figure 3.8(a,b) [see also Figure 3.14(a)]. For the refocusing program, RESIDS.BAS, the feed-phase pattern as computed by HORN4.FOR, for example, must be available. The approximate location of the horn apex with respect to the subreflector focus can be calculated from the computed or measured phase versus θ data. Once this "best phase center" is found, the residual phase errors after refocusing can be computed and represented by a power series. To find the best location for the horn and the residual phase errors, we can use RESIDS.BAS.

The horn phase errors Φ as computed by HORN4 or the horn pattern programs of Chapter 2 can be converted to distances by multiplying by $K' = \lambda/2\pi$. With polar coordinate origin at F_1 the distance to the equiphase surface of the horn phase pattern is R_2 at $\theta = 0$, where Φ is also zero. The distance R to the equiphase surface (in the near-zone of the feed) at any θ is given by $R = R_2 + \Phi K'$; Φ is generally negative near the edges of the phase pattern of corrugated horns.

From the geometry [see (Figure 3.14(a) for the variable definitions], the center of phase, C_2, on the $\theta = 0$ line for any $\Phi(\theta)$ is easily found to be

$$C_2 = (R_2^2 - R^2)/[2(R_2 - R\cos\theta)] \tag{3.14}$$

using the law of cosines in the form

$$R = C_2 \cos\theta + \sqrt{(C_2^2 \cos^2\theta + R_3^2 - C_2^2)}$$

and the relation

$$R_3 + C_2 = R_2$$

RESIDS.BAS allows the designer to compute C_2 values for all θ data from the Φ values and to pick a C_2 value designated C_0 (approximate center of phase on the horn axis). C_0 should be chosen in a region of θ where Φ is greatest. Then the residual horn pattern phase function is recomputed using the second part of RESIDS. If desired, a power series representation can be made for the refocused horn phase function for entry into a pattern computing program such as SUBPAT.FOR. More general refocusing methods to correct feed pattern phase errors are given in Chapter 2.

By refocusing the horn and by using more or less of the horn pattern (the horn semiflare angle need not illuminate the edge of the subreflector) rather optimum trade-offs between near-in sidelobes and gain can be attained. Often the "ideal" or desired radial aperture distributions can be closely approximated by choosing or designing the feed radiation patterns.

3.8 SHAPING REFLECTOR SURFACES

We will discuss reflector shaping at some length because it offers an opportunity to approach the "ideal" aperture distributions of Chapter 1, to increase antenna aperture efficiency greatly, and to control far-out sidelobe levels. Methods for checking the accuracies of shaped reflector surfaces are developed in Section 3.13.

3.8.1 Pattern Shaping, $\csc^2\phi$

Since the 1940s, shaped reflector surfaces other than conic sections have been used to shape radiation patterns using geometric optics (GO) methods. In his MIT Radiation Laboratory report [13], R.C. Spencer attributes this surface shaping technique to L.J. Chu. We will here follow the equations of Spencer's report for two-dimensional shaping to achieve $\csc^2\phi$ radiation patterns. The program CSC.FOR generates a shaped surface that when fed by a line source will generate a $\csc^2\phi$ pattern in a given ϕ coverage sector. The geometric arguments of proof of the integral equation relations are given in [4,13]. The principal equations are now discussed using the variables defined in Figure 3.10.

For the example used to illustrate the method a line source feed with a directive pattern in the θ plane

$$P_f(\theta) = \cos^4(\theta - \theta_0) \tag{3.15}$$

is assumed to illuminate the shaped cylindrical surface with feed pattern peak at θ_0. The shaped reflecting surface to be synthesized will subtend a $\theta = 90°$ arc of the feed pattern.

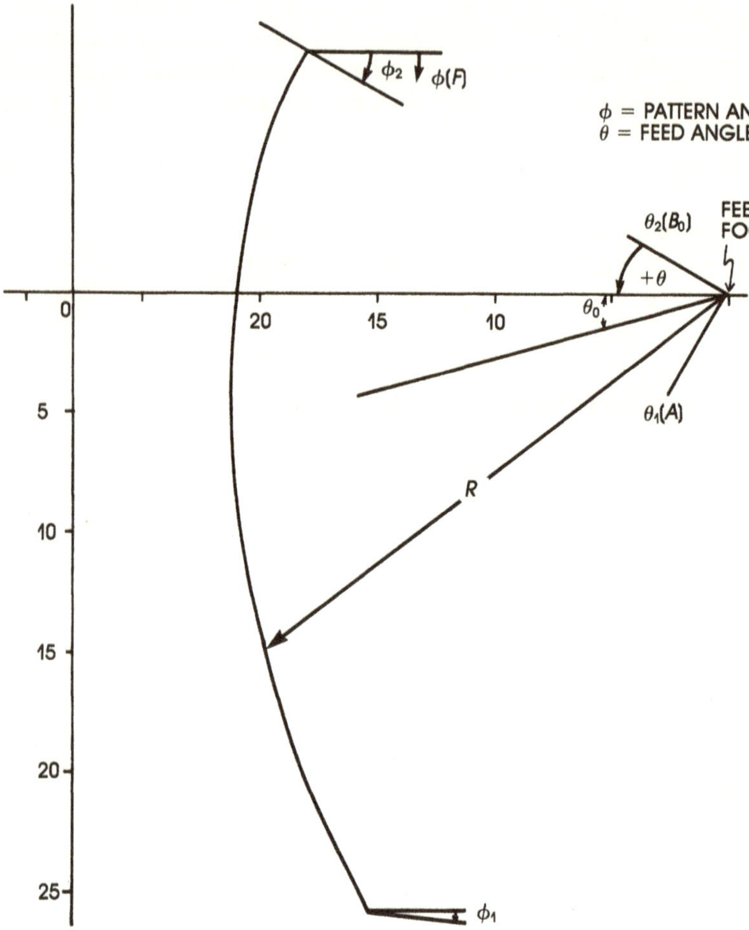

Figure 3.10 Profile of Shaped Cylindrical Reflector for $\csc^2\phi$ Pattern

The desired antenna coverage pattern, $P_p(\phi)$, extends from 5° to 25°, where the ϕ origin lies to the left of the surface in Figure 3.10, with angles measured in a clockwise sense from the horizontal of the figure,

$$P_p(\phi) = \csc^2\phi \tag{3.16}$$

As can be seen in the figure, the angle boundaries are

$$\theta_1 = -60°;\ \theta_2 = 30°;\ \theta_0 = -15°;\ \phi_1 = 5°;\ \phi_2 = 25°$$

The angles θ and ϕ are related to the power integrals

$$\int_{\theta_1}^{\theta} P_f(\theta)\, d\theta = K \int_{\phi_1}^{\phi} P_p(\phi)\, d\phi \qquad (3.17)$$

where K is a normalizing constant obtained by equating the total feed power with the total pattern power (spillover is neglected), or

$$K = \int_{\theta_1}^{\theta_2} P_f(\theta)\, d\theta \Big/ \int_{\phi_1}^{\phi_2} P_p(\phi)\, d\phi$$

Integrating Equation (3.17) gives

$$\cot\phi = C_1 - f(\theta - \theta_0) C_2$$

where $f(\theta - \theta_0) = 3/8(\theta - \theta_0) + 3/16 \sin[2(\theta - \theta_0)] + \cos^3(\theta - \theta_0)\sin(\theta - \theta_0)/4$, and the constants of integration are $C_1 = 6.78728$ and $C_2 = 8.52629$ for the example problem where $C_2 = 1/K$ and C_1 is obtained from fixed lower limits on (3.17).

For the cases where (3.15) and (3.16) can be integrated to obtain closed form solutions then simple equations can be written relating θ and ϕ. To shape the reflector surface we need to find the radial distance, r, from the origin to the surface as a function of θ and ϕ using the relations obtained from the power flow relations above.

Program CSC.FOR uses the explicit relation between θ and ϕ provided by (3.17) and performs numerical integration of the differential relation in polar coordinates to find r from the expression:

$$dr/r = \tan[(\theta - \phi)/2]$$

giving

$$\ln(r/r_0) = \int_{\theta_1}^{\theta} \tan[(\theta - \phi)/2]\, d\theta \qquad (3.18)$$

The equations and constants are seen in the program listing of CSC.FOR. The curve shown in Figure 3.10 and the numbers can be listed by running CSC linked to *Numerical Recipes* programs QTRAP and TRAPZD for numerical integration of (3.18). The tolerance in QTRAP should be set at E-4.

The sign in the argument of the $\tan[(\theta \pm \phi)/2]$ in (3.18) depends on whether angles are increased together in the coordinate frame in the same sense or not.

A fan-shaped $\csc^2\phi$ beam (rather than a cylindrical pattern as calculated above) can be generated by attaching parabolic "ribs" to the central section computed. The focal lengths of the parabolas are different at each point, Y_c, Z_c, and these parabolic rib curves with focal lengths F_c can be computed by using

$$4(F_c)^2 - 4F_cZ_c - Y_c^2 = 0$$

$$X^2 + Y_c^2 = 4(F_c)^2 - 4F_cZ$$

where Z_c and Y_c are a point on the central cut obtained from CSC.FOR and F_c is the focal length of the parabolic rib attached to that point parallel to the Z-axis at Y_c.

Two kinds of central curves, one concave and the other convex, can be found for $\csc^2\phi$ or other coverage patterns [5,13].

3.9 SYMMETRIC CASSEGRAIN SHAPING

The integral method of shaping is easily extended to shaping of Cassegrain surfaces for control of the dual-reflector aperture distributions for axially symmetric antennas. This is an important capability when maximum gain is desired requiring a nearly uniform aperture power density. Because of the rotational symmetry of the subreflector and main reflector, both surfaces can be shaped using what is essentially a two-dimensional formulation. Profiles (two-dimensional curves) of both the subreflector and main reflector can be rotated to generate the reflectors. It is necessary to shape the main reflector to correct for aperture phase errors caused by the shaping for power control of the subreflector.

The synthesis of the central section by integration of power in angular zones proceeds as in the $\csc^2\phi$ example, except that now energy in tubes rather than in planar zones must be considered. Also following papers [13,14,15], we will rotate the θ and β angles in opposite senses. Figure 3.11 defines the variables used in the equations to follow. The spherical angle integration of power in θ cones for feed power, $P(\theta)$, in watts-solid angle is

$$P_c(\theta) = K \int_0^{\theta} P(\theta) \sin\theta \, d\theta \tag{3.19}$$

$P(\theta)$ can be a corrugated horn pattern which we will for this example assume to be a $\cos^N\theta$ function. We will also assume that uniform power

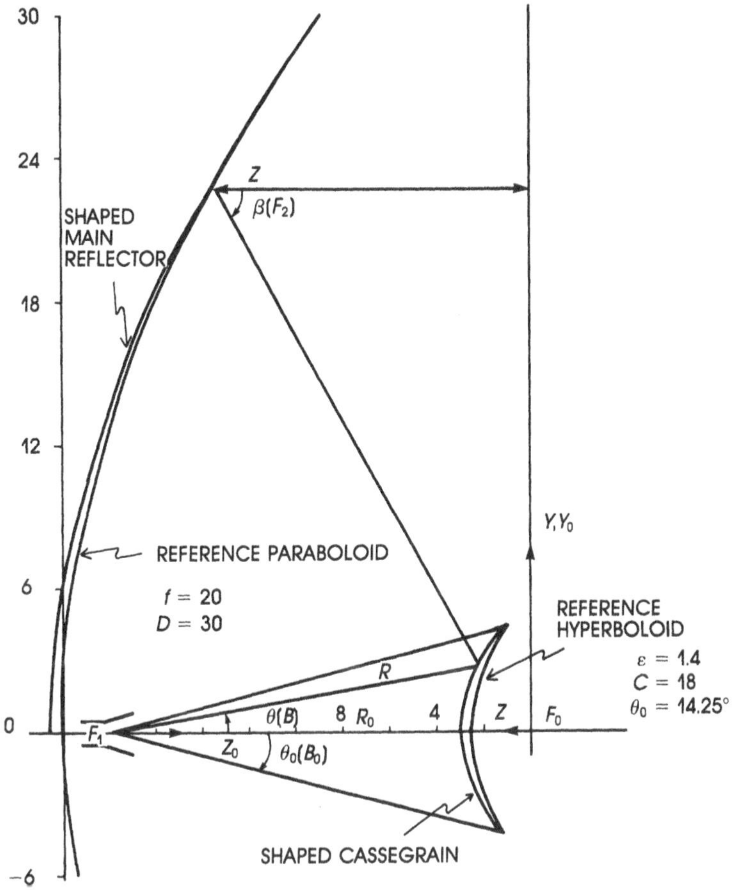

Figure 3.11 Profile of Shaped Subreflectors and Main Reflector for Uniform Aperture Power, Symmetric Cassegrain

density in the aperture is desired or

$$P(r) = \int_0^r r\,\mathrm{d}r \tag{3.20}$$

A numerical example to illustrate the shape synthesis technique uses the antenna configuration shown in Figure 3.6 for the unshaped symmetric Cassegrain geometry. The same paraboloid is used with $F = 20$ and $R = 30$ and the feed horn illuminates the 14.25° subreflector cone with a pattern $P(\theta) = \cos^N\theta$ where $N = 72.6849$ to produce a −10-dB taper on the circular rim of the dish. The shaped subreflector and the parameters used in the shaping procedure to produce uniform amplitude and phase on the main reflector are shown in Figure 3.11.

The integral relations connecting feed radiated power and aperture power for finding the normalization constant K is

$$K = \int_0^{R_{max}} y \, dy \Big/ \int_0^{\theta_{max}} P(\theta) \sin \theta \, d\theta \tag{3.21}$$

The y-axis is used for generating the subreflector-shaped profile, although the x-axis or any radial axis can be used because of the rotational symmetry about the z-axis.

The equations (3.19, 3.20, and 3.21) for this case are easily integrated in closed form where the limits of θ are from 0 to 14.25 and of y from 0 to 30. We relate y to θ by the relation

$$\int_0^R y \, dy = K \int_0^\theta P(\theta) \sin(\theta) \, d\theta$$

to get the equation

$$y^2 = K(1 - \cos^{(N+1)}\theta)/(N + 1) \tag{3.22}$$

or

$$y = 31.62234665\sqrt{(1 - \cos^M\theta)}$$

where $M = N + 1 = C_2$.

In the program ICAS.FOR developed to shape the rotationally symmetrical Cassegrain antenna surfaces, CV is defined as the total distance from the feed center of phase to points on the main reflector. $C_0 = CV$ for the dish rim points. Using again the differential relations of surface slope to angles we write

$$(1/r) \, dr/d\theta = \tan[(\theta + \beta)/2]$$

where β as seen in Figure 3.11 is obtained from the paraboloid surface or

$$dz/dy = -y/(2F) = -\tan(\beta/2)$$

The example in program ICAS.FOR illustrates the shaping method and results in a profile curve for the Cassegrain subreflector and main reflector as shown in Figure 3.11. Some trial and error work is needed to find the relation between the hyperboloidal reference surface obtained from the configuration program CASS.BAS (with $Y_c = 0.001$) and the shaped subreflector profile. We can for example make the rim of the shaped subreflector and its end points, Y_0, Z_0 lie on the reference hyperboloid rim. Of the many subreflectors that will give uniform dish aperture distributions, this is the one chosen in the computer example given. The z-axis crossing point of the shaped reflector is moved toward the main reflector anticipating the steeper slope of the shaped surface that must divert the higher power levels of the horn pattern near the axis to bring the outer portions of the aperture up to a power density equal to the central portion. A trial value of R_0 is chosen less than for the unshaped Cassegrain. Running ICAS, which must be linked with QTRAP and TRAPZD to do the numerical integration, gives an end value of R_{max} that lies on the ray $\theta = \theta_{max} = 14.25°$. Then R_0 can be changed to put the endpoint (rim) of the shaped subreflector at the Y_0, Z_0 values of the original hyperboloid endpoint.

A first order shaping correction of the main reflector to bring all rays in phase on the aperture ($z = 0$ plane) is also included in ICAS. The ray paths in the vertex region of the paraboloid are shorter than at the dish rim by amount DEL. If the z coordinates are all increased by amount DEL/$(1 + \cos \beta)$ then aperture phase errors are reduced.

The procedure can be generalized by using a more general horn pattern represented perhaps by a power series. Also the aperture distribution need not be constant and could for example be one of the ideal aperture distributions of Chapter 1. The blocking of the aperture by the subreflector and feed, however, tend to prevent good control of sidelobe levels. A rather simple variation in the formulation, setting the power from 0 to R_a, where R_a is the subreflector radius, to zero for example, reduces the power reflected back into the horn and also the level of reflected power on the subreflector. The apodized "hole" in the aperture distribution still causes a rise in near-in sidelobes, however. When near maximum gain is demanded from an aperture, this symmetric dish Cassegrain shaping technique is quite useful. This integral shaping technique has been extended [16,17] to offset dual-reflector configurations. In the next sections an alternate technique using ray tubes is developed for offset dual-reflector antennas to control amplitude and phase aperture distributions [18]. The axially symmetric Cassegrain case given in the example above is redone by the ray-tube method in Section 3.11.

3.10 RAY-TUBE NUMERICAL TECHNIQUES FOR REFLECTOR SHAPING

In the focusing reflector-optical systems used for microwave-millimeter wave antennas, energy flows (approximately) in ray tubes which diverge from the feed and upon reflection from subreflectors converge or diverge to focal zones or on to other surfaces. The idea of diverging rays with power varying (decreasing) as $1/R^2$ and converging rays with R^2 increasing in power density can be formulated in numerical differential equations that can be handled efficiently by computers [18]. This shaping method is called the ray-ratio method because power control is accomplished by controlling the ratio of converging and diverging ray pathlengths. This rather general approach is applied to offset and symmetric subreflector configurations.

Aperture phase too can be controlled by numerical algorithms. Phase errors due to nonspherical phase fronts from feeds or phase errors caused by amplitude shaping can be corrected by numerical techniques that invert Snell's law of ray reflection to find the normals and slopes of phase correcting shaped reflectors.

3.10.1 Fundamental Relations in Numerical Shaping of Surfaces

The basic equations in the ray-ratio method for controlling aperture amplitude relate the field strength of the feed pattern on the subreflector to the desired field strength on the antenna aperture. Using the example antenna of Figure 3.12 the three rays carrying power from the source at F_1 to the subreflector surface, R_1, and from the subreflector surface to the focal point at F_0, R_2, and finally from F_0 to the paraboloid surface, R_3, can be related to the feed power density at the subreflector, $G_f(\theta,\phi)$ and the aperture power density at the paraboloid surface $P_a(x,y)$. For the Gregorian dual-reflector antenna this relation for feed pattern rays in a direction (θ,ϕ) is

$$K_0[R_2/(R_3R_1)]^2 \doteq P_a(x,y)/G_f(\theta,\phi)$$

where K_0 is a constant normalized at $\theta = 0$ and $\phi = 0$ where $K_0 = (R_3R_1/R_2)^2$.

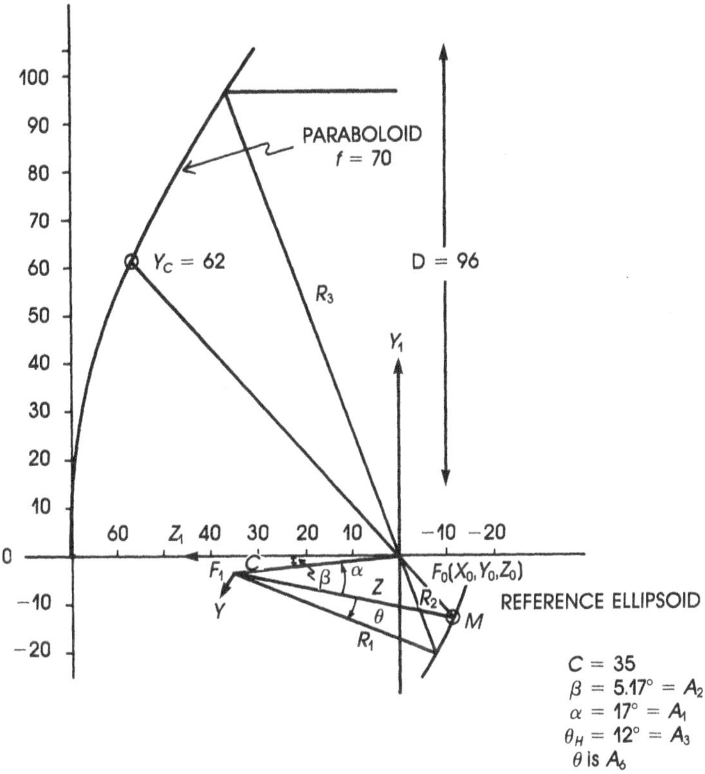

Figure 3.12 Offset Gregorian Subreflector Antenna

The converging ray R_2 is in the numerator and the two diverging rays R_1 and R_3 are in the denominator. We can make the antenna aperture distribution P_a into a function of θ by using the transform relation $\sqrt{x^2 + (y - y_c)^2} = B' \tan(\theta/2)$ for conic sections (see Section 1.4) and always relate the aperture points to the subreflector points (as angle functions) during the shape synthesis process. Working with fields rather than power we can for example set $E_a = 1$ everywhere on the aperture for uniform aperture distributions, often desired to maximize antenna gain. The path attenuation factor for E-fields, $\cos^2(\theta/2)$, must be entered into the aperture amplitude synthesis equation to compensate for path loss. In the program to follow using the ray-ratio technique the field densities at the aperture, E_a, and the feed pattern field at the subreflector surface, $E_f = \sqrt{G_f}$, are designated by the variable F_6 where

$$F_6 = E_a/E_f$$

The ray lengths R_1, R_2, R_3 then must be adjusted (synthesized) for all feed angles to satisfy the condition

$$\sqrt{K_0}R_2/(R_1R_3) = \cos^2(\theta/2)F_6$$

This function is used to control the incremental surface shaping by introducing the correct values of the surface tangents into Adam's projector [19].

The basic equations for controlling the aperture phase function relates the feed rays R_1 to the R_2 rays reflected from the subreflector. For example, if the reflected rays from all subreflector points X_1, Y_1, Z_1 must be focused to some point X_0, Y_0, Z_0, then the direction cosines, L, M, N, of the R_2 rays must be

$$L = (X_0 - X_1)/R_2; \; M = (Y_0 - Y_1)/R_2; \; N = (Z_0 - Z_1)/R_2$$

and

$$R_2 = \sqrt{[(X_0 - X_1)^2 + (Y_0 - Y_1)^2 + (Z_0 - Z_1)^2]}$$

This or a similar phase control function on the reflected rays R_2 is used to construct the slope of the shaped subreflector in the incremental shaping process. Often the incident ray vector R_1 is the most difficult to find. The "true" incident ray directions for a corrugated conical horn feed pattern can be calculated for a finite number of incidence angles and the true ray direction cosines at any subreflector point obtained by interpolation among the set of discrete incident rays. For the corrugated horn it is also possible to compute the incident ray directions from series representations of the feed pattern. It is also possible in some cases to compute the incident ray direction on the subreflector surface by solving a set of nonlinear equations based on Snell's law of reflection from another surface (this is done in Chapter 7).

For both phase and amplitude shaping for control of aperture field distributions the two rays (vectors) R_1 and R_2 are calculated at each new point extrapolated by Adam's projector using the concepts just described. Then Snell's law must be inverted to obtain the slopes or partial derivative of the shape synthesized surface. This numerical step is accomplished by writing R_1, R_2 and the normal to the shaped surface $\mathbf{n}$ as unit vectors

$$\mathbf{r}_2 = \mathbf{r}_1 + 2(\mathbf{r}_1 \cdot \mathbf{n})\mathbf{n}$$

$$\mathbf{r}_1 = a_1\mathbf{x} + b_1\mathbf{y} + c_1\mathbf{z}$$

$$\mathbf{r}_2 = a_2\mathbf{x} + b_2\mathbf{y} + c_2\mathbf{z}$$

with a_1, b_1, c_1 and a_2, b_2, c_2 the direction cosines of the unit vectors $\mathbf{r}_1$ and $\mathbf{r}_2$ respectively. To find the subreflector normal $\mathbf{n}$ given $\mathbf{r}_1$ and $\mathbf{r}_2$ we write

$$\mathbf{n} = a_n\mathbf{x} + b_n\mathbf{y} + c_n\mathbf{z}$$

Inversion of Snell's law to calculate a_n, b_n, c_n is accomplished by writing

$$D = [2(1 - (a_1a_2 + b_1b_2 + c_1c_2))]^{1/2}$$

and

$$a_n = (a_1 - a_2)/D; \; b_n = (b_1 - b_2)/D; \; c_n = (c_1 - c_2)/D$$

from which the surface slopes $\partial z/\partial x = -a_n/c_n = K_8$ and $\partial z/\partial y = -b_n/c_n = K_9$ can be calculated. The variables K_8 and K_9 are used in shaping programs listed in Appendix 3A for these slopes.

Most of the algebraic difficulties in these numerical shaping programs are in solving quadratic equations to bring the coordinates of the main reflector and subreflector into a common coordinate system and to find the position of the moving focal point X_0, Y_0, Z_0 in amplitude shaping. For this latter calculation the lengths R_1, R_2, R_3 can be calculated from the ray-ratio relationship and the moving focal point must line on the R_2 vector. The proper ordering of the calculations is crucial in these numerical methods. Start-up slopes and points must be calculated for use in the Adam's projector from the reference surface which is, for the examples in this chapter, computed for the subreflector surface area near the midpoint (the point M of Figure 3.12).

The numerical shaping procedures are best understood by examining and running the examples programmed in RAYRAE.BAS (Gregorian surface shaping), RAYRAH.BAS (Cassegrain surface shaping), 4XSTUP.BAS (Gregorian surface shaping), and PARBUP.BAS (paraboloid surface shaping). These programs are listed in Appendix 3A and treat both amplitude and phase shaping problems.

3.11 NUMERICAL ALGORITHMS FOR POWER CONTROL

The offset antenna fed by a horn and the ellipsoid subreflector shown in Figure 3.12 produce an aperture power distribution $P(x, y)$ which depends on the horn feed pattern $G(\theta,\phi)$ and antenna configuration constants. The transformation of the feed pattern to the aperture distribution in dual-reflector antennas is discussed in Chapter 1. We will assume in the tutorial examples that the horn pattern has θ dependence only. The purpose of the demonstration of RAYRAE.BAS is to illustrate that (1) when the feed pattern is uniform in power and phase, the reference ellipsoidal subreflector is synthesized when uniform paraboloid aperture distribution is required; and (2) when a directive feed is programmed into the software a shaped subreflector results in order to produce a uniform aperture field on the main reflector.

We begin the shape synthesis procedure by finding the "reference" surface of the subreflector using the conic section solution for optimally tilted feed axis and ellipsoid axis. We enter the constants F, Y_c, R for the main reflector ($F = 70$, $Y_c = 62$, and $R = 48$ for the illustrating example) and the subreflector constants $E = 0.538$, and $C = 35$ into our Gregorian subreflector shaping program RAYRAE.BAS in lines 70 and 80. This program first computes (as in GREG.BAS) the tilt angle α, β for the reference (unshaped) ellipsoid and also computes and prints the center point and an edge point of the subreflector in X_1, Y_1, Z_1 coordinates with F_0 as focus and X_0, Y_0, X_0 with F_1 as focus (see Figure 3.12). These preliminary size data on the reference subreflector are for user orientation and for later comparison with a shaped subreflector. The coordinates used in the computation after the program pauses at line 545 for data check all have the F_1 point as origin.

The differential shape synthesis procedure requires start-up points in a small region about the midpoint M of the subreflector and these values are computed from the conic section in lines 570–710 of RAYRAE. These start-up values, X, Y, Z, and back slopes for three points are located on a rectangular grid, with spacing $D_1 = 0.1$ in both X and Y near the central zone of the subreflector.

When the feed pattern is uniform the aperture field is also uniform except for the ray-path attenuation function given in Table 3.1 for the offset Gregorian antenna. We will first show that for the above condition the shaping algorithm gives the ellipsoidal coordinates. Then the subreflector shape needed to convert the tapered horn pattern, $E = \cos^{MM}\theta$ of line 995, into a uniform main reflector aperture distribution will be computed by RAYRAE.BAS.

The procedure can best be understood by following the steps of RAYRAE.BAS. In lines 70–330 the antenna configuration constants are computed as in GREG.BAS. Then in lines 350–530 the subreflector center and the edge point at $\theta = A_3$, $\phi = 0$ for the unshaped subreflector are computed and displayed. The central ray path distant from the feed-phase center to the aperture plane, $Z_1 = 0$, is computed and saved as T_0. A normalizing constant, K_0, is also found for the central rays based on this ray-ratio expression for the central ray (see Figure 3.12 for ray definitions):

$$K_0 = R_3^2 R_1^2 / R_2^2 \text{ (see lines 500 and 520)} \tag{3.23}$$

As mentioned the numerical shaping algorithm requires start-up data. In lines 570–710 the small central surface of the conic section subreflector is computed and the three back slopes are calculated using the ellipsoid equation

$$X^2 + Y^2 + Z^2 = E^2(Z \cos \alpha - Y \sin \alpha + D)^2 \tag{3.24}$$

where in the program $\alpha = A_2$.

To find the slopes of this surface of revolution we need the partial derivatives $\partial Z/\partial X$ and $\partial Z/\partial Y$. These derivatives are found and programmed in lines 640–700. These values and the Z coordinates are stored in array S_0. $Z = S_0(I, 1)$ and $\partial Z/\partial X = S_0(I, 0)$ and $\partial Z/\partial Y = S_0(I, 2)$.

The variables X_0, Y_0, Z_0 (here) give the position of the second focal point of the subreflector in coordinates from origin F_1 of Figure 3.13. The direction cosines for the ray R_1 are given by L_1, M_1, and N_1. For R_2 they are L_2, M_2, and N_2. Lines 800–830 invert Snell's law to find the direction cosines of the surface normal, L_3, M_3, N_3. To find the pathlength R_3 we must find where a line lying along R_2 beginning at the subreflector point (X, Y, Z) intersects the paraboloid. This is complicated algebraically as we must write the paraboloid

$$x'^2 + y'^2 = F^2 + 4Fz'$$

in the same coordinates as the subreflector by the transform equations

$$\begin{aligned} y' &= Y_2 \cos A_4 + Z_2 \sin A_4 + C \sin A_1 \\ z' &= -Y_2 \sin A_4 + Z_2 \cos A_4 - C \cos A_1 \end{aligned} \tag{3.25}$$

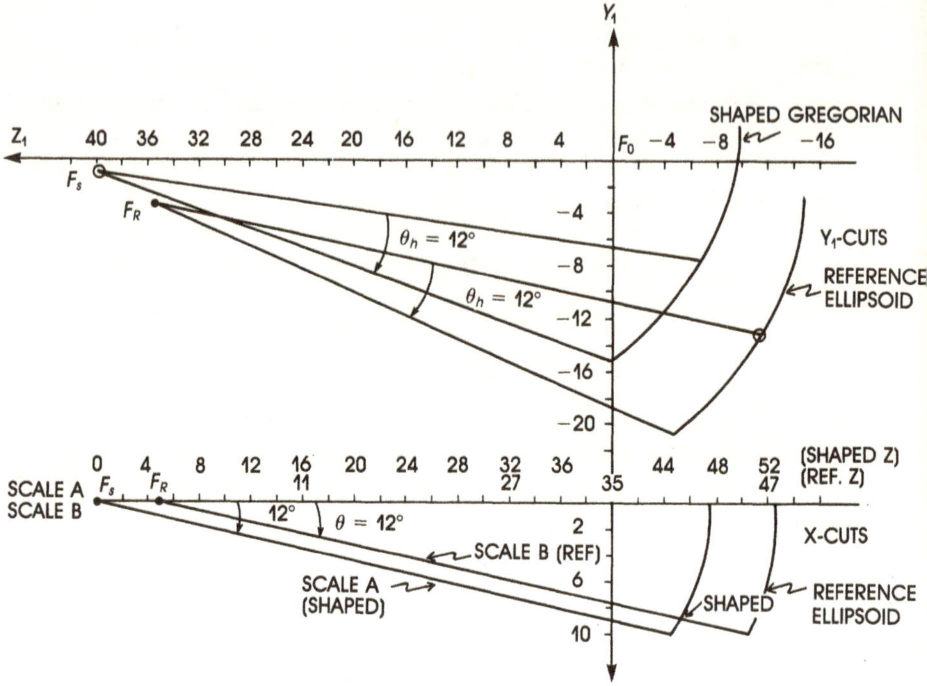

Figure 3.13 Gregorian Subreflector Contours for Uniform Aperture Power

The line equation along R_2 gives these relations

$$X_2 = L_2 Z_2/N_2 + L$$

$$Y_2 = M_2 Z_2/N_2 + M \tag{3.26}$$

with $L = X - L_2 Z/N_2$ and $M = Y - M_2 Z/N_2$. These equations are solved simultaneously in lines 840–940.

With this task done we have all the ray lengths $T = R_2 + R_3$, R_2, and R_1 needed for computing the ray ratios that relate the aperture power $P(x',y')$ to the feed-radiation pattern $P(\theta)$ (where $\theta = A_6$) by the equation discussed earlier

$$(K_0 R_2^2 G(\theta))/(R_3^2 R_1^2) = P(x',y') \tag{3.27}$$

A function $G = R_2/R_3$ is defined from (3.27) and $R_2 = GT/(1 + G)$ follows as shown in lines 990 and 1000. $T = R_2 + R_3$.

The function $P = \cos^4(\theta/2)$ accounting for ray-path attenuation on the unshaped system is included in line 990. The expressions are for fields, not power. Line 995 (when the REM command is deleted in a later example) introduces the feed pattern which is again assumed to be a $\cos^{MM}\theta$ power function.

Given the length of R_2 from line 1000 and using the fact that it lies along L_2, M_2, N_2, we compute the new location of the second (moving) focal point, X_0, Y_0, Z_0 in lines 1010 to 1060. The moving focal coordinates referred to the F_1 focus are displayed as X_F, Y_F, Z_F. The partials for the reflector slopes are $\partial Z/\partial X = K_8$ and $\partial Z/\partial Y = K_9$. We are working on the X-cut across the subreflector surface so K_8 only is used in the Adam's numerical projector equation [26] of line 1130:

$$Z_{(n+1)} = Z_n + D_1/24(55K_8 - 59K_5 + 37K_6 - 9K_7) \tag{3.28}$$

X is incremented by $D_1 = 0.1$ and we have only to repeat the process to generate an X-cut and the FOR loop on Y, line 570, continues the process to calculate more X-cuts. By storing the K_9 values and the X_0, Y_0, Z_0 (these are the focal point coordinates referred to F_1 which for in-power pattern synthesis is a moving point) values for the four X-cuts generated here, cuts in Y can be made and the entire subreflector shaped.

To get confidence in the method it is instructive first to omit line 995 with the REM command and compute the subreflector surface for uniform horn and aperture functions to get the conic section ellipsoid curve.

The subroutine 1220 prints the DELZ errors, difference between the synthesized Z values and the reference ellipsoid Z_E values in line 1420, and also an error E_1 which is the total of the path-length errors from F_1 to the aperture compared to T_0 for the central ray. The A_6 or feed θ angle and J, the array index, are also listed for every third point computed along the X-cuts. This subroutine was originally written to produce the back-up data for Y-cut programs that shape the entire subreflector.

The next exercise with this software is done by deleting the REM statements in lines 90 and 995. This makes operative the feed pattern $E = \cos^{88.85}\theta$. Now the program shapes the subreflector so that again uniform power density is maintained across the dish aperture (where the rays from this X-cut on the subreflector strike the paraboloid). By varying E, the reference ellipsoid eccentricity, and C, the expansion constant, a solution for the horn with $\theta_h = 12°$ can be found that spreads the power uniformly over the offset parabolic section. There is no unique solution, but by choosing $E = 0.69$ and $C = 40$ the subreflector is about the same size as for the reference unshaped ellipsoidal subreflector. It lies with different tilt angles α and β, however, as shown in Figure 3.13. Running this case with RAYRAE shows that the pathlength error E_1 is about 0.3 (length units) for this shaped subreflector when the main reflector is not shaped. Means for shaping the paraboloid are discussed in Section 3.12.2 for the numerically shaped Cassegrain.

3.11.1 Shaping Offset Cassegrain Antennas

The program RAYRAH.BAS performs the same task of shaping for power control using an offset Cassegrain configuration. The program is very similar to RAYRAE.BAS except for changes needed to convert from Gregorian to Cassegrain geometry. The hyperboloidal constants are from the CASS.BAS equations and are first computed and displayed as in RAYRAE. Again we compute the reference subreflector points at the center and for one X-cut edge point. Constants T_0 and K_0 are found for the offset Cassegrain geometry. The center point of the subreflector and the edge point in both X_0, Y_0, Z_0 (focus at F_1) and X_1, Y_1, Z_1 (focus at F_0) coordinates for the reference hyperboloid are also printed to aid in establishing reference subreflector size.

The changes to convert RAYRAE to RAYRAH consist mainly in sign changes. In addition the ray-ratio relation now has the form for Cassegrain geometry [*cf.* (3.27)]:

$$K_0 T^2 G(\theta)/(R_1^2(R_2 + T)^2 = P(x',y') \tag{3.29}$$

This leads to a change in R_2,

$$R_2 = GT/(1 - G)$$

The program subroutine has been enlarged to compute the reflected rays as line parameters from the subreflector and store these in the array $S(\ )$. These rays will be used (in program RAYRSH.BAS) to shape the main reflector to correct for aperture phase errors caused by subreflector shaping.

The BASICA interpreter will not give double precision on trigonometric functions. So only 10^{-5} or 10^{-6} accuracy is computed when the synthesized curve is subtracted from the ellipsoid curve (DELZ values) for the case of uniform feed power used to synthesize the subreflector for uniform aperture power. With double precision available from an MS-DOS compiler 10^{-8} or better is actually obtained. This can be checked by running the compiled program RAYRAH.EXE.

3.12 NUMERICAL SHAPING FOR PHASE CONTROL

Aperture phase errors can result from the phase pattern of the horn feed. HORN4 provides radiated phase versus θ data, which indicates the corrugated horn does not have a true center of phase or radiate spherical phase surfaces. We have seen how by refocusing some improvement in aperture phase errors can be realized. The subreflector surface and particularly the edges of the subreflector, however, can be shaped to eliminate, in an optical (GO) sense, the horn phase errors. We first convert the near-field horn phase function to a set of rays normal to the equiphase surface. The incident rays are reflected by the shaped subreflector to focus at the paraboloid focal point F_0.

For shaping to correct for phase-front deviations from spherical-wave front we can use program HORNF0.BAS with program 4XSTUP. BAS. An illustrating example of the shaping method for an offset Gregorian subreflector is shown in X-Z profile through its midpoint M in Figure 3.14(b). The reference or unshaped subreflector has eccentricity, $E = 0.4$, illuminating a paraboloid section of focus, $F = 22$, center at $Y_c = 24.5$ and dish radius of 18 units.

The technique used in HORNF0.BAS represents the equiphase surface of the horn phase pattern by a family of discrete rays normal to it. The ray striking any point on the shaped subreflector surface can be found by interpolation of these rays which is done in 4XSTUP.BAS. (It is sometimes possible to express explicitly the ray parameters at a given point in the numerical synthesis process, and this will be done in Chapter 8.)

3.12.1 Shaping Gregorian Subreflector Edges to Correct for Horn Phase Errors

The horn-phase pattern for a conical corrugated horn of semiflare angle $\theta_h = 21.5°$ computed by HORN4 on a sphere of radius R_2 referred to the apex of the horn is expressed in electrical degrees of phase error as a function of θ. This phase function is shown tabulated in the insert of Figure 3.14(b). It is written in the program HORNF0.BAS as a power series obtained from HORN5. Program HORNF0.BAS converts this horn phase information into parameters of a discrete set of rays (lines) to be used in a method that reshapes the subreflector so that all rays pass through the paraboloid focus.

Figure 3.14(b) illustrates the geometry and defines the variables for computing a discrete family of rays normal to the near-zone equiphase surface of the conical corrugated horn radiation near-field pattern at a radius $R_2 = 26$ units as computed using HORN4. We begin with the horn

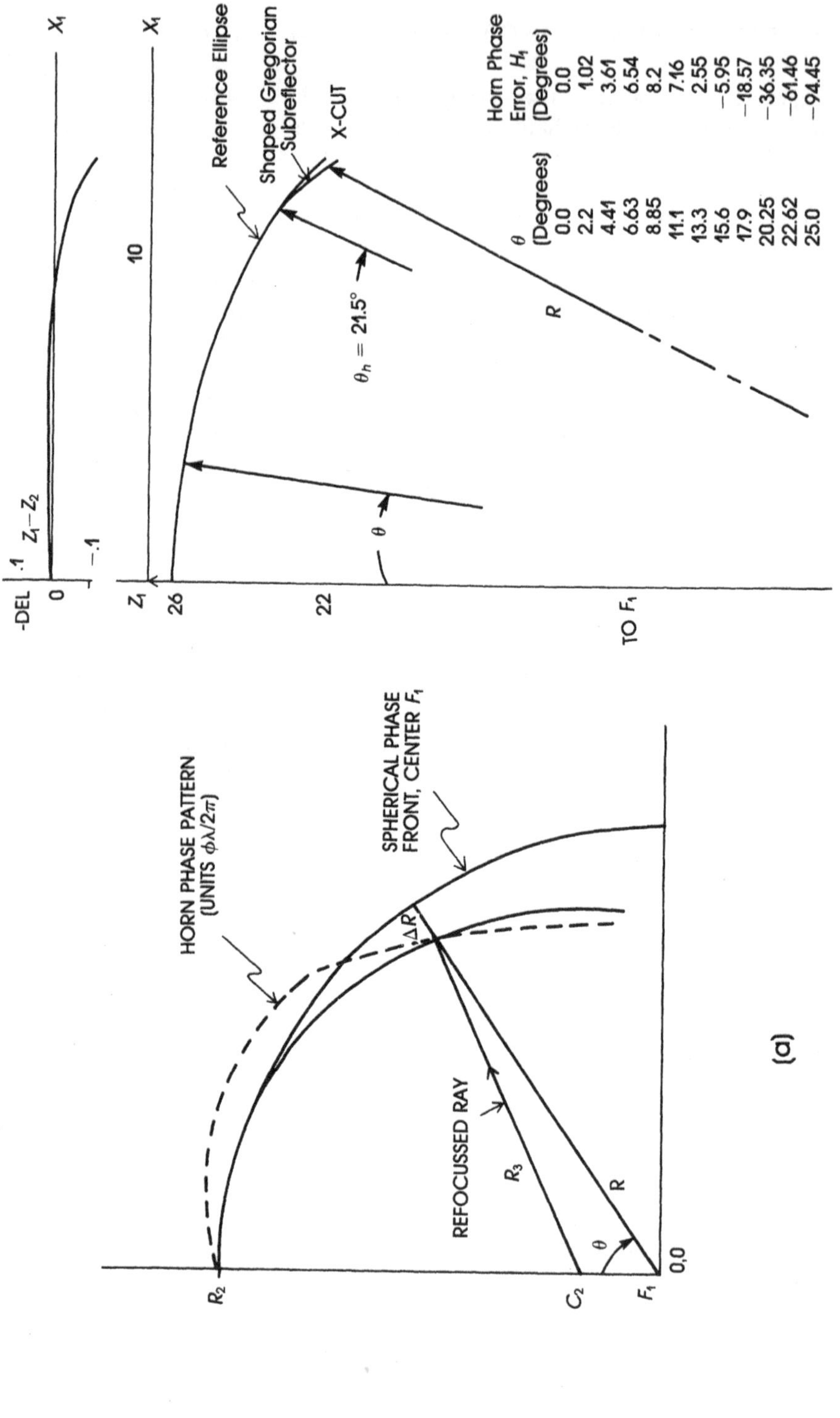

θ (Degrees)	Horn Phase Error, H_1 (Degrees)
0.0	0.0
2.2	1.02
4.41	3.61
6.63	6.54
8.85	8.2
11.1	7.16
13.3	2.55
15.6	−5.95
17.9	−18.57
20.25	−36.35
22.62	−61.46
25.0	−94.45

Figure 3.14 (a) Method for Correcting Horn Phase Errors; (b) Subreflector Shaping for Correcting Horn Phase Errors

phase function $\Phi(\theta)$ which is assumed to be independent of ϕ. Phase is referred to a spherical phase front of radius R_2 passing through the subreflector focus F_1. To change radian phase deviations from the spherical phase front to distances Δr, we write,

$$\Phi = H_1 = B_1\theta^2 + B_2\theta^4 + B_3\theta^6 + B_4\theta^8 + B_5\theta^{10}$$
$$k'\Phi(\theta) = \Delta r(\theta) = r - R_2 \tag{3.30}$$

where $k' = (\lambda)/2\pi$.

To find rays normal to the surface $r(\theta)$, we take the gradient in spherical coordinates of (3.30)

$$f = r - R_2 - k'\Phi$$

$$\mathbf{N} = \nabla(f) = \mathbf{r} - 1/r\ \partial(k'\Phi)/\partial\theta\boldsymbol{\theta}$$

There are no ϕ components; $\mathbf{r}$ and $\boldsymbol{\theta}$ are unit vectors.

$$k'\partial\Phi/\partial\theta = \partial r/\partial\theta = R_1 \text{ from (3.30)}$$

or

$$\mathbf{n} = O_1\mathbf{r} + O_2\boldsymbol{\theta}$$

where $\mathbf{n}$ is a unit vector normal to the horn equiphase front, and

$$O_1 = 1/\sqrt{(1 + (R_1/r)^2)} \text{ and } O_2 = -(R_1/r)/\sqrt{(1 + (R_1/r)^2)}$$

These variables, where ($r = R$), can be identified in program HORNF0.BAS. The horn phase function has been represented by a power series with coefficients B_1, B_2, B_3, B_4, B_5 identified by the DATA lines 60 and 80. The constants R_2 and K are in DATA line 70. The partial derivative with respect to $\theta = A_4 = A_1$ is done in line 210.

To return to rectangular coordinates the spherical unit vectors are transformed to direction cosines L_1, M_1, N_1. Rays normal to the surface are written as lines passing through the points X_1, X_2, X_3 (these are rectangular coordinates) at the end of R. The line equations for each ray are

$$X = k_xZ + e_x$$
$$Y = k_yZ + e_y \tag{3.31}$$

where $k_x = L_1/N_1$ and $k_y = M_1/N_1$ and $E_x = e_x = L_1/N_1(-Z + C_2) + X$ and $E_y = e_y = M_1/N_1(-Z + C_2)$. C_2 can be used to represent the center of phase of the horn if one exists. These line parameters are stored in the array $S()$ for use in the shaping program 4XSTUP.BAS.

Program HORNF0.BAS runs without interactive data inputs. It prints out THETA and HORN PHASE for the sequence of array index numbers I. Also printed on the screen are the discrete line parameters e_x, e_y, k_x, and k_y. The program 4XSTUP.BAS is chained in by HORNF0.BAS, so that it must be available in the BASIC diskette or directory in use.

The set of lines from HORNF0.BAS carry the phase error data from the horn. Typically these phase errors are largest at the edges of the subreflector for larger values of θ. This means that by shaping only the edges of the subreflector the horn phase errors are corrected.

Unlike shaping for control of antenna aperture amplitude, shaping the entire subreflector surface (when necessary) for phase control does not require shaping of the main reflector. Phase shaping to correct horn phase nonsphericity causes all the rays to pass through the paraboloid focus. Note from the HORN4 data that the (near-field) horn phase function changes with wavelength so that the wideband characteristics of the antenna need to be studied when subreflector shaping is used to correct horn phase errors. A compromise shaping correction can be done at some middle frequency when two narrow but separated frequency bands need to be accommodated.

In 4XSTUP.BAS, four X_1 cuts are computed across a central strip of the subreflector in the $+X_1$ direction starting again, as for the amplitude shaping programs, with a small area or section at the middle of a reference subreflector (an optimal tilted subreflector surface). The constants for the example reference subreflector are seen in lines 60–80. The variables have the same symbols as in GREG.BAS except that $A = A_2 = \alpha$. In lines 110–260, the central start-up surface points and slopes from the reference ellipsoid are computed. The rays or lines from HORNF0 are stacked in rows with 22 points on a row so that the increase of array index by 23 brings one to the next higher row. The interpolation procedure in lines 310–420 substitutes sequentially the subreflector point X_1, Y_1, Z_1 (these coordinates have origin at F_1) into equation,

$$\begin{aligned} G_1 &= k_x Z_1 + e_x - X_1 \\ G_3 &= k_y Z_1 + e_y - Y_1 \end{aligned} \tag{3.32}$$

where k_x, e_x, K_y, e_y are the line parameters of the discrete rays carrying the horn phase error. First the program finds an error, G_1, in the x-z plane

($G_1 = 0$ and $G_3 = 0$ if a discrete ray (line) is incident on the point X_1, Y_1, Z_1) and notes when this function changes signs as line parameters are called from the array $S()$. The error G_2 before the sign change is saved and used in lines 340 and 350 to find the slope of the surface at the ray incidence point of the subreflector being synthesized (which generally is not at one of the line intersection points of the finite set of HORNF0 rays). The process is repeated in the y-z plane with error function G_3 in lines 370–420. This interpolation effort gives us the direction cosines A_6, B_6, C_6 of the "true" incident ray from the horn at the point X_1, Y_1, Z_1. We require that the reflected ray pass through the focus of the paraboloid ($X_0 = 0, Y_0, Z_0$) which in the subreflector coordinates is given in line 90 and is used to compute the direction cosines of the desired reflected ray in lines 520–540. Again Snell's law is inverted to find the normal with direction cosines A_3, B_3, C_3 in lines 550–580. The surface slopes as partial derivatives are K_8 (with respect to X_1) and K_9 (in Y_1). Adam's projector in the X_1 direction is used in lines 610–630 to find the next point and the process is repeated until the end of the X_1 cut is reached at $X_1 = 10.8$ for this example.

In later application the Z_1 and K_9 data from 4-X_1 cuts is stored in an array to be used as start-up data for the Y-up and Y-down cuts to synthesize the entire subreflector surface. Only half of the subreflector surface is needed due to left-right symmetry of the subreflector about the $X_1 = 0$ plane.

3.12.2 Shaping Cassegrain Antennas for Control of Amplitude and Phase

The shaping of an axially symmetric Cassegrain to achieve uniform dish aperture distribution, as carried out using ICAS.FOR in Section 3.8 for the antenna of Figure 3.11 by the integral technique, can also be done using the ray-ratio technique. This is done using program RAYRSH.BAS chained to the program PARBUP.BAS. RAYRSH.BAS (which is very similar to RAYRAH.BAS) computes the coordinates of the shaped Cassegrain subreflector. PARBUP.BAS computes the shaped main reflector profile to correct for the phase errors introduced by amplitude control shaping by RAYRSH.BAS.

To shape the symmetric main dish we first demonstrate with programs PARBF0.BAS and PARBUP.BAS that when the feed pattern at the focus of a paraboloid is free of phase errors [fed by a perfect horn, for example, in which e_x and e_y of the horn phase pattern discrete rays (3.31) are always zero], then the numerical shaping process synthesizes a parabolic curve.

To shape the paraboloid surface to correct for aperture phase errors caused by amplitude shaping of the subreflector the program 4XSTUP is modified to give PARBUP.BAS. Now the rays incident on the main reflector are reflected so that they are parallel to the Z axis (not passing through a focal point as in 4XSTUP). The family of incident rays for the example are saved from program RAYRSH.BAS, the Cassegrain shaping program modified for the symmetric antenna.

To demonstrate that the main reflector shaping technique gives the reference paraboloid, the HORNF0 program is modified to make a truly spherical phase front, i.e., the discrete family of lines all pass through the prime focus, F_0. This modification is done in PARBF0.BAS. When PARBF0.BAS is run and chained to PARBUP.BAS, the original profile of the paraboloid is obtained indicating that when the illuminating rays all pass through the prime focus the synthesis procedure produces the parabolic curve.

To compare these numerical shaping methods with the integral methods discussed earlier, the offset geometry for the Cassegrain geometry has been specialized to the symmetric case treated in Section 3.8. Program RAYRSH.BAS is edited to include a feed pattern, $E_h = \cos^N \theta$, where $N = 72.6849/2$ for the horn fields on the subreflector over a 14.25° angular sector subtended by the shaped Cassegrain. The selection of initial ($\theta = 0$) conditions and reference hyperboloid parameters for the small start-up area is done so that when $\theta = 14.25°$ the 30-unit paraboloidal aperture radius (X_2) is illuminated by the horn rays reflected from the subreflector. When the eccentricity $E = 1.275$ and $C = 17$ for the start-up area, then the subreflector curve is similar to that obtained by the integral shaping method of ICAS.FOR. Again, as in the integral method, there are many subreflectors that will produce uniform dish amplitude distributions. The reference subreflector parameters E and C can be chosen for control of the start-up central section, and they also determine with the feed function the shape of the subreflector surface.

Shaping of the main reflector to compensate for phase errors due to shaping for uniform aperture power can now be done by chaining the program RAYRSH.BAS to PARBUP.BAS. Now the original paraboloid surface is obtained in the vertex region and the shaped main reflector departs from the reference paraboloid as Y increases. The Z values are longer, that is, DELZs are positive in the PARBUP.BAS printouts. We see then that the two surface shaping methods to achieve uniform aperture distribution with the same feed illumination lead to different solutions. The ICAS program moves the vertex section of the paraboloid back from the subreflector to compensate for phase errors due to amplitude shaping of the subreflector. The programs RAYRSH and PARABUP leave the vertex region of the paraboloid unchanged, but shape the dish to correct phase errors.

As RAYRSH.BAS can be used for offset Cassegrains, it is evident that by making the array $S()$ large enough to contain a set of discrete rays for the entire subreflector surface the main reflector can be shaped for offset main reflectors. By using 4XSTUP.BAS and similar Y-cut programs, the entire shaped offset subreflector surface can be synthesized.

The subreflector and main reflector shapes obtained by the integral method and the ray-ratio methods are not identical. The ICAS.FOR programs modify the parabolodial shape of the main reflector in the vertex region, whereas the ray-ratio numerical method is controlled by $\cos^N\theta$ which is 1 at $\theta = 0$, and so the true paraboloid is constructed near the vertex. The main reflector is, however, shaped at larger θ angles for both methods. Either shape synthesis technique can be extended and programmed to shape offset dual-reflector antennas for control of aperture field amplitudes.

3.13 METHODS FOR CHECKING REFLECTOR SURFACE ACCURACY

Numerical surface shaping techniques as discussed above lead to point defined surfaces. These three-dimensional surfaces usually have a square grid in two coordinates, here x and y with the z coordinate defining the shaped surface. It is useful to have techniques for measuring the surfaces after they are constructed and for analytically checking the reflection properties of the shaped surfaces. The program TRACE.BAS is provided for analytical testing of point defined surfaces and the program MEASE.BAS is useful for measuring surface tolerances.

In [20] the equations for first finding where an illuminating ray strikes a point defined surface and then computing the reflected ray direction are derived. The procedure is based on finite-difference approximations to terms in a Taylor series expansion for the reflector surface. The steps in the finite-difference calculus [20] are outlined to identify the variable and terms in the program TRACE.BAS. In the example given, a small section of an ellipsoidal reflector surface is computed by the program at a finite set of points with X and Y spacings both T. To demonstrate the method a trial incident ray is entered to see if the program can find the impact point on a known, but point-defined, surface and calculate the correct direction cosines of the reflected ray. For general application, which will be done in Chapter 8, the points defining the trial (usually shaped) reflector surface are entered into the program from an array in a data file. The target rays incident on the point-defined reflector are entered into the program, also with a data file, using their four-line parameters.

TRACE.BAS is listed in Appendix 3A. The reflector surfaces are defined on a square grid by a one-dimensional array, $S_2()$. Because the grid of points are regular, only the Z values are stored in the array and the X and Y coordinates are obtained from the index K of the array. (See lines 490–510. This is a memory-saving technique to handle large surfaces in BASIC.)

The constants E, D, C_S, and $A = \alpha$ for a strip of ellipsoidal point-defined reflector are given in line 80 of TRACE.BAS and the surface coordinates computed in lines 120–230. The first exercise is to demonstrate that, by entering a ray defined by its line parameters that intercepts a known point on the elliptical surface, the program finds that point. When the incident ray passes through one ellipsoid focus, the reflected ray passes approximately through the other focus.

3.13.1 Ray Tracing Point-Defined Surfaces

$\mathbf{S}_1 = a\mathbf{x} + b\mathbf{x} + c\mathbf{z}$ can be written as a line referred to coordinates of Figure 3.15 as

$$x = k_x z + x_0$$

$$y = k_y z + y_0$$

where in TRACE.BAS $k_x = a/c = K_1$ and $k_y = b/c = K_2$. Also $x_0 = E_1$ and $y_0 = E_2$.

The search process for finding the intercept of the ray (line) on the surface is done in steps. First, "miss" functions G_1 and G_2 are defined in lines 350–360, and the absolute amount of miss calculated in line 370. The increments Δx and Δy are both equal to T, which is taken in the demonstration to be 0.5.

The normal mode of using the program is to enter K_1, K_2, E_1, and E_2 in line 280 defining the incident ray. For the demonstration, however, we enter the true intercept points on the ellipse X_3 and Y_3, and compute the Z_3 value from the ellipsoid equation. For this known case we check to see if the finite-difference calculus of [20] again computes the values X_3 and Y_3, which are identified in the program as X_9 and Y_9.

The computer search for the grid point P_A of Figure 3.15 which is only an approximate solution is done in steps 285–470. The eight other points can be identified in the grid in relation to P_A. The first order solution using two terms of the Taylor series only uses five points: P_A, P_B, P_D, P_E, and P_F. The more accurate solution uses all nine grid points surrounding and including P_A.

The calculations in lines 550–720 are omitted, as they are not needed. (The lines illustrate some algebraic steps in [20].) The first-order (two-term) finite solution using five points is done in lines 730–1060. The quadratic or second-order solution of the finite Taylor series is done in lines 1080–1190. The direction cosine of the reflected ray is computed and compared with a true ray through the second focus of the ellipse in lines 1200–1370.

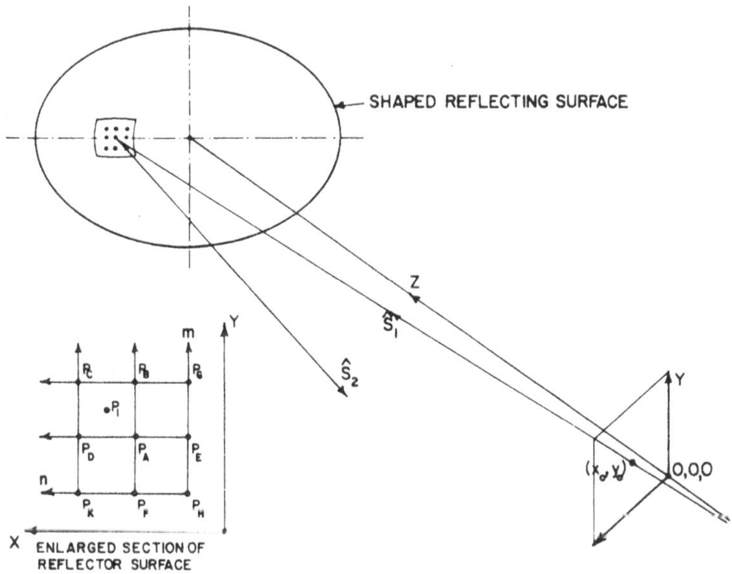

Figure 3.15 Ray Tracing on Shaped Reflector Antennas (From Sletten, *Proc. IEEE* [20]. Reprinted with permission.)

It is helpful to know the variable symbols used in TRACE to represent the partial derivatives in the Taylor series of [20]. The values of Z and the locations X, Y for the nine grid points are obtained from the array $S_2()$ by lines 730–860. With these values we can define the partials using the following variables:

$$U_1 = (Z_D - Z_A)/T \text{ and } U_2 = (Z_A - Z_E)/T$$

$$V_1 = (Z_B - Z_A)/T \text{ and } V_2 = (Z_A - Z_F)/T$$

Then, taking the partial derivatives, we write:

$$\partial^2 z/\partial x^2 = (U_1 - U_2)/T = W_1 \text{ and } \partial^2 z/\partial y^2 = (V_1 - V_2)/T = W_2$$

$$\partial z/\partial x = U_1 - W_1 T/2 \text{ and } \partial z/\partial y = V_1 - W_2 T/2$$

$$P_5 = (Z_C - Z_G)/(2T) \text{ and } P_6 = (Z_K - Z_H)/(2T)$$

$$\partial^2 z/\partial x \partial y = (P_5 - P_6)/(2T)$$

To demonstrate the program's capability to find the intercept point on the ellipsoidal strip, the trial points X_3, Y_3 must not be too near the borders of the strip in order that a complete set of nine surrounding points can be used by the program. The strip is ten units tall (Y) and only three units wide (X). Numbers like 7.111 and 1.111 are acceptable for X_3 and Y_3. Also, the error values E_1 and E_2 must be set to zero if the reflected ray is to pass through the second focus at X_0, Y_0, Z_0. For the demonstration in this tutorial program X_0 and Y_0 should always be set to zero and the trial impact points X_3, Y_3 chosen not too near to the strip boundaries in order that the eight surrounding points are available.

3.13.2 Measuring Reflector Surface Errors

Surface deviations from computed values on reflectors generally produce gain losses or increases in sidelobes [21,22]. Computer techniques are useful aids in measuring the mechanical surface errors and for relating the errors to antenna performance.

For the statistical method for relating gain and sidelobe levels to surface deviations only two parameters, the rms surface errors and the correlation distances typical of these surface deviations from the desired reflector surface, which is usually a paraboloid, are needed. We can define the root mean squared error, rms,

$$\text{rms} = [(1/n)\Sigma\varepsilon_i^2]^{1/2}$$

where ε is the surface deviation from the desired surface. The correlation length is generally rather vaguely determined by relating it to the average size "dents" or to the panel structure of the surface.

A more accurate way to estimate the effects of surface errors and distortion analytically is to deterministically enter the measured values of surface errors (or departures from the wanted surface) into the numerical computation of the antenna radiation pattern using a program like

PATT.FOR. Because it is not practical to measure enough points for the computation grids, some method for interpolation from the measured points is needed.

There are several methods for accurately measuring the position of points on a surface: using optical theodolites, laser ranging, and milling tools measuring probe distances from a reference plane. To demonstrate the value of software in surface accuracy measurements the last method is developed in the program MEASE.BAS.

Referring to Figures 3.16(a) and 3.16(b), an offset paraboloid surface of given focal length F has four accurately located points A, B, C, and D, on its surface. The dish rim describes a circle in the aperture plane. The measurement plane of the probing instrument is located in front of the four reference points, and it is approximately aligned relative to the antenna reference plane (which is tangent to the paraboloid at the midpoint ($X_c = 0, Y_c, Z_c$). There is, however, some error in aligning the measuring plane such that the probe measurements perpendicular from the measurement plane to the points D_1, D_2, D_3 and D_4 are not equal. We need to relate points on the paraboloid X_1, Y_1, Z_1 with probe points, X_3, Y_3, Z_3, on the measurement plane. The origin for points in the measurement plane is located above and perpendicular to the check point A on the dish. Given F, Y_c, D_1, D_2, D_3, and D_4, and the locations of the points A, B, C, and D, we compute the midpoint tangent plane of the offset paraboloidal aperture plane from

$$Z_1 = F - (X_1^2 + Y_1^2)/(4F)$$

and

$$\tan \theta_0 = 2F/Y_c$$

for the slope of the paraboloid.

The intercepts of points, X_1, Y_1, Z_1 on the paraboloid, made by lines normal to the measurement plane, are given labels X_2, Y_2, Z_2. For the tilt angles defined by Figure 3.16(a), unit vectors along the axis of the measurement plane in terms of the x_1, y_1, z_1 unit vectors (the coordinate frame of the paraboloid) are,

$$\mathbf{x}_3 = \cos O_2\mathbf{x}_1 + \sin O_2 \sin A_2\mathbf{y}_1 + \sin O_2 \cos A_2\mathbf{z}_1$$

$$\mathbf{y}_3 = \cos A_2\mathbf{y}_1 - \sin A_2\mathbf{z}_1$$

$$\begin{aligned}\mathbf{z}_3 &= -\sin O_2\mathbf{x}_1 + \sin A_2 \cos O_2\mathbf{y}_1 + \cos A_2 \cos O_2\mathbf{z}_1 \\ &= C_1\mathbf{x}_1 + C_2\mathbf{y}_1 + C_3\mathbf{z}_1\end{aligned}$$

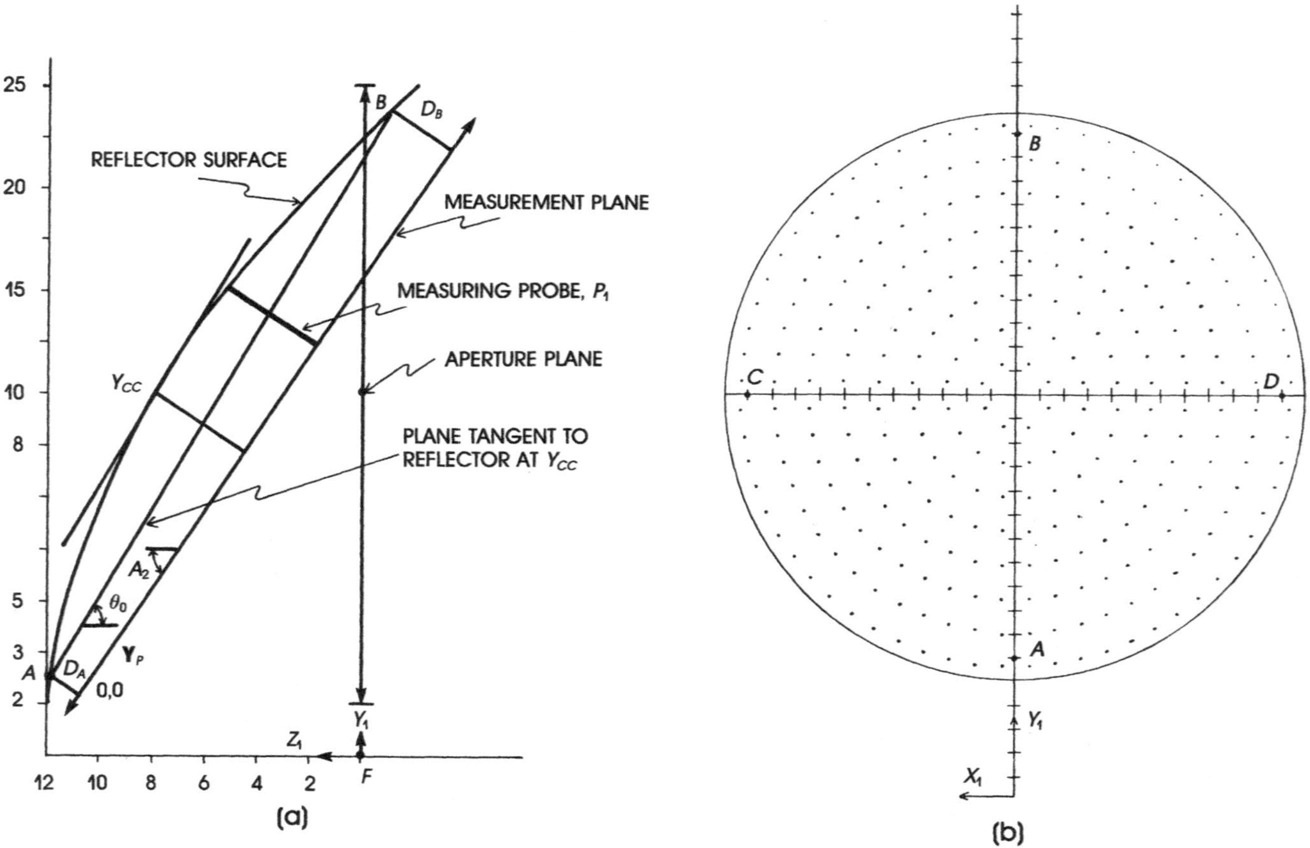

Figure 3.16 (a) Profile of Probe Method for Measuring Surface Errors; (b) Aperture Plane Diagram Locating Reference Points A, B, C, D and Probe Measurement Points

where O_2 is the x-plane tilt angle and A_2 is the y-plane tilt angle of the measurement plane as computed in the program MEASE.BAS. The probe lies along $\mathbf{z}_3$ and the measurement plane is normal to $\mathbf{z}_3$. From the definition of distances from a plane to a line we get the probe depths to any point on the reflector surface. The coordinate transform equations are seen in lines 450–470.

The selection of points on the dish surface is done to make the points nearly equally distributed over the aperture. The method for doing this is due to L.I. Schwab, consultant to GTE, and is included with his permission.

When interpolation from measured probe points to general aperture points closely spaced and on a regular grid is desired r,θ coordinates on the dish aperture plane can be used and linear range and angle interpolation done to get error values for the grid points used in pattern integration. It is well to select the measured points on the edge of panels where discontinuities in surface shape and gaps between panels are likely to occur. In [23] such an interpolation method was programmed for smooth reflector panels having r,θ geometry to compute the radiation patterns for measured surface errors.

The program MEASE.BAS is meant for entering probe measurements, P1. To illustrate the method of computing rms values, however, a random error function is introduced where experimental measurements would normally be entered. The last printout gives K, the number of points measured, V, the sum of errors squared, and rms, the root mean squared error for the reflector.

3.14 COMMENTS ON THE TECHNIQUES

The comparison using computer examples of symmetric, offset, prime-focus and subreflector feeding tend to show that each configuration can give similar antenna gains and near-in sidelobe performance. The design choices are often based on construction considerations. The trade-offs with offset subreflectors to achieve lower cross polarization and less aperture blocking are discussed.

Some important performance improvements can be realized with shaped dual-reflector antennas. The very high aperture efficiencies attained by shaped Cassegrains (or Gregorians) are of some economic importance. Subreflectors can also provide means for obtaining low near-in and far-out sidelobes. Subreflector shaping to synthesize ideal aperture distributions in a gain versus sidelobe sense is possible. Also phase errors caused by horn phase errors can be corrected by subreflector shaping.

The radiation from subreflector edges is critical for control of both near-in and far-out sidelobes. Good data on edge diffraction can be obtained using the UGTD methods described in Chapter 4. Wilkinson [24] used "undersized" Gregorian subreflectors with absorbing shields around both the main and subreflector edges to achieve very good sidelobe performance. In reference [25] and in Chapter 8 "oversized" subreflectors are suggested with shaping to control or reduce both near-in and far-out sidelobes.

Subreflector and especially main reflector shaping for offset subreflector antennas is costly because surfaces lack axial symmetry. Shaping for feed phase errors only [26] can be done without shaping the main reflector. Subreflector shaping only may be justified, if the feed refocusing techniques are inadequate.

By using computer simulation to investigate the performance of several antennas over the frequency bands of interest, the best configuration can be selected using the methods of this chapter. Fine tuning can then be done, using the diffraction techniques of Chapter 4. Finally, experimental models of reflectors and feeds can be developed for many applications, using techniques and suggestions from Chapters 2, 3, 7, and 8.

References

[1] Clarricoats, P.J.B., and G.T. Poulton, "High Efficiency Microwave Reflector Antenna—a Review," Proc. IEEE, Vol. 65, No. 10, Oct. 1977.

[2] Rudge, A.W., and N.A. Adatia, "Offset-Parabolic-Reflector Antennas: a Review," Proc. IEEE, Vol. 66, No. 12, Dec. 1978.

[3] Jones, E.M.T., "Paraboloid Reflector and Hyperboloid Lens Antennas," IRE Trans., Antennas and Propagation, Vol. AP-2, pp. 119–127, July 1954.

[4] Collin, R., and F.J. Zucker, *Antenna Theory,* Part II, Chapters 16 and 17. New York: McGraw Hill, 1969.

[5] Silver, S., *Microwave Antenna Theory and Design,* pp. 169–199, New York: McGraw Hill, 1949.

[6] Spencer, R.C., "Fourier Integral Methods of Pattern Analysis," MIT Rad. Lab. Rep. No. 762-1, Jan 1946.

[7] Spencer, R.C., "Tables and Methods of Calculation for Line Sources," MIT Rad. Lab. Rep. No. 762-2.
[8] Lee, S.H., and R.C. Rudduck, "Aperture Integration and GTD Techniques Used in the NEC Reflector Antenna Code," IEEE Trans., Vol. AP-33, No. 2, pp. 189–194, Feb. 1985.
[9] Ruze, J., "Feed Support Blockage Loss in Paraboloid Antennas," Microwave Journal, pp. 76–80, Dec. 1968.
[9A] Ko, W.L., R. Mittra, and S.W. Lee, "Aperture Blockage in Reflector Antennas," IEEE Trans., Vol. AP-32, No. 3. pp. 282–286, Mar. 1984.
[10] Collin, R., and F.J. Zucker, *Antenna Theory,* Part I, pp. 460–563, New York: McGraw Hill 1969.
[11] Ludwig, A.C., "The Definition of Cross Polarization," IEEE Trans., Vol. AP-21, pp. 116–119, Jan. 1973.
[12] Adatia, N.A., and A.W. Rudge, "Bean-squint in Circularly Polarised Offset Reflector Antennas," Electronics Letters, pp. 513–515, Oct. 1975.
[13] Spencer, R.C., "Synthesis of Microwave Diffraction Patterns with Application to Csc$^2\theta$ Patterns," MIT Rad. Lab. Rept. 54-24, June 23, 1943.
[14] Williams, W.F., "High Efficiency Antenna Reflector," Microwave Journal, Vol. 8, pp. 79–82, July 1965.
[15] Galindo, V., "Design of Dual Reflector Antennas with Arbitrary Phase and Amplitude Distributions," PTGAP Digest, Boulder, CO, July 1963.
[16] Lee, J.J., L.I. Parad, and R.S. Chu, "A Shaped Offset-fed Dual Reflector Antenna," IEEE Trans., Vol. AP-27, No. 2, pp. 165–171, Mar. 1979.
[17] Galindo-Isreal, V., R. Mittra, and A.G. Cha, "Aperture Amplitude and Phase Control of Offset Dual Reflectors," IEEE Trans., Vol. AP-27, No. 2, pp. 154–164, Mar. 1979.
[18] Sletten, C.J., "Numerical Technique for Shaping Reflector Surfaces to Synthesize Antenna Patterns," Proc. IEEE, Vol. 69, No. 3, pp. 392–394, Mar. 1981.
[19] Press, W.H., B.P. Flannery, S.A. Teukolsky, and W.T. Vetterling, *Numerical Recipes,* "The Art of Scientific Computing," pp. 569–572, Cambridge Univ. Press, 1986.
[20] Sletten, C.J., "Ray Tracing Method for Doubly Curved Reflector Surfaces," Proc. IEEE, Vol. 69, No. 6, June 1981.
[21] Meeks, M.L., and J. Ruze, "Evaluation of Haystack Antenna and Radome," IEEE Trans., Vol. AP-19, No. 6, Nov. 1971.

[22] Bao, Vu The, "Influence of Correlation Interval and Illumination Taper in Antenna Tolerance Theory," Proc. IEEE, Vol. 116, No. 2, Feb. 1969.

[23] 30/20 GHz Satellite Ground Terminals, Design Plan Phase II, Part 2, GTE-CSD Report for US Army Sat. Com. Agency, Contr. DAAB07-82-C-J227/00004, June 1985.

[24] Burdine, B.H., and E.J. Wilkinson, "A Low Sidelobe Earth Station Antenna for 4/6 GHz Band," Microwave Journal, pp. 53–58, Nov. 1980.

[25] Sletten, C.J., W. Payne, and W. Shillue, "Offset Dual Reflector Antennas for Very Low Sidelobes," Microwave Journal, May 1986.

[26] Faigen, I.M., C.F. Reichert, C.J. Sletten, and R.A. Shore, "Reflector Antennas with Low Sidelobes, Low Cross Polarization, and High Aperture Efficiency," NASA CR 174670, March 1984.

Appendix 3A
List of Chapter 3 Computer Programs and Source Code Listings of Selected Programs

In this appendix the names of computer programs developed especially for Chapter 3 are listed. The source code for some of the simpler, basic programs are included. These tutorial programs are discussed in the text, but by studying and running the code, detailed information on the design software is clearly illustrated. All these programs are available on 5¼-inch diskettes in Microsoft BASIC and FORTRAN for IBM PC and compatible computers in the optional software package with user's manual and example booklet.

The source code printouts attached have the program names in the first few lines near the top of the listings. The suffix .FOR indicates a FORTRAN program and .BAS a BASIC program.

SOFTWARE LIST, CHAPTER 3

PROGRAM NAME; ASSOCIATED PROGAMS, FILES AND SUBROUTINES:

*SYMPF.FOR; DISH.DAT, PHAMP.DAT, ALAM.DAT, PATT.FOR, CJS.DAT, PLPATT.BAS
*SLOTS.BAS
*DIPOL.BAS
SYMPF2.FOR; DISH3.DAT
SYMPFH.FOR; DISH3.DAT
*ASYMM.BAS
OFFSET.FOR; DISH1.DAT, PHAMP.DAT, ALAM.DAT, PATT.FOR, CJS.DAT, PLPATT.BAS
*APRX.BAS
*PFCAS.BAS

SUBPAT.FOR; DISH2.DAT, PHAMP.DAT, ALAM.DAT, PATT.FOR, CJS.DAT, PLPATT.BAS
RESIDS.BAS
*CSC.FOR; CSC.DAT, QTRAP.FOR, TRAPZD.FOR
ICAS.FOR; QTRP.FOR, TRAPZD.FOR
*RAYRAE.BAS
*HORNF0.BAS
*4XSTUP.BAS
RAYRAH.BAS
PARBF0.BAS
PARBUP.BAS
RAYRSH.BAS
*TRACE.BAS
*MEASE.BAS

* These programs are printed in this appendix.

QTRAP.FOR and TRAPZD.FOR are copyrighted Numerical Recipes subroutines that must be linked to CSC.FOR and ICAS.FOR. The output data obtained by running CSC for the example problem is listed with its source code. The pattern programs SYMPF.FOR, OFFSET.FOR, and SUBPAT.FOR must be run to termination to avoid deleting the data files PHAMP.DAT and ALAM.DAT. Programs (mentioned in the text) SPILL.BAS, CROSS.BAS, and PATT.FOR are developed and described in Chapter 1.

```
      PROGRAM SYMPF
C     Computes amplitude distribution on Symmetric Prime Focus
C     fed Parboloids.
      EXTERNAL PHSAMP
      COMMON A,F,NN
      REAL LAM,NN
      DATA PI/3.141592654/

      OPEN(3,FILE='DISH.DAT',STATUS='OLD')
      REWIND 3
      READ(3,'(4F11.6)') A,LAM,NN,F
      READ(3,'(2I4)') NRAD,NANG

      WRITE(*,8) A,LAM,NN,F
8     FORMAT(' A=',F11.6,' LAM=',F11.6,' NN=',F11.6,' F=',F11.6)
      WRITE(*,11) NRAD,NANG
11    FORMAT(' NRAD=',I4,' NANG=',I4)
      ANRAD=PI/180.0
      TWOPI=2.0*PI
      OPEN(4,FILE='PHAMP.DAT',STATUS='NEW',FORM='UNFORMATTED')
      OPEN(7,FILE='ALAM.DAT',STATUS='NEW',FORM='UNFORMATTED')
      WRITE(7) A,LAM
      WRITE(7) NRAD,NANG
      PAUSE'FILES OK?'

C     Calculates aperture gird points
      DELPHI=TWOPI/FLOAT(NANG-1)
      DELRAD=1./FLOAT(NRAD-1)
      PHI0= -PI
      R0=0.0

      DO 101 I=1,NANG

      PHI=PHI0+(I-1)*DELPHI
      DO 100 J=1, NRAD
      R=R0+(J-1)*DELRAD
      X0=A*R*COS(PHI)
      Y0=A*R*SIN(PHI)

      CALL PHSAMP(AMP,PHAS,X0,Y0,LAM)

      WRITE(4) AMP,PHAS
      WRITE(*,98) I,AMP,PHAS
98    FORMAT(' I=',I4,' AMP=',F12.6,' PHAS='E13.4)

100   CONTINUE
101   CONTINUE
      STOP
      END

C     Computes Aperture Illumination from Feed Pattern
      SUBROUTINE PHSAMP(AMP,PHAS,X0,Y0,LAM)
      COMMON A,F,NN
      REAL LAM,NN

      Z0=F-(X0*X0+Y0*Y0)/4./F
      R1=SQRT(X0*X0+Y0*Y0+Z0*Z0)
      TH=ACOS(Z0/R1)
      AMP=COS(TH)**(NN/2.)*COS(TH/2.)**2.
      PHAS=0.0

      RETURN
      END
```

```
10 REM Demonstration program to show polarized fields for
20 REM a Dipole or a pair of Dipoles as primary feeds.
30 PRINT "DIPOL.BAS"
40 PI=3.14159265359# : LC=.4342944819#
50 E=.000001
60 INPUT "F=focal length of Symm. Parab. ",F
70 INPUT "Aperture point, x,y ",X,Y
80 Z=F-(X*X+Y*Y)/4/F
90 R=SQR(X*X+Y*Y+Z*Z)
100 T=Z/R
110 A=ATN(SQR(1-T*T)/T)
115 XE=X+E
120 B=ATN(Y/XE)
130 IF XE<=0 THEN B=B+PI
140 REM Slot or dipole scalar pattern
150 F0=COS(PI/2*SIN(B)*SIN(A))/(1-(SIN(B)*SIN(A))^2)
160 REM Slot pair array pattern
170 F7=COS(PI/2*SIN(A)*COS(B))
180 REM Feed polarization components, F1 is theta, F2 is phi.
190 F1=F0*F7
195 REM Transformed unit vector theta in x,y,z unit vectors.
200 T1=SIN(A)^2*SIN(B)*COS(B)
202 T2=-(1-(SIN(A)*SIN(B))^2)
204 T3=SIN(A)*SIN(B)*COS(A)
250 REM Computing unit vector normal to paraboloid surface.
260 R1=SQR(X*X+Y*Y+4*F*F)
270 L=X/R1 : M=Y/R1 : N=2*F/R1
280 EPX=T1*F1
290 EPY=T2*F1
300 EPZ=T3*F1
310 REM Using Snell's law for E components on parab. surface.
320 B3=2*(L*EPX+M*EPY+N*EPZ)
330 EXA=B3*L-EPX
340 EYA=B3*M-EPY
350 EZA=B3*N-EPZ
360 P=10*LC*LOG((EYA+E)^2)
370 PX=10*LC*LOG((EXA+E)^2)
380 PRINT "dB=";P
390 PRINT "dBx=";PX
400 PRINT "Theta=";A/PI*180;"Phi=";B/PI*180
410 PRINT "F0=";F0;"Fpair=";F7;"F1=";F1
420 PRINT "EPX=";EPX;"EPY=";EPY;"EPZ=";EPZ
430 PRINT "EXA=";EXA;"EYA=";EYA;"EZA=";EZA
440 GOTO 70
450 END
```

```
10 REM Demonstration program to show polarized fields for
20 REM slots or dipoles as primary feeds.
30 PRINT "SLOTS.BAS"
40 PI=3.14159265359# : LC=.4342944819#
50 E=.000001
60 INPUT "F=focal length of Symm. Parab. ",F
70 INPUT "Aperture point, x,y ",X,Y
80 Z=F-(X*X+Y*Y)/4/F
90 R=SQR(X*X+Y*Y+Z*Z)
100 T=Z/R
110 A=ATN(SQR(1-T*T)/T)
120 B=ATN(Y/(X+E))
130 IF X<=0 THEN B=B+PI
140 REM Slot or dipole scalar pattern
150 F0=COS(PI/2*SIN(B)*SIN(A))/(1-(SIN(B)*SIN(A))^2)
160 REM Slot pair array pattern
170 F7=COS(PI/2*SIN(A)*COS(B))
180 REM Feed polarization components, F1 is theta, F2 is phi.
190 F1=F0*F7*COS(B)
200 F2=-F0*F7*SIN(B)*COS(A)
210 REM unit vector theta in x,y,z unit vectors
220 T1=COS(A)*COS(B) : T2=COS(A)*SIN(B) : T3=-SIN(A)
230 REM Unit vector phi in x,y unit vectors
240 P1=-SIN(B)   : P2=COS(B)   : P3=0
250 REM Computing unit vecto normal to paraboloid surface.
260 R1=SQR(X*X+Y*Y+4*F*F)
270 L=X/R1 : M=Y/R1 : N=2*F/R1
280 EPX=T1*F1+P1*F2
290 EPY=T2*F1+P2*F2
300 EPZ=T3*F1
310 REM Using Snell's law for E components on parab. surface.
320 B3=2*(L*EPX+M*EPY+N*EPZ)
330 EXA=B3*L-EPX
340 EYA=B3*M-EPY
350 EZA=B3*N-EPZ
360 PX=10*LC*LOG((EYA+E)^2)
370 P=10*LC*LOG(EXA^2)
380 PRINT "dB=";P
390 PRINT "dBx=";PX
400 PRINT "Theta=";A/PI*180;"Phi=";B/PI*180
410 PRINT "F0=";F0;"Fpair=";F7;"F1=";F1;"F2=";F2
420 PRINT "EPX=";EPX;"EPY=";EPY;"EPZ=";EPZ
430 PRINT "EXA=";EXA;"EYA=";EYA;"EZA=";EZA
440 GOTO 70
450 END
```

```
10 REM *Calculates the aperture fields for an Offset*
20 REM *paraboloid fed by horn with cos(theta)^MM pattern.*
30 PRINT "ASYMM.BAS"
40 PI=3.14159265359# : Y3=35 : A0=60.51287 : J=1
50 F=30 : EP=.00001 : LC=.4342944819# : MM=11
60 INPUT "Tilt angle of feed, A0, Degs.=",A0
70 A0=A0*PI/180
80 B0=0 : RA=0 : GOTO 110
90 REM The Feed is a Y-polarized Huygen's Source Horn.
100 PRINT : INPUT "Enter PHI in degs., RA: ",B0,RA
110 B0=B0*PI/180
120 X2=RA*COS(B0) : Y2=RA*SIN(B0)
130 X=X2+EP : Y=Y2+Y3
140 Z=F-(X*X+Y*Y)/4/F
150 R1=SQR(X*X+Y*Y+4*F*F)
160 L=X/R1 : M=Y/R1 : N=2*F/R1
170 REM Rotation of coordinates to tilt, Z1-axis
180 Y1=Y*COS(A0)-Z*SIN(A0)
190 Z1=Y*SIN(A0)+Z*COS(A0)
200 X1=X
210 REM PRINT "X1=";X1;"Y1=";Y1;"Z1=";Z1
220 REM PRINT "L,M,N=";L;M;N
230 R=SQR(X*X+Y*Y+Z*Z)
240 T=(Z1)/R - EP
250 A=ATN(SQR(1-T*T)/T)
260 B=ATN(Y1/X1)
270      IF X1<=0 THEN B=B+PI
280 IF J=1 THEN R0=R : J=0
290 E0=COS(A)^(MM/2)*R0/R
300 REM Huygen source polarization
310 L1=-(1-COS(A))*SIN(B)*COS(B)
320 M1=1-SIN(B)^2*(1-COS(A))
330 N1=-SIN(A)*SIN(B)
340 L2=L1*E0
350 REM Rotation of polarization vectors back to Z-axis.
360 M2=E0*(M1*COS(A0)+N1*SIN(A0))
370 N2=E0*(N1*COS(A0)-M1*SIN(A0))
380 B2=2*(L*L2+M*M2+N*N2)
390 L3=B2*L-L2
400 M3=B2*M-M2
410 N3=B2*N-N2
420 PRINT "A=";A*180/PI
430 REM PRINT "L1,M1,N1=";L1;M1;N1
440 PRINT "E2,E2,E2=";L2;M2;N2
450 PRINT "E3,E3,E3=";L3;M3;N3
460 P0=10*LC*LOG(M3*M3)
470 P00=10*LC*LOG(E0*E0)
480 PRINT "X=";X2;"Y=";Y;"Z=";Z;"R=";R
490 P1=10*LC*LOG(L3*L3)
500 PRINT "Poo=";P00;"Po=";P0;"Px=";P1
510 GOTO 100
520 END
```

```
10 REM First approximations to symmetric paraboloid performance.
20 PRINT "APRX.BAS"
30 PI=3.14159265359# : LC=.4342944819#
40 INPUT "Enter frequency, gHz ",FQ
50 INPUT "Enter 1 for centimeters, 2 for inches ", M
60 L=30/FQ
70 IF M=1 THEN GOTO 90
80 L=30/FQ/2.54
90 PRINT "Wavelength is ";L
100 INPUT "Enter diameter of dish ",D
110 BW=1.1*L/D*180/PI
120 PRINT "Beamwidth, -3dB, is approx.";BW;"degrees"
130 G0=(2*PI*D/2/L)^2
140 DB0=10*LC*LOG(G0)
150 PRINT "Maximum antenna gain is ";DB0;"dB
160 PRINT
170 INPUT "Enter paraboloid focal length ",F
180 Z=F-(D/2)^2/4/F
190 AE=ATN(D/2/Z)
200 INPUT "Enter dish edge taper in dB (neg.) ",DBE
210 N=(DBE/10-4*LC*LOG(COS(AE/2)))/(LC*LOG(COS(AE)))
220 PRINT "Horn pattern approx. by Cos(theta)^N is N=";N
230 PRINT "Edge angle for symmetric dish is ";AE*180/PI;"degrees
240 PRINT "The f/D of antenna is ";F/D
250 END
```

```
10 REM Prime focus vs. Cassegrain feeding compared.
20 PRINT "PFCAS.BAS"
30 PRINT "Prime focus-fed, axially symmetric dish"
40 PI=3.14159265359#
50 N=1.1076 : F=20
60 FOR Y=0 TO 30 STEP 2
70 Z=F-Y*Y/4/F
80 A=ATN(Y/Z)
90 E1=COS(A) (N/2)*COS(A/2)^2
100 PRINT "Y=";Y;"A=";A*180/PI;"E1=";E1
110 NEXT Y
120 INPUT "PRESS RETURN TO CONTINUE ",A$
130 PRINT "Cassegrain-fed, axially symmetric dish"
140 MM=72.6849
150 E=1.4
160 M=(E+1)/(E-1)
170 B=2*F*M
180 PRINT "M=";M
190 PRINT "B=";B
200 FOR A2=.25 TO 14.25
210 A1=A2*PI/180
220 R=B*TAN(A1/2)
230 E2=COS(A1)^(MM/2)*COS(A1/2)^2
240 PRINT "R=";R;"A1=";A2;"E2=";E2
250 NEXT A2
260 END
```

```
      PROGRAM CSC
C     Computes csc^2 pattern from Holt
      PARAMETER(PIO2=1.5707963)
      EXTERNAL FUNC
      COMMON C1,C2,A0
      ANRAD=2.*PIO2/180.0
      OPEN(3,FILE='CSC.DAT',STATUS='OLD')
      REWIND 3
      READ(3,'(2F15.9)') C1,C2
      READ(3,'(4F11.5,I4)') A,B0,A0,R0,M

      WRITE(*,'(A,2F15.9)') ' C1,C2= ', C1,C2
      WRITE(*,'(A,4F11.5,I4)') ' A,B0,A0,R0,M= ',A,B0,A0,R0,M
      A=A*ANRAD
      B0=B0*ANRAD
      A0=A0*ANRAD

      DO 10 I=1,M
      B=FLOAT(I)/FLOAT(M)*(B0-A) + A
      CALL QTRAP(FUNC,A,B,S)
      BA=B/ANRAD
      R=EXP(S)*R0
      WRITE(*,'(1X,A,F12.6,A,F12.6)') ' THETA=',BA,' R=',R
10    CONTINUE

      END

      FUNCTION FUNC(X)
      COMMON C1,C2,A0
      F=ATAN(1./(C1-C2*(3./8.*(X-A0)+3./16.*SIN(2.*(X-A0))+
     *COS(X-A0)**3*SIN(X-A0)/4.)))
C     WRITE(*,'(A,F12.6)') ' F=',F
      FUNC=TAN((X-F)/2.)
      END
```

```
C>CSC
C1,C2=      6.787280000    8.526291000
A,B0,A0,R0,M=    -60.00000   30.00000  -15.00000   30.00000  20
 THETA=  -55.500000 R=   28.595710
 THETA=  -51.000000 R=   27.367670
 THETA=  -46.500000 R=   26.293300
 THETA=  -42.000000 R=   25.353850
 THETA=  -37.500000 R=   24.533630
 THETA=  -33.000000 R=   23.819470
 THETA=  -28.500000 R=   23.200150
 THETA=  -24.000000 R=   22.666110
 THETA=  -19.500000 R=   22.209060
 THETA=  -15.000000 R=   21.821750
 THETA=  -10.500000 R=   21.497740
 THETA=   -6.000000 R=   21.231160
 THETA=   -1.500000 R=   21.016610
 THETA=    3.000000 R=   20.849080
 THETA=    7.500000 R=   20.723990
 THETA=   12.000000 R=   20.637320
 THETA=   16.500000 R=   20.585980
 THETA=   21.000000 R=   20.568230
 THETA=   25.500000 R=   20.584120
 THETA=   30.000000 R=   20.635780

C>
```

```
10 REM "For ellipsoidal offset subreflector shaping for amplitude
20 REM "control using ray tube method.
30 PRINT "INPUT CONFIGURATION CONSTANTS"
40 PI=3.141592654#
50 PRINT "RAYRAE.BAS"
60 RANG=180/PI
70 I=0 : J=0 : XE=10 : MM=88.5822
80 E=.538 : Y3=62 : F=70 : C=35 : R=48
85 REM INPUT "Y3,F,R ";Y3,F,R
90 REM INPUT "ENTER E,C "; E,C
100 PRINT "E=";E;"YC=";Y3;"F=";F;"C=";C;"R=";R
110 PRINT
120 T1=Y3/F : A4=T1^2/4 +1
130 B4=-T1/(2*E)*(1+E^2)
140 C4=(T1/(4*E)*(1+E^2))^2-T1^2/4
150 S3=(-B4-SQR(B4^2-4*A4*C4))/(2*A4)
160 A1=ATN(S3/SQR(1-S3^2))
170 S4=SIN(A1)*(1-E^2)/((1+E^2)*COS(A1)-2*E)
180 A2=ATN(S4)
190 D=C*(1-E^2)/(2*E^2)
200 B=2*F*(1-E^2)/(1+E^2-2*E*COS(A1))
210 S5=R/B : S6=ATN(S5) : A3=S6*2
220 AB4=A2-A1 : M1=B/(2*F) : A4=AB4
230 AA1=A1*RANG : AA2=A2*RANG : AA3=A3*RANG : AA4=AB4*RANG
240 PRINT "FOR THE ELLIPSOID:" : PRINT
250 PRINT "BETA=";AA1
260 PRINT "ALPHA=";AA2
270 PRINT "HORN SEMIFLARE ANGLE=";AA3
280 PRINT "ELLIPSOID CONSTANT, D=";D
290 PRINT "MAGNIFICATION=";M1
300 PRINT "ALPHA-BETA=";AA4
310 PRINT "TRANSFORM CONSTANT, B=";B
320 PRINT "ECCENTRICITY, E=";E
330 REM
340 ANGR=PI/180
350 REM  Computes subreflector center and edge point.
360 A6=0 :A7=0
370 R1=E*D/(1-E*COS(A2)*COS(A6)+E*SIN(A2)*SIN(A6)*SIN(A7))
380 X0=R1*SIN(A6)*COS(A7)
390 Y0=R1*SIN(A6)*SIN(A7)
400 Z0=R1*COS(A6)
410 PRINT "X0=";X0;"Y0=";Y0;"Z0=";Z0
420 X1=X0 : Y1=Y0*COS(A4)+Z0*SIN(A4)+C*SIN(A1)
430 Z1=-Y0*SIN(A4)+Z0*COS(A4)-C*COS(A1)
440 PRINT "X1=";X1;"Y1=";-Y1;"Z1=";-Z1
450 R2=2*D*E/(1-E^2)-R1
460 X=-B*TAN(A6/2)*COS(A7)
470 Y=-Y3-B*TAN(A6/2)*SIN(A7)
480 Z=(X*X+Y*Y)/4/F-F
490 R3=SQR(X*X+Y*Y+Z*Z)
500 KL=R3^2*R1^2/R2^2
510 TL=R1+R2+R3-Z
520 IF I=0 THEN I=1 : A6=A3 : T0=TL : K0=KL : GOTO 370
530 PRINT "T0=";T0;"K0=";K0
540 PRINT : PRINT : INPUT "Check values, Press return ",L$
```

```
550 DIM S0(12,2),S1(143,4)
560 D1=.1 : F6=1
570 FOR Y=0 TO 3*D1 STEP D1
580 FOR X=0 TO -3*D1 STEP -D1
590 I=-X/D1
600 A5=COS(A2)^2-1/E^2
610 B5=2*(D*COS(A2)-Y*COS(A2)*SIN(A2))
620 C5=D^2-(X^2+Y^2)/E^2+Y^2*SIN(A2)^2-2*D*Y*SIN(A2)
630 S0(I,1)=(-B5-SQR(B5^2-4*A5*C5))/(2*A5)
640 P1=X/E^2
650 RS=COS(A2)*S0(I,1)-Y*SIN(A2)+D
660 F0=Y/E^2-COS(A2)*RS
670 F3=S0(I,1)/E^2-COS(A2)*RS
680 S0(I,0)=-P1/F3
690 S0(I,2)=-F0/F3
700 PRINT S0(I,1),S0(I,0)
710 NEXT X
720 PRINT
730 Z=S0(0,1) : K5=S0(1,0) : K6=S0(2,0) : K7=S0(3,0)
740 X0=0 : Y0=-C*SIN(A2) : Z0=C*COS(A2)
750 X=.00001
760 R1=SQR(X^2+Y^2+Z^2)
770 L1=X/R1 : M1=Y/R1 : N1=Z/R1
780 R2=SQR((X0-X)^2+(Y0-Y)^2+(Z0-Z)^2)
790 L2=(X0-X)/R2 : M2=(Y0-Y)/R2 : N2=(Z0-Z)/R2
800 D4=SQR(2*(1-(L1*L2+M1*M2+N1*N2)))
810 L3=(L1-L2)/D4
820 M3=(M1-M2)/D4
830 N3=(N1-N2)/D4
840 L=X-L2/N2*Z
850 M=Y-M2/N2*Z
860 AQ=L2^2/N2^2+(M2/N2*COS(A4)+SIN(A4))^2
870 B7=2*(COS(A4)*(COS(A4)*M*M2/N2+M2/N2*C*SIN(A1)-2*F+M*SIN(A4)))
880 B8=2*(SIN(A4)*(C*SIN(A1)+2*F*M2/N2)+L*L2/N2)
890 B6=B7+B8
900 C6=L^2+COS(A4)^2*M^2+C^2*SIN(A1)^2+2*C*M*COS(A4)*SIN(A1)-4*F^2+4*F*SIN(A4)
+4*F*C*COS(A1)
910 Z2=(-B6-SQR(B6^2-4*AQ*C6))/(2*AQ)
920 X2=L2/N2*Z2+L
930 Y2=M2/N2*Z2+M
935 REM Prints paraboloid coordinates in common coordinates.
940  REM PRINT "X2=";X2;"Y2=";Y2;"Z2=";Z2
950 T=SQR((X2-X)^2+(Y2-Y)^2+(Z2-Z)^2)
960 T1=Y/X : A7=ATN(T1)
970 T2=X/(Z*COS(A7))
980 A6=ATN(T2)
985 REM Shaping process starts with theta=A6.
990 G=R1*COS(A6/2)^2/SQR(K0)*F6
995 REM G=G/COS(A6)^MM
1000 R2=G*T/(1+G)
1010 A9=(L2/N2)^2+(M2/N2)^2+1
1020 B9=2*(L2/N2*(L-X)+M2/N2*(M-Y)-Z)
1030 C9=R1^2-R2^2+L^2+M^2-2*X*L-2*Y*M
1040 Z0=(-B9-SQR(B9^2-4*A9*C9))/(2*A9)
1050 X0=Z0*L2/N2+L
1060 Y0=Z0*M2/N2+M
```

```
1065 REM Prints moving focal coordinates near Fo.
1070 PRINT "XF=";X0;"YF=";Y0;"ZF=";Z0
1080 K8=-L3/N3
1090 K9=-M3/N3
1100 IF I>=3 THEN GOSUB 1220
1110 I=I+1
1120 X=X+D1
1130 Z=Z+D1/24*(55*K8-59*K5+37*K6-9*K7)
1140 K7=K6 : K6=K5 : K5=K8
1150 PRINT "X=";X;"Y=";Y;"Z=";Z : PRINT
1160 IF X>=XE THEN GOTO 1200
1170 REM INPUT "STOP";A$
1180 GOTO 760
1190 J=360*Y
1200 NEXT Y
1210 END
1220 L4=L2
1230 M4=M2*COS(A4)+N2*SIN(A4)
1240 N4=N2*COS(A4)-M2*SIN(A4)
1250 X3=X2
1260 Y3=Y2*COS(A4)+Z2*SIN(A4)+C*SIN(A1)
1270 Z3=-Y2*SIN(A4)+Z2*COS(A4)-C*COS(A1)
1280 T3=SQR(X3^2+Y3^2+4*F^2)
1290 L5=X3/T3 : M5=Y3/T3 : N5=-2*F/T3
1300 T4=-2*(L4*L5+M4*M5+N4*N5)
1310 L6=L4+T4*L5
1320 M6=M4+T4*M5
1330 N6=N4+T4*N5
1340 X4=-L6/N6*Z3+X3
1350 Y4=-M6/N6*Z3+Y3
1360 T5=SQR((X4-X3)^2+(Y4-Y3)^2+Z3^2)
1370 E1=R1+T+T5-T0
1380 A5=COS(A2)^2-1/E^2
1390 B5=2*(D*COS(A2)-Y*COS(A2)*SIN(A2))
1400 C5=D^2-(X^2+Y^2)/E^2+Y^2*SIN(A2)^2-2*D*Y*SIN(A2)
1410 ZE=(-B5-SQR(B5^2-4*A5*C5))/(2*A5)
1420 DELZ=Z-ZE
1430 PRINT "J=";J;"E1=";E1;"DELZ=";DELZ;"A6=";A6*RANG
1440 REM S1(J,0)=Z : S1(J,1)=K9 : S1(J,2)=X0 : S1(J,3)=Y0 : S1(J,4)=Z0
1450 I=0 : J=J+1
1460 RETURN
1470 END
```

```
10 REM Find kx,ex,ky,ey for Horn phase functions.
20 PRINT "HORNFO.BAS"
30 DIM S(550,3)
40 COMMON S()
50 PI=3.14159265359#
55 REM Horn phase pattern data.
60 READ R2,L,C2,E1,K,B1,B2,B3,B4,B5
70 DATA 26,.59,0,.000001,.0939891396#
80 DATA 12.5074,-332.003,2774.04,-12773.3,22589
85 REM Grid of X,Y points on spherical cap.
90 FOR Y=0 TO 4 STEP .5
100 FOR X=0 TO 11 STEP .5
110 X1=X : X2=Y
120 I=2*X1+46*X2
130 IF X1=0 THEN X1=E1
135 REM Computes A1=theta for grid points.
140 A2=ATN(X2/X1)
150 T1=X1/(R2*COS(A2))
160 A1=ATN(T1/SQR(1-T1^2))
170 A4=A1
175 REM Finds Phase errror H1 from power series.
180 H1=B1*A4^2+B2*A4^4+B3*A4^6+B4*A4^8+B5*A4^10
190 IF X=0 THEN H1=0
200 R=R2+K*H1
205 REM Finds derivative of horn phase curve.
210 R1=K*(2*B1*A4+4*B2*A4^3+6*B3*A4^5+8*B4*A4^7+10*B5*A4^9)
215 REM Computes gradients in spherical and rectangular coordinates.
220 H2=R1/R
230 H3=SQR(1+H2^2)
240 O1=1/H3
250 O2=-H2/H3
260 X1=R*SIN(A1)*COS(A2)
270 X2=R*SIN(A1)*SIN(A2)
280 X3=R*COS(A1)
290 L1=O1*SIN(A1)*COS(A2)+O2*COS(A1)*COS(A2)
300 M1=O1*SIN(A1)*SIN(A2)+O2*COS(A1)*SIN(A2)
310 N1=O1*COS(A1)-O2*SIN(A1)
320 S(I,0)=L1/N1
330 S(I,1)=S(I,0)*(-X3+C2)+X1
340 S(I,2)=M1/N1
350 S(I,3)=S(I,2)*(-X3+C2)+X2
360 PRINT "I=";I;"THETA,A1=";A1*180/PI;"HORN PHASE,H1=";H1*180/PI
370 PRINT "Ex=";S(I,1);"Ey=";S(I,3);"kx=";S(I,0);"ky=";S(I,2)
380 PRINT
390 IF A1>.46  THEN GOTO 410
400 NEXT X
410 NEXT Y
420      CHAIN "4XSTUP.BAS"
430 END
```

```
10 REM For shaping offset Gregorian subreflector
20 REM using HORNFO ray data.  Shapes surface to correct for
25 REM horn phase errors. Must be CHAINed, can't run alone.
30 PRINT "4XSTUP.BAS"
40 DIM S0(4,2)
50 PI=3.14159265359#
55 REM Data for example Gregorian subreflector.
60 D1=.1 : D0=.1 : F=22
70 E=.4 : YC=24.5 : C=17 : R=18
80 A1=15.752354# : A=35.77891 : D=44.625
85 REM INPUT "E,YC,F,C,R,BETA,ALPHA,D ",E,YC,F,C,R,A1,A,D
88 A1=A1*PI/180 : A=A*PI/180
90 X0=0 : Y0=-C*SIN(A) : Z0=C*COS(A)
100 COMMON S()
105 REM Computes start-up points and slopes from middle
107 REM zone of reference ellipsoid.
110 FOR Y=0 TO .3 STEP .1
120 FOR X=0 TO -.2 STEP -.1
130 I=-10*X
140 AQ=COS(A)^2-1/E^2
150 BQ=2*(D*COS(A)-Y*COS(A)*SIN(A))
160 CQ=D^2-(X^2+Y^2)/E^2+Y^2*SIN(A)^2-2*D*Y*SIN(A)
170 Z=(-BQ-SQR(BQ^2-4*AQ*CQ))/(2*AQ)
180 S0(I,1)=Z
190 P1=X/E^2
200 R=COS(A)*S0(I,1)-Y*SIN(A)+D
210 F0=Y/E^2+SIN(A)*R
220 F3=Z/E^2-COS(A)*R
230 S0(I,0)=-P1/F3
240 S0(I,2)=-F0/F3
250 PRINT "Z=";Z;"dz/dx=";S0(I,0);"dz/dy=";S0(I,2)
260 NEXT X
270 INPUT "PAUSE, PRESS RETURN ",A$
280 PRINT S0(1,1);S0(2,0)
290 I=0 : G2=0 : G1=0 : Z1=S0(1,1) : Y1=Y : K5=S0(0,0)
300 X1=.1 : K6=S0(1,0) : K7=S0(2,0) : G4=0 : G3=0
310 G2=G1
320 G1=S(I,0)*Z1+S(I,1)-X1
330 IF G1<=0 THEN I=I+1 : GOTO 310
340 F1=G1/(G1-G2)
350 K3=S(I,0)-F1*(S(I,0)-S(I-1,0))
360 I=I-1
365 REM y-cut interpolation for true ray intercept.
370 G4=G3
380 G3=S(I,2)*Z1+S(I,3)-Y1
390 IF G3<=0 THEN I=I+23 : GOTO 370
400 F2=G3/(G3-G4)
```

```
410 K4=S(I,2)-F2*(S(I,2)-S(I-23,2))
420 I=I-23
430 AQ=COS(A)^2-1/E^2
440 BQ=2*(D*COS(A)-Y*COS(A)*SIN(A))
450 CQ=D^2-(X1^2+Y^2)/E^2+Y^2*SIN(A)^2-2*D*Y*SIN(A)
460 Z2=(-BQ-SQR(BQ^2-4*AQ*CQ))/(2*AQ)
470 PRINT "X1=";X1;"Z1=";Z1;"Z2=";Z2;"DEL=";Z1-Z2
475 REM Inverts Snell's law for shaped normal.
480 T1=SQR(K3^2+K4^2+1)
490 C6=1/T1
500 A6=K3/T1
510 B6=K4/T1
520 T2=SQR(X1^2+(Y0-Y1)^2+(Z0-Z1)^2)
530 A7=-X1/T2 : B7=(Y0-Y1)/T2
540 C7=(Z0-Z1)/T2
550 T3=SQR(2*(1-(A6*A7+B6*B7 +C6*C7)))
560 A3=(A6-A7)/T3
570 B3=(B6-B7)/T3
580 C3=(C6-C7)/T3
590 K8=-A3/C3
600 K9=-B3/C3
605 REM Uses Adams projector for shaped Z1 points.
610 X1=X1+D0
620 Z1=Z1+D0/24*(55*K8-59*K5+37*K6-9*K7)
630 K7=K6 : K6=K5 : K5=K8
640 IF X1>=10.8 THEN PRINT X1,Z1 : GOTO 660
650 GOTO 310
660 NEXT Y
670 END
```

```
10 REM For tracing rays on a point defined surface.
20 REM In this example of the technique a vertical strip of
30 REM an Ellipsoid is target surface.
40 PRINT "TRACE.BAS"
50 DIM S2(980),S0(10,2)
60 PI=3.141593 : N=0
70 REM Constants of Ellipsoid
80 E=.538# : A=17.00461*PI/180 : CS=35 : D=42.9607454292#
90 U0=0  : U1=3  : V0=0  : V1=10
100     INPUT "Point spacing, T= ",T
110 K3=INT((V1-V0)/T+1.5)
115 REM Computes target surface strip.
120 FOR X=U0 TO U1 STEP T
130 FOR Y=V0 TO V1 STEP T
140 K=INT(((Y-V0)+K3*(X-U0))/T +.5)
150 T1=COS(A)^2-1/E^2
160 T2=2*(D*COS(A)-Y*COS(A)*SIN(A))
170 T3=D^2-(X^2+Y^2)/E^2+Y^2*SIN(A)^2-2*D*Y*SIN(A)
180 Z=(-T2-SQR(T2^2-4*T1*T3))/(2*T1)
190 IF N=1 THEN Z3=Z : GOTO 260
200 S2(K)=Z
210 PRINT K;Z;X;Y
220 NEXT Y
230 NEXT X
240 INPUT "ENTER X0,Y0,X3,Y3= ",E1,E2,X3,Y3
250 N=1 : X=X3 : Y=Y3 : GOTO 150
260 K1=(X3-E1)/Z3
270 K2=(Y3-E2)/Z3
280 PRINT "E1=";E1;"E2=";E2;"K1=";K1;"K2=";K2
285 REM Search for point closest to X3,Y3.
290 K=0 :          N=0
300 C1=99999! :C2=99999! : L2=0
310 I=INT(K/K3)
320 X=I*T + U0
330 Y=(K-I*K3)*T + V0
340 Z=S2(K)
350 G1=Z*K1+E1-X
360 G2=Z*K2+E2-Y
370 L=ABS(G1)+ABS(G2)
380 IF L>=C1 THEN K5=K : L2=L3 :N=N+1 : GOTO 410
390 C1=L : K=K+1 : L3=L
400 GOTO 310
410 K=N*K3
420 C1=99999!
430 IF L2>=C2 THEN GOTO 460
440 K6=K5 : C2=L3
450 GOTO 310
460 K=K6-1
470 PRINT "K=";K
475 REM Starts first approx. to find the impact point.
480 I=INT(K/K3)
490 X1=I*T + U0
500 Y1=(K-I*K3)*T +V0
510 Z1=S2(K)
```

```
520 T4=X1/E^2
530 C3=(X1-E1-K1*Z1)/K1
540 C4=(Y1-E2-K2*Z1)/K2
550 GOTO 730
555 REM Skips over another approx. method to find impact point.
560 T5=COS(A)*Z1-Y1*SIN(A)+D
570 F0=Y1/E^2+SIN(A)*T5
580 F3=Z1/E^2-COS(A)*T5
590 P1=-T4/F3
600 P2=-F0/F3
610 PRINT "P1=";P1;"P2=";P2
620 REM C3=(X1-E1-K1*Z1)/K1
630 REM C4=(Y1-E2-K2*Z1)/K2
640 R=(1/K1-P1)/P2
650 M=C3/P2
660 S=(1/K2-P2)/P1
670 P=C4/P1
680 D1=(M*S+P)/(1-R*S)
690 D2=(P*R+M)/(1-R*S)
700 PRINT "D1=";D1;"D2=";D2
710 X9=X1+D1 : Y9=Y1+D2
720 PRINT "X9=";X9;"Y9=";Y9
730 J=1 : K=K6-1
740 I2=INT(K/K3)
750 S0(J,0)=I2*T
760 S0(J,1)=(K-I2*K3)*T
770 S0(J,2)=S2(K)
780 IF J=1 THEN J=J+1 : GOTO 830
790 IF J=2 THEN J=J+1 : GOTO 840
800 IF J=3 THEN J=J+1 : GOTO 850
810 IF J=4 THEN J=J+1 : GOTO 860
820 IF J=5 THEN GOTO 870
830 K=K6 : GOTO 740
840 K=K6-2 : GOTO 740
850 K=K6+K3-1 : GOTO 740
860 K=K6-K3-1 : GOTO 740
870 U1=(S0(4,2)-S0(1,2))/T
880 U2=(S0(1,2)-S0(5,2))/T
890 W1=(U1-U2)/T
900 P1=U1-W1*T/2
910 V1=(S0(2,2)-S0(1,2))/T
920 V2=(S0(1,2)-S0(3,2))/T
930 W2=(V1-V2)/T
940 P2=V1-W2*T/2
950 P5=(S2(K6+K3)-S2(K6-K3))/(2*T)
960 P6=(S2(K6+K3-2) -S2(K6-K3-2))/(2*T)
970 P7=(P5-P6)/(2*T)
980 R=(1/K1-P1)/P2
990 M=C3/P2
1000 S=(1/K2-P2)/P1
1010 P=C4/P1
1020 D1=(M*S+P)/(1-R*S)
1030 D2=(P*R+M)/(1-R*S)
1040 PRINT "D1=";D1;"D2=";D2
```

```
1050 X9=X1+D1 : Y9=Y1+D2
1060 PRINT "X9=";X9;"Y9=";Y9
1070 INPUT "PAUSE, PRESS RETURN ",A$
1075 REM Starts second appox. for impact point using 9 gird points.
1080 C5=K2/K1
1090 C6=(C3-C4)*K2
1100 Q1=W1/2+W2*C5^2/2+P7*C5
1110 Q2=P1-1/K1+P2*C5+W2*C5*C6+P7*C6
1120 Q3=-C3+P2*C6+W2*C6^2/2
1130 D7=(-Q2-SQR(Q2^2-4*Q1*Q3))/(2*Q1)
1140 IF ABS(D7-D1)<.2 THEN D1=D7 : GOTO 1160
1150 D1=(-Q2-SQR(Q2^2-4*Q1*Q3))/(2*Q1)
1160 D2=C5*D1+C6
1170 PRINT "D1=";D1;"D2=";D2
1180 X9=X1+D1 : Y9=Y1+D2
1190 PRINT "X9=";X9;"Y9=";Y9
1200 F1=D1/T
1210 P3=U1+(U2-U1)*(.5-F1)
1220 F2=D2/T
1230 P4=V1+(V2-V1)*(.5-F2)
1240 R1=SQR(P3^2+P4^2+1)
1250 A1=-P3/R1 : A2=-P4/R1 : A3=1/R1
1260 R2=SQR(K1^2+K2^2+1)
1270 B1=-K1/R2 : B2=-K2/R2 : B3=-1/R2
1280 R3=2*(A1*B1+A2*B2+A3*B3)
1290 H1=B1-R3*A1
1300 H2=B2-R3*A2
1310 H3=B3-R3*A3
1320 PRINT "l=";H1;"m=";H2;"n=";H3
1330 Z00=CS*COS(A) : Y00=-CS*SIN(A)
1340 Z9=(X9-E1)/K1
1350 D5=SQR(X9^2+(Y9-Y00)^2+(Z9-Z00)^2)
1360 IL=X9/D5 : IM=(Y9-Y00)/D5 : IN=(Z9-Z00)/D5
1370 PRINT "IL=";IL;"IM=";IM;"IN=";IN
1380 GOTO 240
1390 END
```

```
10 REM Program to control a surface probe to measure the
20 REM RMS surface errors on an offset paraboloid.
30 PRINT "MEASE.BAS"
35 PI=3.141592654#
40 REM Data defining offset section of paraboloid.
50 F=12 : YCC=14 : XA=0 : YA=3 : YB=0 : YB=25 : XC=11 : YC=14
60 XD=-11 : YD=14 : K=0 : I=1 : J=1 : V=0 : R0=12
65 DA=.6 : DB=.7 : DC=.6 : DD=.7
70 REM INPUT "ENTER DA,DB,DC,DD= ",DA,DB,DC,DD
80 F1=DB-DA : F2=DD-DC
90 ZA=F-YA^2/4/F : ZB=F-YB^2/4/F
100 S=SQR((YB-YA)^2+(ZB-ZA)^2)
110 A0=ATN(2*F/YCC)
120 T1=F1/S : O1=ATN(T1/SQR(1-T1^2))
130 A1=A0-O1 : A2=PI/2-A1
140 T2=F2/(XC-XD)
150 O2=ATN(T2/SQR(1-T2^2))
160 REM PRINT S,A2,O2
170 REM Equation of measurement plane
180 C1=-SIN(O2)
190 C2=SIN(A2)*COS(O2)
200 C3=COS(A2)*COS(O2)
210 D=-(C2*YA+C3*ZA-DA)
220 REM Origin of first point
230 X1=0 : Y1=YA
240 GOTO 340
250 PRINT "R1=0"
260 X1=0 : Y1=YCC
265 GOTO 340
270 REM Select dish aperture points on concentric circles
280 REM and nearly equally spaced.
290 FOR N=1 TO 8
300 U1=PI/8/N : R1=N*SQR(2)*R0/12
310 PRINT : PRINT "R1= ";R1
320 FOR U=U1 TO 4*N*U1 STEP 2*U1
330 X1=-(R1*COS(U)) : Y1=R1*SIN(U)+YCC
340 Z1=F-(X1^2+Y1^2)/4/F
350 N1=-(C1*X1+C3*Z1-Y1*(C3^2/C2+C1^2/C2)+D)
360 M1=C1^2/C2+C3^2/C2+C2
370 Y2=N1/M1
380 X2=X1+C1/C2*(Y2-Y1)
390 Z2=Z1+C3/C2*(Y2-Y1)
400 P=C1*X1+C2*Y1+C3*Z1+D
410 P1=SQR((X1-X2)^2+(Y1-Y2)^2+(Z1-Z2)^2) + RND/10
420 REM PRINT P,P1
430 REM Coordinate transform to get probe position
440 REM above the X1,Y1,Z1 points on the dish.
450 X3=COS(O2)*X2+SIN(O2)*SIN(A2)*Y2+SIN(O2)*COS(A2)*Z2
460 Y3=COS(A2)*Y2-SIN(A2)*Z2
```

```
470 Z3=C1*X2+C2*Y2+C3*Z2
480 IF I=1 THEN Y4=Y3 : X4=X3 : I=0 : GOTO 250
490 Z3=Z3+D
500 Y3=Y3-Y4 : X3=X3-X4
510 K=K+1
520 E=P1-P : V1=E^2
530 V=V+V1
540 PRINT "XP=";-X3;"YP=";Y3;"P=";P
550 PRINT "X=";-X1;"Y=";Y1-YCC;"E=";E
560 IF J=1 THEN J=0 : GOTO 290
570 NEXT U
580 NEXT N
590 REM Computes RMS for dish
600 W=SQR(V/K)
610 PRINT "K=";K;"V=";V;"RMS=";W
620 END
```

Chapter 4
COMPUTER TECHNIQUES FOR DIFFRACTION PATTERNS
Y. C. Chang

4.1 INTRODUCTION

Several analytical techniques have been applied to calculate the diffraction patterns of reflector antennas. In the forward region, which includes the main beam and the near-in sidelobes, aperture field integration or surface current integration [1,2] have been widely used for many decades. These methods, however, fail to predict accurately the far-out sidelobes and backlobe. For the calculation of these wide-angle sidelobes, geometrical theory of diffraction (GTD) has been found very efficient and accurate in recent years [3,4,5]. The combination of GTD and aperture field or surface current integration has almost become a standard method for the analysis of reflector antennas. Numerical approaches to carry out these techniques efficiently may vary, however, depending on the antenna configuration, area of interest, and, sometimes, personal preference.

The blockage and scattering due to the feed or subreflector and its support struts usually have significant effects on the antenna gain and the sidelobe level [6]. The treatment of feed-subreflector blockage and strut scattering becomes an essential part of the analysis, unless the offset configuration is chosen to avoid this problem.

Sections 4.1 to 4.3 of this chapter describe the main approaches employed in the computer program REF.FOR, which calculates the far-field patterns of a circular reflector. The reflector can be either the main reflector or a subreflector, and either center-fed or offset. Section 4.4 introduces the capabilities of the program and presents an example for the purpose of demonstration.

4.2 APERTURE FIELD INTEGRATION

Aperture field integration (AI) has been applied to compute the radiation patterns of reflector antennas for at least three decades. It states that the radiated field of an aperture antenna can be obtained by integrating the field distribution over the antenna aperture. The radiated electric field can be expressed by [1,7,8]

$$\bar{E} = (jk/2\pi) \iint [\bar{F}_x E_x^a + \bar{F}_y E_y^a](e^{-jks}/s)\ dx\ dy \tag{4.1}$$

for a reflector with its boresight axis coincident with the Z-axis and with its aperture defined in the $X - Y$ plane as shown in Figure 4.1. E_x^a and E_y^a are the x- and y-components of the aperture field. $\bar{F}_x$ and $\bar{F}_y$ are the modified vector element patterns associated with two Huygens sources each having its electric field vector parallel to the X-axis and the Y-axis, respectively. These vector element patterns are given by [7]

$$\bar{F}_x = [\boldsymbol{\theta} \cos \phi - \boldsymbol{\phi} \sin \phi] \cos(\theta/2) \tag{4.2}$$

$$\bar{F}_y = [\boldsymbol{\theta} \sin \phi + \boldsymbol{\phi} \cos \phi] \cos(\theta/2) \tag{4.3}$$

where θ and ϕ are the spherical coordinates of the observation point as shown in Figure 4.1.

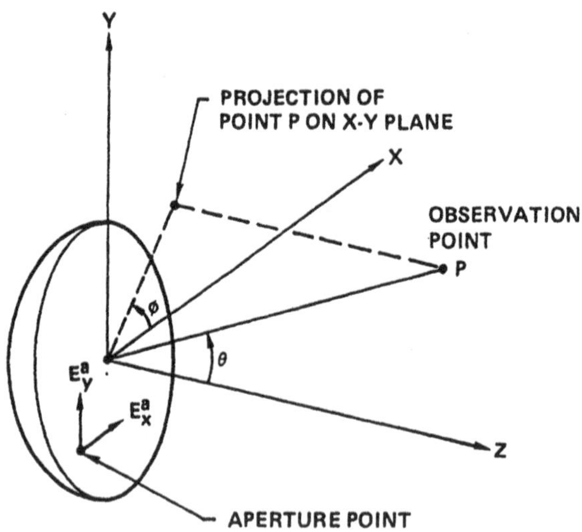

Figure 4.1 Coordinate System for the Aperture Field Integration

Aperture integration is performed over the portion of the aperture plane which is perpendicular to the reflector front axis (Z-axis). A rectangular grid size (D_x and D_y) is chosen so that the aperture can be divided into a principal rectangular grid as shown in Figure 4.2. Using the approach of overlapping subapertures [7], the aperture is treated like a collection of overlapping subapertures as shown in Figure 4.3. Each subaperture is rectangular in shape and consists of four adjacent grid rectangles. The aperture distribution for each subaperture is triangular. The use of overlapping, rectangular subapertures with triangular distribution permits a piece-wise linear approximation to the overall aperture distribution of the reflector. Furthermore, the grid spacings D_x and D_y can be electrically large, i.e., several wavelengths in size. This further minimizes the computation time. Thus, aperture integration results in a sum of the pattern functions of the rectangular subapertures weighted by the aperture field $\bar{E}^a$ and their respective areas.

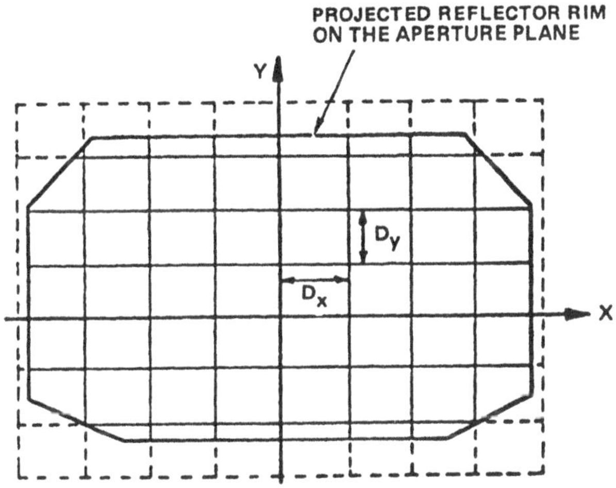

Figure 4.2 Geometry for Principal Rectangular Grid

The aperture integration method requires a two-dimensional integration for each pattern point as indicated in Equation (4.1). For far-field computations, however, an appropriate rotation of the grids [7] as shown in Figure 4.4 can greatly improve efficiency in carrying out integration. Note that the y-axis is the new axis after rotation of the original Y-axis. With this approach, y-integration is carried out for each column of the aperture parallel to the y-axis and each one-dimensional integration result is stored. The stored values of the y-integration are then used for each

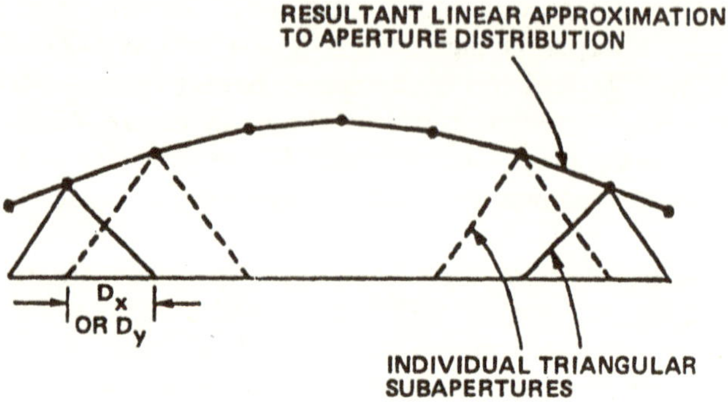

Figure 4.3 Numerical Integration Using Overlapping Subapertures with Triangular Distributions

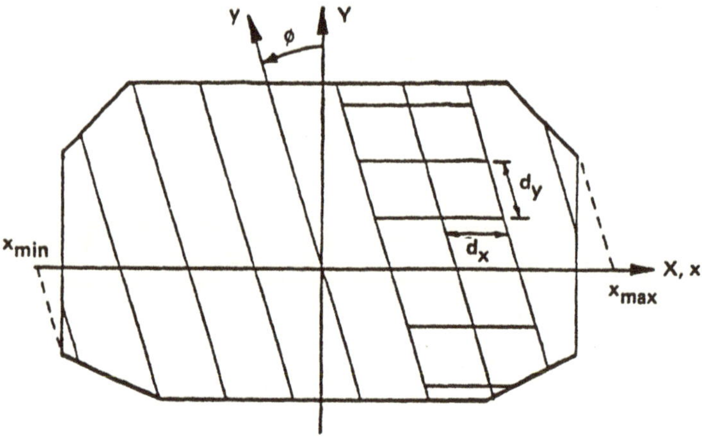

Figure 4.4 Geometry for Rotated Grid

pattern angle in the plane perpendicular to the y-axis. As a result, the efficiency of the calculation approaches that of a one-dimensional integration.

Consider the far-field expression of Equation (4.1) which is given by

$$\bar{E} = (\mathrm{j}k/2\pi)(\mathrm{e}^{-\mathrm{j}kR}/R) \int\int [\bar{F}_x E_x^a + \bar{F}_y E_y^a]\mathrm{e}^{\mathrm{j}k(x \sin\theta \cos\phi + y \sin\theta \sin\phi)}\, dx\, dy \tag{4.4}$$

where (θ,ϕ) represent the angular coordinates of a far-field point. By rotating only the Y-axis in the rectangular grid system, an angle ϕ, a nonorthogonal grid is set up as shown in Figure 4.4. After the coordinate transformation, Equation (4.4) becomes

$$\bar{E} = (jk/2\pi)(e^{-jkR}/R) \sum_{k=x,y} \bar{F}_k(\theta,\phi) \int I_k(x')e^{jkx'\sin\theta\cos\phi} \cos\phi \, dx' \quad (4.5)$$

where

$$I_k(x') = \int E_k^a(x',y') \, dy' \quad (4.6)$$

and $\bar{F}_{x,y}$ are defined in Equations (4.2) and (4.3). It can be seen in Equation (4.6) that the y integrations denoted as $I_x(x')$ and $I_y(x')$ are independent of θ and can be stored. Consequently, the far-field pattern in the plane perpendicular to the rotated y-axis is reduced to one-dimensional integrations; this provides greatly improved efficiency over the many two-dimensional integrations that would otherwise be required.

4.3 GEOMETRICAL THEORY OF DIFFRACTION (GTD)

Geometrical theory of diffraction (GTD) has been found very effective in analyzing the wide-angle sidelobes and near-field patterns of both main reflectors and subreflectors [9,10,11,12,13]. A two-point GTD has been employed successfully by Ratnasiri, *et al.* [5] to calculate the patterns for a circular reflector. Their method is restricted to reflectors with focused feeds such that the two diffraction points on the reflector rim can be found easily. For reflectors with general rim shapes, the NEC Reflector Antenna Code [7,8] uses a segmented-rim GTD which divides the reflector rim into many straight segments and calculates the contributions from all the rim segments. Since each rim segment needs to be small, the diffractions from its two endpoints are significant. These diffractions are included, using corner diffraction [14]. Segmented-rim GTD is more time-consuming, because usually at least 36 segments are needed.

In the computer program REF.FOR, two-point GTD is employed in most of the wide-angle region except for the backlobe region, where segmented-rim GTD is used.

4.3.1 Two-Point GTD

According to Keller's extension of Fermat's principle, the maximum and minimum paths from the feed to the reflector rim and then to the observation point P as shown in Figure 4.5 are the trajectories of the two

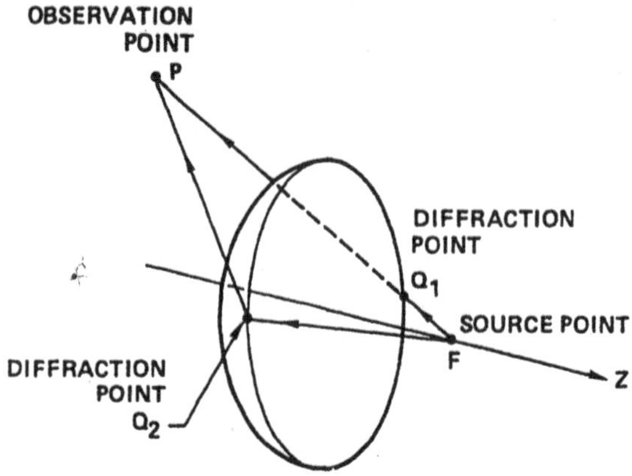

Figure 4.5 The Two Edge-Diffracted Rays which Contribute to the Far-Out Sidelobes of the Reflector

edge-diffracted rays of the reflector antenna. The diffracted field at P is the sum of the fields which are diffracted from diffraction points Q_1 and Q_2, respectively. Note that over a certain range of angles (θ), the contributions to the diffracted field come only from one of these two points, as the other gets shadowed by the reflector itself.

Since high frequency diffraction is essentially a local phenomenon, the edge diffracted field $\bar{E}^d(P)$ from either diffraction point Q can be determined in terms of the reflector geometry and ray geometry at Q on the edge. Two orthogonal components of $\bar{E}^d(P)$ can be expressed in terms of the incident field $\bar{E}^i$, the diffraction coefficients D_s and D_h, the spread factor $A(s)$, and the phase term e^{-jks} as [4]:

$$E^d_{\beta_0}(P) = -E^i_{\beta_0'}(Q)\, D_s(\phi,\phi';\beta_0')\, A(s)\, e^{-jks} \tag{4.7}$$

$$E^d_{\phi}(P) = -E^i_{\phi'}(Q)\, D_h(\phi,\phi';\beta_0')\, A(s)\, e^{-jks} \tag{4.8}$$

where the directions of β_0, ϕ, β_0', and ϕ' are shown in Figure 4.6 and

$$D_{h,s}(\phi,\phi';\beta_0') = -\frac{e^{-j\pi/4}}{2\sqrt{2\pi k}\,\sin\beta_0'}\left\{\frac{F[kL^i a(\phi-\phi')]}{\cos[(\phi-\phi')/2]} \pm \frac{F[kL^r a(\phi+\phi')]}{\cos[(\phi+\phi')/2]}\right\}$$

$$= \mathrm{DI}^- \pm \mathrm{DI}^+ \tag{4.9}$$

$$L^i = [ss'/(s+s')]\,\sin^2\beta_0' \tag{4.10}$$

$$L^r = [s\rho_1^r/(s + \rho_1^r)] \sin^2\beta_0 \tag{4.11}$$

$$a(\phi \pm \phi') = 2 \cos^2[(\phi \pm \phi')/2] \tag{4.12}$$

and

$$F[x] = 2j \sqrt{x}\, e^{jx} \int_{\sqrt{x}}^{\infty} e^{-j\tau^2}\, d\tau \tag{4.13}$$

which is the transition function.

In Equation (4.11), ρ_1^r represents the principal radius of curvature of the reflected wavefront at Q in the plane of incidence. It can be obtained by

$$\frac{1}{\rho_1^r} = \frac{1}{\rho_1^i} + \frac{2}{\rho_{g1} \cos \theta^i} \tag{4.14}$$

where

ρ_1^i is the principal radius of the incident wavefront at Q in the plane of incidence, which is the distance s' from the source to Q,
ρ_{g1} is the radius of curvature of the surface at Q in the plane of incidence, and
θ^i is the angle of incidence at Q.

The spread factor $A(s)$ describes how the amplitude of the field varies along the diffracted ray. It can be expressed as

$$A(s) = \sqrt{\rho/[s(\rho + s)]} \tag{4.15}$$

where the caustic distance ρ from the diffraction point Q on a curved edge has been derived [15] using differential geometry. It can be expressed as

$$\frac{1}{\rho} = \frac{1}{\rho_e^i} - \frac{\mathbf{n}_e \cdot (\mathbf{s}' - \mathbf{s})}{a_e \sin^2\beta_0'} \tag{4.16}$$

where

ρ_e^i is the radius of curvature of the incident wavefront at Q in the plane containing $\mathbf{e}$ and $\mathbf{s}'$,

a_e is the radius of curvature of the edge at Q, and
$\mathbf{n}_e$ is the unit normal vector of the edge at Q directed outward from the center of curvature.

All other terms associated with the edge diffraction are defined as:

D_s = the acoustic soft diffraction coefficient,
D_h = the acoustic hard diffraction coefficient,
s' = the distance from the source to Q,
s = the distance from Q to observation point P,
$\mathbf{e}$ = the unit vector tangent to the edge at Q,
$\mathbf{s}'$ = the unit vector in the incident-ray direction at Q,
$\mathbf{s}$ = the unit vector in the diffraction-ray direction at Q,
$\boldsymbol{\phi}', \boldsymbol{\phi}$ = the unit vector in the plane perpendicular to the edge at Q as shown in Figure 4.6,
$\boldsymbol{\beta}_0'$ = the unit vector in the plane containing $\mathbf{e}$ and $\mathbf{s}'$,
$\boldsymbol{\beta}_0$ = the unit vector in the plane containing $\mathbf{e}$ and $\mathbf{s}$, and
β_0' = the angle between the incident and edge tangent directions.

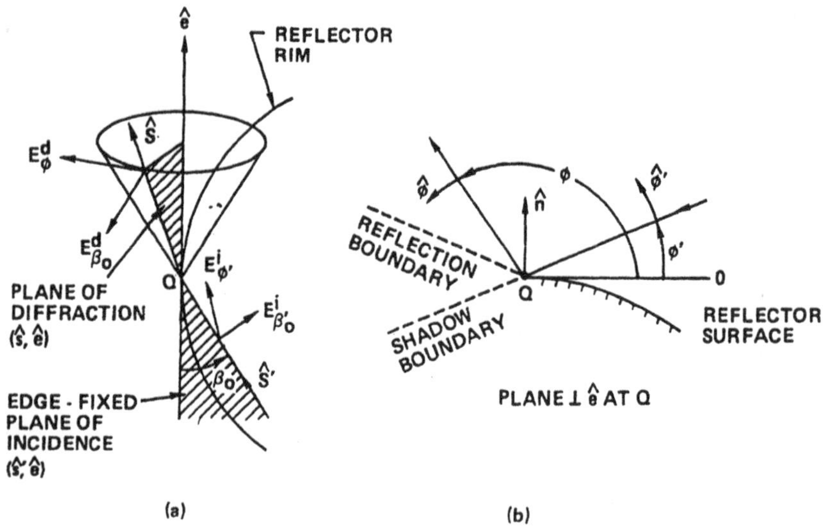

Figure 4.6 Geometry for Three-Dimensional Diffraction of Reflector Rim

Note that

$$\boldsymbol{\beta}_0' = \mathbf{s}' \times \boldsymbol{\phi}' \tag{4.17}$$

$$\boldsymbol{\beta}_0 = \mathbf{s} \times \boldsymbol{\phi} \tag{4.18}$$

$$\mathbf{e} \cdot \mathbf{s}' = \mathbf{e} \cdot \mathbf{s} = \cos \beta_0' = \cos \beta_0 \tag{4.19}$$

Equation (4.19) expresses the law of diffraction at the edge diffraction point Q. As a result, all the diffracted rays emanating from the same diffraction point form a diffraction cone with semiflare angle β_0.

Note that for the far-field cases, the distance s approaches infinity. Consequently, the distance parameters L^i and L^r become $s' \sin^2 \beta_0$ and $\rho_1^r \sin^2 \beta_0$, respectively, and $A(s)$ can be expressed as $\sqrt{\rho}/s$.

In the computer program REF.FOR, the caustic distance ρ is calculated using a numerical approach other than Equation (4.16) so that it can be applied to either center-fed or offset main reflectors or subreflectors.

Slope diffraction can be used to model variations in the incident fields at the diffraction points [16,17]. For reflectors using a tapered feed pattern with a steep slope at the reflector rim, the slope diffraction term must be included in the edge diffraction part. The slope diffracted fields are calculated in a way similar to that in which edge diffraction is calculated, except that the slope diffraction coefficients $\partial D_s/\partial \phi'$ and $\partial D_h/\partial \phi'$, and the slope of the incident field $\partial E^i/\partial n'$ at the edge are used. Thus, the respective β_0 and ϕ components of the slope diffracted field are given by

$$E_{\beta_0}^{sd} = -\frac{1}{jk \sin \beta_0} \frac{\partial E_{\beta_0'}^i}{\partial n'} \frac{\partial D_s}{\partial \phi'} A(s)\, \mathrm{e}^{-jks} \tag{4.20}$$

$$E_{\phi}^{sd} = -\frac{1}{jk \sin \beta_0} \frac{\partial E_{\phi'}^i}{\partial n'} \frac{\partial D_h}{\partial \phi'} A(s)\, \mathrm{e}^{-jks} \tag{4.21}$$

where n' is the distance in the direction of $\boldsymbol{\phi}'$ normal to the plane of incidence of the edge. The slope diffraction coefficients $\partial D_s/\partial \phi'$ and $\partial D_h/\partial \phi'$ can be expressed as

$$\frac{\partial D_{s,h}}{\partial \phi'} = -j \sqrt{\frac{k}{2\pi}} \frac{\mathrm{e}^{-j\pi/4}}{\sin \beta_0} \left\{ L^i \sin\left(\frac{\phi - \phi'}{2}\right) [1 - F[kL^i a(\phi - \phi')] \right.$$
$$\left. \pm L^r \sin\left(\frac{\phi + \phi'}{2}\right) [1 - F[kL^r a(\phi + \phi')] \right\} \tag{4.22}$$

4.3.2 Segmented-Rim GTD

This section summarizes the segmented-rim GTD approach which is similar to that of diffraction by a flat plate, except that the curvature of the reflector surface must be taken into account. Although this approach is applied only to the backlobe region of the reflector in the program REF .FOR, it was originally developed for the whole GTD region in the NEC Code for reflectors with general rim shape [18]. In this segmented-rim approach, the reflector rim is subdivided into many straight segments. A suitable criterion to decide the number of segments is that each segment of the reflector rim must be small enough so that the focus lies in the far field of the rim segment.

The segmented-rim GTD approach focuses on each individual rim segment and determines whether there is a significant diffraction contribution from that segment. If the diffraction for that segment is not significant, the code skips the edge diffraction part and checks the next rim segment. If the diffraction is significant, the diffraction point and the unit vectors and angles associated with the edge diffraction are calculated. This procedure is the same as that used for flat plate scattering calculation, except that the geometrical information associated with the parabolic reflector surface is changed. Note that for the backlobe region, all the segments contribute significantly to diffraction. As a result, the diffracted field includes at least 36 straight edge diffraction points and possibly slope diffraction as well, if desired.

The calculations of the edge diffraction and slope diffraction of each segment are similar to those of Equations (4.7), (4.8), (4.20), and (4.21), except that some of the parameters can be simplified because of the straight edges. Since both ρ_{g1} and a_e are infinite for flate plate and straight edge, both ρ_1^r and ρ become s'. Consequently, the far field expression of Equations (4.10), (4.11), and (4.15) can be written as

$$L^i = L^r = s' \sin^2\beta_0 \tag{4.23}$$

$$A(s) = \sqrt{s'}/s \tag{4.24}$$

Since each rim segment is finite, the corner effect at the endpoints of the segment must be taken into account. A corner diffraction solution has been developed by Burnside [14] which is based on the asymptotic evaluation of the radiation integral which employs the equivalent edge currents that would exist in the absence of the corners. The corner diffraction term is then found by appropriately modifying the asymptotic result for the radiation integral which is characterized by a saddle point near an endpoint. The corner diffraction compensates for the discontinuity which occurs when the diffraction point moves off of the rim segment. The corner diffraction field is given by

$$E^c_{\beta 0}(P) = -E^i_{\beta 0'}(Q)\, C' C_s \tag{4.25}$$

$$E^c_{\phi}(P) = -E^i_{\phi'}(Q)\, C' C_h \tag{4.26}$$

where

$$C' = -\frac{\sin \beta_c e^{-j\pi/4}}{2\pi(\cos \beta_{0c} + \cos \beta_c)} F[kL_c a(\beta_{0c} + \beta_c)] \sqrt{\frac{s'}{s_c}}\, e^{-jk(s_c - s')} \tag{4.27}$$

$$C_{h,s} = \mathrm{DI}^- \cdot \mathrm{AFC}^- \pm \mathrm{DI}^+ \cdot \mathrm{AFC}^+ \tag{4.28}$$

The Heuristic function $\mathrm{AFC}^{\pm}$ is expressed as

$$\mathrm{AFC}^{\pm} = \left| F\left[\frac{L^{r,i}a(\phi \pm \phi')/\lambda}{kL_c a(\beta_{0c} + \beta_c)}\right]\right| \tag{4.29}$$

in which

$$L_c = s_c s/(s_c + s) \tag{4.30}$$

The other variables associated with this geometry are shown in Figure 4.7.

In the program REF.FOR the diffracted fields from both corners ME and ME + 1 of a rim segment ME are combined in the following way

$$E^c_{\beta 0}(P) = -E^i_{\beta 0'}(Q) B_s \tag{4.31}$$

$$E^c_{\phi}(P) = -E^i_{\phi'}(Q) B_h \tag{4.32}$$

where the coefficients B_s and B_h are given by

$$B_{h,s} = \mathrm{DI}^- \cdot \mathrm{CC}^- \pm \mathrm{DI}^+ \cdot \mathrm{CC}^+ \tag{4.33}$$

and

$$\mathrm{CC}^{\pm} = C'_{\mathrm{ME}} \cdot \mathrm{AFC}^{\pm}_{\mathrm{ME}} + C'_{\mathrm{ME}+1} \cdot \mathrm{AFC}^{\pm}_{\mathrm{ME}+1} \tag{4.34}$$

Similarly, the coefficients for slope corner diffraction are given by

$$\frac{\partial B_{s,h}}{\partial \phi'} = \frac{\partial D_{s,h}}{\partial \phi'} (C'_{\text{ME}} + C'_{\text{ME}+1}) \tag{4.35}$$

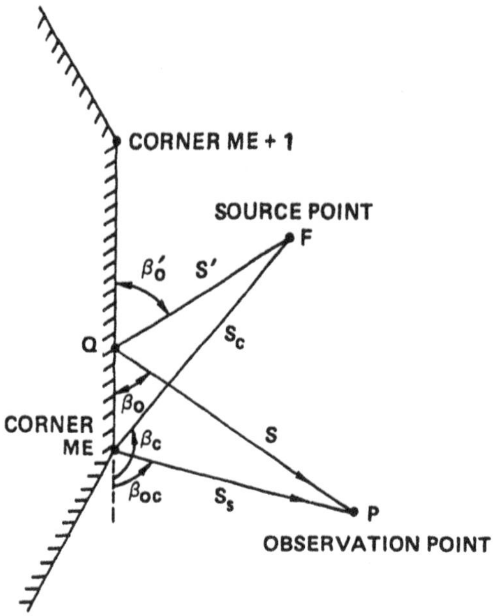

Figure 4.7 Geometry for Corner Diffraction Problem

4.4 FEED-SUBREFLECTOR BLOCKAGE AND STRUT SCATTERING

A hybrid approach [19] is employed in the program REF.FOR for the treatment of feed-subreflector blockage and strut scattering. The hybrid approach is a combination of the null-field approach [6,20] and the induced current approach [21,22,23].

Using the null-field approach, the optical shadows of the struts and feed-subreflector are located as shown in Figure 4.8. Note that the pie-shaped shadow is caused by the spherical wave emanating from the focus of the reflector and hence can be called the spherical-wave component. The narrow rectangular shadow is produced by the plane wave reflected from the reflector and can be called the plane-wave component. After these shadows are defined geometrically, either surface current or aperture field integration can be carried out over the lit areas. The null-field approach provides accurate results only for the main beam region and only if the strut width is much greater than a wavelength. Furthermore,

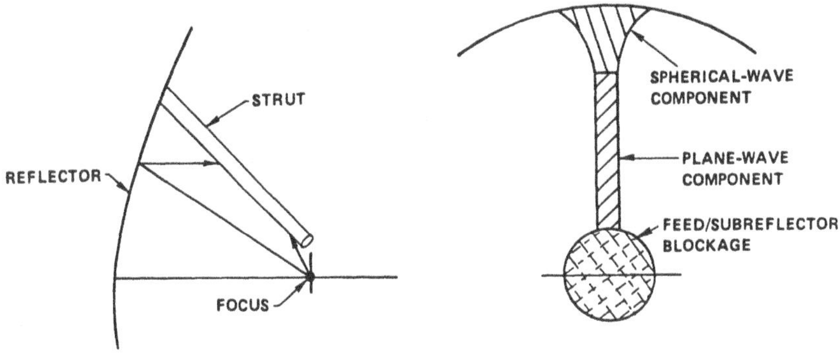

Figure 4.8 Geometry for the Null-Field Approach

the null-field approach fails to account for the effects caused by frequency, polarization, and the depth, cross section, or tilt of the struts.

In the induced current approach, the incident field on the strut is the reflected field from the reflector as shown in Figure 4.9. The induced current on each strut is calculated assuming an infinitely long strut. In the program REF.FOR, only perfectly conducting strut sections with circular cross section are considered. The strut axes are specified as linear with arbitrary locations for the strut endpoints. Eigenfunction solutions are used to calculate the diffraction coefficients of the circular cylinder. For the TE case, the diffraction coefficient is given by [24]

$$D_s(\gamma,\beta_0) = \frac{1}{\sin \beta_0} \sqrt{\frac{2j}{\pi k}} \sum_{n=0}^{\infty} \varepsilon_n a_n \cos n\gamma \tag{4.36}$$

Whereas for the TM case,

$$D_h(\gamma,\beta_0) = \frac{1}{\sin \beta_0} \sqrt{\frac{2j}{\pi k}} \sum_{n=0}^{\infty} \varepsilon_n b_n \cos n\gamma \tag{4.37}$$

where

$$a_n = \frac{J_n(ka \sin \beta_0)}{H_n^2(ka \sin \beta_0)} \tag{4.38}$$

$$b_n = \frac{J_n'(ka \sin \beta_0)}{H_n^{2'}(ka \sin \beta_0)} \tag{4.39}$$

$$\gamma = \tan^{-1}\left(\frac{\mathbf{V}_N \cdot \mathbf{d}}{\mathbf{V}_P \cdot \mathbf{d}}\right) \tag{4.40}$$

β_0 is the incident angle,
a is the radius of the strut, and

$$\varepsilon_n = \begin{cases} 1, & n = 0 \\ 2, & \text{otherwise} \end{cases}$$

Note that J_n, H_n^2, and J_2', $H_n^{2'}$ are Bessel functions and their corresponding derivatives, respectively.

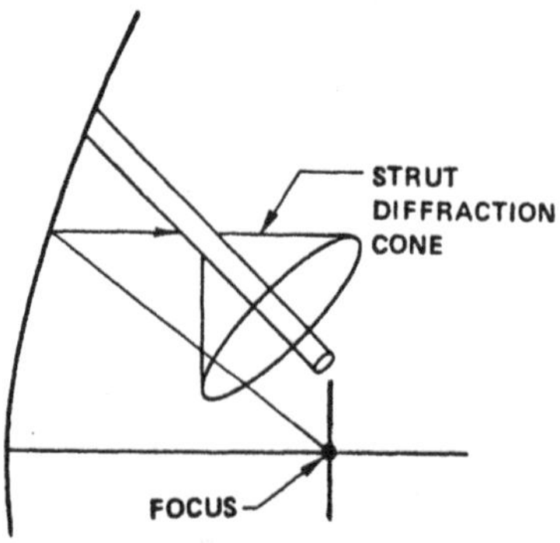

Figure 4.9 Geometry for the Induced Current Approach

Also note that **d** is the unit vector in the direction of diffraction, $\mathbf{V}_N$ and $\mathbf{V}_P$ are the unit normal vector and unit binormal vector to the strut, respectively, as shown in Figure 4.10. The strut scattered fields are assumed to be shadowed by the reflector for scattering directions **d** beyond the strut boundary angle (θ_{ST}), which is set at 100° in the default data in the program REF.FOR.

Using the method of equivalent currents, the scattered field from each strut is given by:

$$E_\parallel^s = -\frac{\sqrt{jk}}{2\pi} D_s(\gamma,\beta_0) \sin\alpha \int_{\xi_1}^{\xi_2} E_\parallel^i \, e^{jk\xi\cos\beta_0} \frac{e^{-jks}}{s} \, d\xi \tag{4.41}$$

and

$$E_{\perp}^{2} = -\frac{\sqrt{jk}}{2\pi} D_h(\gamma,\beta_0) \sin\alpha \int_{\xi_1}^{\xi_2} E_{\perp}^{i} e^{jk\xi \cos\beta_0} \frac{e^{-jks}}{s} d\xi \qquad (4.42)$$

where (s, α, γ) are the spherical coordinates of the observation point in the strut coordinate system.

The induced current approach predicts the plane-wave component of the strut scattered field very well. There is, however, no treatment for the spherical-wave component by the induced current approach. Hence the solution is not complete unless the outside end of the strut is attached to the reflector rim in which the strut does not cast a shadow on the reflector from the focus.

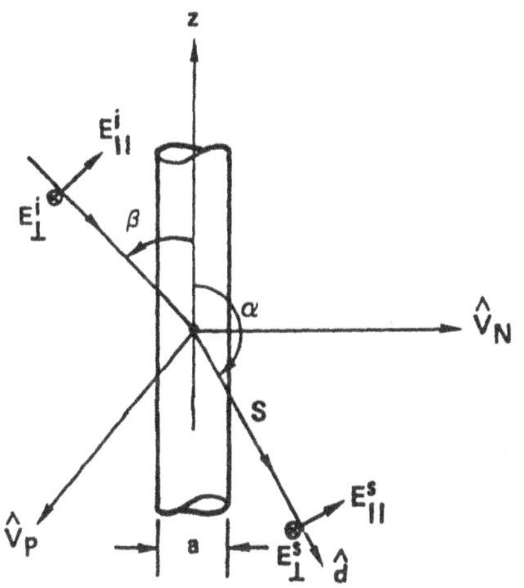

Figure 4.10 Geometry for Strut Scattering

In the hybrid approach, the null-field approach is used for the spherical-wave component of the strut blockage and the feed-subreflector blockage, and the induced current approach is used for the plane-wave component of the strut scattering as shown in Figure 4.11. Although this hybrid approach is not a complete solution either, it has been found very useful in predicting the antenna gain, first sidelobe level, and some of the wide-angle sidelobes. A more complete solution can be obtained by calculating the diffracted fields from the struts for the spherical-wave component. This approach will yield more accurate surface current of aperture fields, and the diffraction cones of the struts can be included in the backlobe regions. Much more work is required for this approach, however, and the improvement over the hybrid approach is usually insignificant.

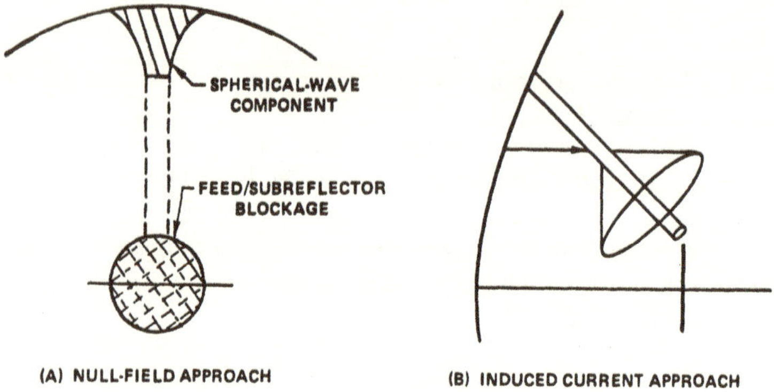

Figure 4.11 Geometry for the Hybrid Approach

4.5 INTRODUCTION OF COMPUTER PROGRAM REF.FOR

This section is a brief introduction to the computer program REF.FOR. Section 4.5.1 summarizes the capabilities of the program to provide a general picture of REF.FOR. Section 4.5.2 presents certain calculated patterns as an example of a typical circular Cassegrain reflector to demonstrate some of the program's capabilities.

4.5.1 Summary of Capabilities

Reflector surface: NTYPE = 1 for parabolic surface,
NTYPE = 10 for hyperbolic surface,
NTYPE = 11 for elliptical surface.
Reflector configuration: Center-fed or offset with focused feed.
Reflector rim: Circular for center-fed,
Elliptical for offset.
Feed pattern input: Linearly piece-wise.
Reflector pattern output: Far-field patterns with arbitrary ϕ-plane.
Analytical techniques: Aperture field integration (AI),
GTD and geometrical optics.
Feed-Subreflector blockage: Included only in AI region.
Strut scattering: Included in both AI and GTD regions.

To calculate the antenna patterns of a Cassegrain or Gregorian antenna, a two-step procedure can be applied. That is, to calculate the subreflector patterns as the first step, then use that result as part of the input for the second step to calculate the main reflector patterns. This procedure is demonstrated in the following example.

4.5.2 Example

The example chosen here is a circular Cassegrain reflector with a 24″-diameter parabolic main reflector and a 4″-diameter hyperbolic subreflector as shown in Figure 4.12. The focal length of the paraboloid is 8″ and the distance between the two foci of the hyperboloid is 5.412″. The distance from the phase center of the feed (which is also one of the foci) to the subreflector is 4.277″. The feed is focused and assumed to have a vertical polarization with a $\cos^{18}\theta$ field pattern, which produces a 40.6°, 10-dB beamwidth. The subreflector is supported by three equally spaced struts with circular cross section and 0.5″ diameter. Note that one of the struts is lying in the vertical plane (*E*-plane) and that the position of the struts can be obtained from the side view of that vertical strut in Figure 4.12.

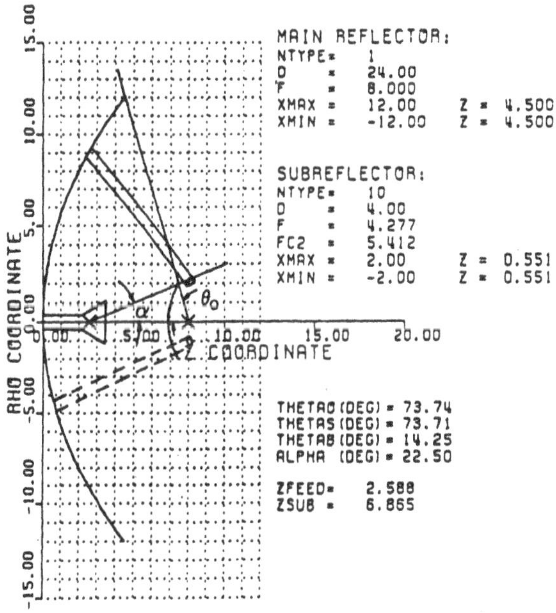

Figure 4.12 Side View Geometry for a 24-inch Cassegrain Antenna with Three Struts

The far-field patterns of the subreflector at 20 GHz are calculated by using GTD and geometrical optics, and shown in Figure 4.13. Both amplitude and phase patterns are shown for both *H*-plane and *E*-plane. These subreflector patterns are used as part of the input data to calculate the main reflector patterns. Figure 4.14 shows the far-field patterns without the subreflector blockage and strut scattering, while the effects of the subreflector and the struts are both included in Figure 4.15. It is obvious that the subreflector and the struts reduce the antenna gain and increase most of the sidelobe levels significantly. It should be pointed out that the program does not take into account surface error and feed loss, which sometimes can be very significant. Excellent agreement between measurements and calculations using the same or similar approaches have been reported [19,25].

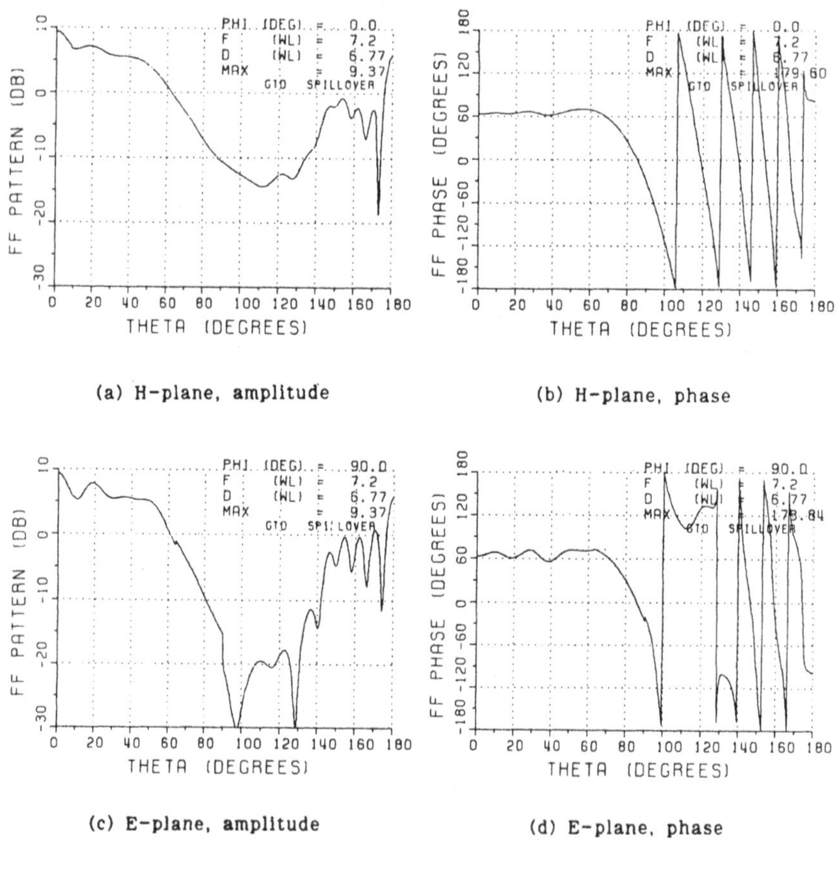

(a) H-plane, amplitude (b) H-plane, phase

(c) E-plane, amplitude (d) E-plane, phase

Figure 4.13 Calculated Far-Field Patterns of the Subreflector

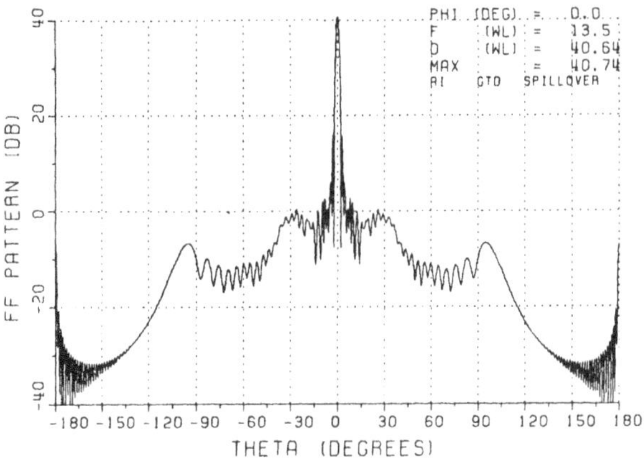

(a) H-plane

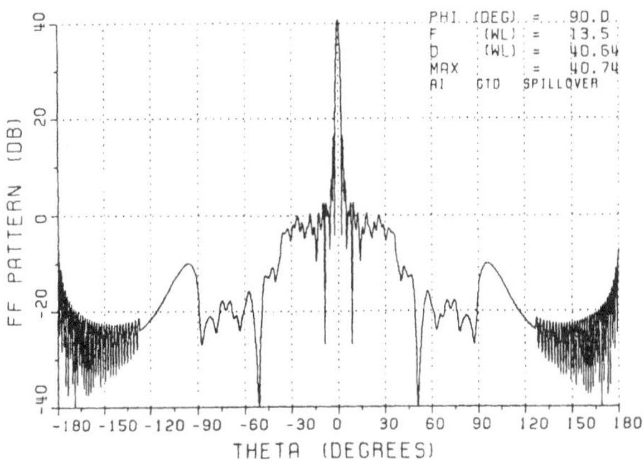

(b) E-plane

Figure 4.14 Calculated Far-Field Patterns of the Main Reflector without Feed Blockage and Strut Scattering

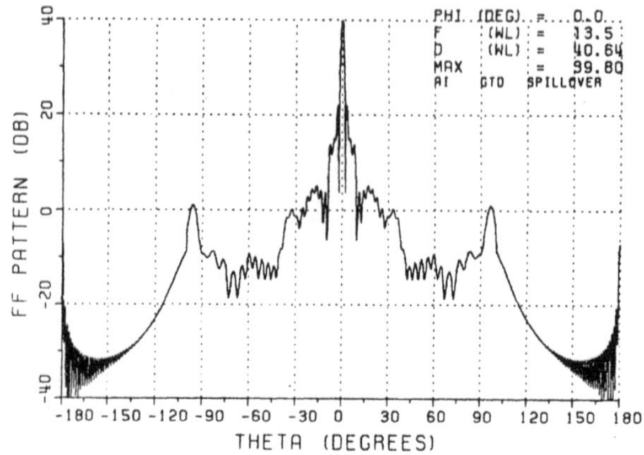

(a) H-plane

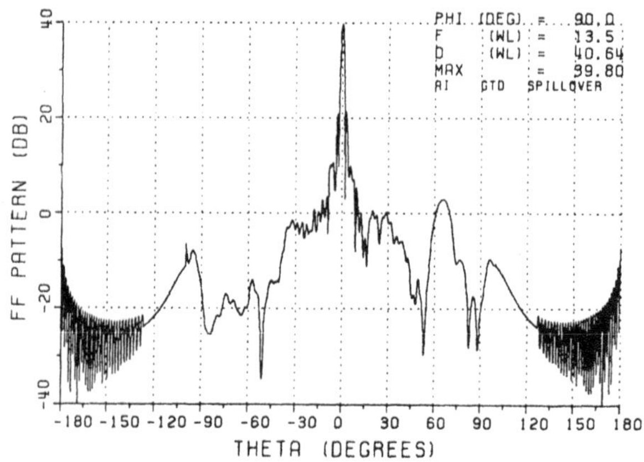

(b) E-plane

Figure 4.15 Calculated Far-Field Patterns of the Main Reflector with Feed Blockage and Strut Scattering

References

[1] Silver, S. (ed.), *Microwave Antenna Theory and Design,* New York: McGraw-Hill, 1949.

[2] Kauffman, J.H., W.F. Croswell, and L.J. Powers, "Analysis of Radiation Patterns of Reflector Antennas," IEEE Trans., Vol. AP-24, Jan. 1976, pp. 53–65.

[3] Keller, J.B., "Geometrical Theory of Diffraction," J. Optics Soc. Amer., Vol. 52, pp. 116–130, 1962.

[4] Kouyoumjian, R.G., and P. Pathak, "A Uniform Geometrical Theory of Diffraction for an Edge of a Perfectly Conducting Surface," Proc. IEEE, Vol. 62, No. 11, Nov. 1974, pp. 1448–1461.

[5] Ratnasiri, P.A.J., R.G. Kouyoumjian, and P.H. Pathak, "The Wide Angle Side Lobes of Reflector Antennas," Rep. No. 2183-1, March 1970, The Ohio State Univ. Electro-Science Lab., Dept. of Elec. Eng.; prepared under Contract AF 19(628)-5929 for Air Force Cambridge Research Lab.

[6] Gray, C.L., "Estimating the Effect of Feed Support Member Blocking on Antenna Gain and Side Lobe Level," Microwave Journal, Vol. 7, March 1964, pp. 88–91.

[7] Lee, S.H., and R.C. Rudduck, "Aperture Integration and GTD Techniques Used in the NEC Reflector Antenna Code," Trans. IEEE, Vol. AP-33, 1985, pp. 189–194.

[8] Chang, Y.C., and R.C. Rudduck, "Numerical Electromagnetic Code—Reflector Antenna Code, NEC-REF (Version 2), Part II: Code Manual," Report 712242-17, Dec. 1982, The Ohio State Univ. Electro-Science Lab., Dept. of Elec. Eng.; prepared under Contract No. N00123-79-C-1469 for Naval Regional Procurement Office.

[9] Rusch, W.V.T., "A Comparison of Geometrical and Integral Fields from High-Frequency Reflectors," IEEE Proc., Vol. 62, Nov. 1974, pp. 1603–1604.

[10] Mentzer, C.A., and L. Peters, Jr., "A GTD Analysis of the Far-Out Sidelobes of Cassegrain Antennas," IEEE Trans., Vol. AP-23, Sep. 1975, pp. 702–709.

[11] Lee, S.W., P. Cramer, Jr., K. Woo, and Y. Rahmat-Samii, "Diffraction by an Arbitrary Subreflector: GTD Solution," IEEE Trans., Vol. AP-27, March 1979, pp. 305–316.

[12] Lee, T.H., "Geometrical Optics and GTD Analysis of Subreflectors in Cassegrain and Gregorian Reflector Antennas," M.S. Thesis, The Ohio State Univ. Electro-Science Lab., Dept. of Elec. Eng., June, 1984.

[13] Chang, Y.C., and H.V. Morrison, "A GTD Analysis of Shaped Cassegrain Antennas," 1985 IEEE AP-S Symposium, Vancouver, Canada, pp. 345–348.

[14] Burnside, W.D., N. Wang, and E.L. Pelton, "Near Field Pattern Analysis of Airborne Antennas," IEEE Trans., Vol. AP-28, May 1980, pp. 318–327.

[15] Pathak, P.H., and R.G. Kouyoumjian, "The Dyadic Diffraction Coefficient for a Perfectly Conducting Wedge," Report 2183-4, June 1970, The Ohio State Univ. Electro-Science Lab., Dept. of Elec. Eng.

[16] Karp, S.N., and J.B. Keller, *Optica Acta,* Vol. 8, 1961, pp. 61.

[17] Rudduck, R.C., and D.C.F. Wu, "Slope Diffraction Analysis of TEM Parallel-Plate Guide Radiation Patterns," IEEE Trans., Vol. AP-17, Nov. 1969, pp. 797–799.

[18] Lee, S.H., and R.C. Rudduck, "Numerical Electromagnetic Code—Reflector Antenna code, NEC-REF, Part II: Code Manual," Report 784508-16, Sep. 1979, The Ohio State Univ. Electro-Science Lab., Dept. of Elec. Eng.; prepared under Contract No. N00123-76-C-1371 for Naval Regional Procurement Office.

[19] Chang, Y.C., H.V. Morrison, and R.C. Rudduck, "A Hybrid Approach for the Strut Analysis of Reflector Antennas," 1986 IEEE AP-S Symposium, Philadelphia, PA, pp. 523–526.

[20] Ruze, J., "Feed Support Blockage Loss in Parabolic Antennas," Microwave Journal, Vol. 11, Dec. 1968, pp. 76–80.

[21] Lee, S.H., R.C. Rudduck, C.A. Klein, and R.G. Kouyoumjian, "A GTD Analysis of the Circular Reflector Antenna Including Feed and Strut Scatter," Report 4381-1, May 25, 1977, The Ohio State Univ. Electro-Science Lab., Dept. of Elec. Eng., prepared under Contract F30602-76-0024 with Rome Air Development Center.

[22] Brachat, P., and P.F. Combes, "Effects of Secondary Diffractions in the Radiation Pattern of the Paraboloid," IEEE Trans., Vol. AP-28, Sept. 1980, pp. 718–721.

[23] Rusch, W.V.T., O. Sorensen, and J.W.M. Baars, "Radiation Comes from Feed-Support Struts of Symmetric Paraboloidal Antennas," IEEE Trans., Vol. AP-30, July 1982, pp. 786–790.

[24] Harrington, R.F., *Time Harmonic Fields,* pp. 232–235. New York: McGraw-Hill, 1961.

[25] Lambert, K.M., R.C. Rudduck, and T.H. Lee, "Far Field Measurements of 8-foot Reflector Antenna in the Compact Range at the Ohio State University," 1987 IEEE AP-S Symposium, Blacksburg, WV, pp. 382–385.

[26] Boersma, J., "Computation of Fresnel Integrals," Math. Comp., Vol. 14, 1960, p. 380.

Appendix 4A
Diffraction Coefficients Subroutines

The FORTRAN program REF.FOR is the only computer program associated with this chapter. It includes, however, many subroutines and functions that can be used for other applications. For the purpose of demonstration, this appendix presents three subroutines used to calculate the diffraction coefficients. Note that, although most of the program was developed by the author, part of it was written by the staff of the Ohio State University ElectroScience Laboratory over a period of years.

The first subroutine HPDC calculates the edge diffraction coefficients $D_{h,s}$ of (4.9), the slope diffraction coefficients $\partial D_{s,h}/\partial\phi'$ of (4.22), the corner diffraction coefficients $B_{h,s}$ of (4.33), and the slope corner diffraction coefficients $\partial B_{s,h}/\partial\phi'$ of (4.35) for a plate or reflector rim. The second subroutine is a complex function FFCT to calculate the transition function $F[X]$ in (4.13). The third subroutine is FRNELS to calculate the Fresnel used for the transition function $F[X]$.

The input variables of the subroutine HPDC are:

DEL: Variable representing the value of $kL_c a(\beta_{0c} + \beta_c)$
CP: Variable representing C' in Equation (4.27)
RL: Distance parameters L^i (RL(1)) and L^r (RL(2)) of (4.10) and (4.11)
PS: Diffraction angle (ϕ) for the diffraction field
PSO: Diffraction angle (ϕ') for the incident field
SBO: Sine of the diffracted cone angle β_0

Note that the variable DEL, the coefficient C', and the distance parameters L^i and L^r are calculated in subroutines GTDSR or GTDTP. All the output coefficients DS (D_s), DH (D_h), DPS ($\partial D_s/\partial\phi'$), DPH ($\partial D_h/\partial\phi'$), BS ($B_s$), BH ($B_h$), BPS ($\partial B_s/\partial\phi'$), and BPH ($\partial B_h/\partial\phi'$) can be obtained through the COMMON block /DSC/.

The complex function FFCT calculates the transition function $F[X]$ in (4.13). For $X \geq 0$, $F[X]$ can be written as

$$F[X] = \mathrm{j}\sqrt{2\pi X}\, e^{\mathrm{j}X}\{(0.5 - \mathrm{j}0.5) - [C(\sqrt{2X/\pi}) - \mathrm{j}S(\sqrt{2X/\pi})]\} \tag{4A.1}$$

where C and S represent the Fresnel integral function defined by

$$f(x) = C(x) - \mathrm{j}S(x) = \int_0^x e^{-\mathrm{j}\pi t^2/2}\, dt \tag{4A.2}$$

For $X < 0$, $F[X]$ is equal to the conjugate of $F[|X|]$.

The Fresnel integral $f(x)$ is evaluated using an approximation by J. Boersma [26] in the subroutine FRNELS. It is approximated as

$$f(x) \approx e^{-\mathrm{j}X} \sqrt{X/4} \sum_{n=0}^{11} (a_n + \mathrm{j}b_n)(X/4)^n \text{ for } 0 \leq X < 4 \tag{4A.3}$$

and

$$f(x) \approx (1 - j)/2 + e^{-\mathrm{j}X} \sqrt{4/X} \sum_{n=0}^{11} (c_n + \mathrm{j}d_n)(4/X)^n \text{ for } X \geq 4 \tag{4A.4}$$

where $X = \pi x^2/2$. The constants a_n, b_n, c_n, and d_n are provided by Boersma and are defined in data statements in the subroutine.

STATEMENT LISTINGS OF THE PROGRAMS

```
      SUBROUTINE HPDC(DEL,CP,RL,PS,PSO,SBO)
C!!!
C!!!  CALCULATES THE DIFFRACTION COEFFICIENTS FOR A PLATE OR
C!!!  REFLECTOR RIM
C!!!
      DIMENSION DEL(2),RL(2)
      COMPLEX CP(2),DI(2),DPI(2),CC(2),TERM,TERMP(2),FFCT,FA,TOP
      COMPLEX CJ,DS,DH,DPS,DPH,BS,BH,BPS,BPH,CIN,CIP,CCP
      COMPLEX CSL(2)
      COMMON /DSC/DS,DH,DPS,DPH,BS,BH,BPS,BPH
      COMMON /LOG1/LSLOPE,LCORNR
      COMMON /PIS/PI,TPI,DPR,CJ
      COMMON /TOPD/TOP
      LOGICAL LCORNR,LSLOPE
      TERM=TOP/(2.*TPI*SBO)
      DO 10 N=1,2
      TERMP(N)=CJ*2.*TPI*TERM*RL(N)
10    CSL(N)=0.5*CSQRT(CMPLX(RL(N)))/SBO
      IF (DEL(1).LT.1.D-20) DEL(1)=1.D-20
      IF (DEL(2).LT.1.D-20) DEL(2)=1.D-20
      BET=PS-PSO
      DO 100 N=1,2
      DI(N)=(0.,0.)
      DPI(N)=(0.,0.)
      ANG=BET/DPR
      SB=SIN(ANG/2.)
      CB=COS(ANG/2.)
      A=2.*CB*CB
      X=TPI*RL(N)*A
      IF ((LSLOPE).OR.(ABS(X).LE.10.)) FA=FFCT(X)
      IF (ABS(X).GT.10.) FA=1.+CJ/(2.*X)-3./(4.*X*X)
      IF (A.GT.1.D-20) GO TO 20
      DI(N)=-CSL(N)+CJ*0.
      GO TO 30
20    CONTINUE
      DI(N)=TERM*FA/CB
30    CONTINUE
      DPI(N)=TERMP(N)*SB*(1.-FA)
      IF (.NOT.LCORNR) GO TO 50
      CC(N)=(0.,0.)
      DO 40 I=1,2
      ARG=RL(N)*A/DEL(I)
      AFC=1.
      IF (ABS(ARG).LE.10.) AFC=BABS(FFCT(ARG))
      CC(N)=CC(N)+CP(I)*AFC
40    CONTINUE
50    BET=PS+PSO
100   CONTINUE
```

```
      DS=DI(1)-DI(2)
      DH=DI(1)+DI(2)
      DPS=DPI(1)+DPI(2)
      DPH=DPI(1)-DPI(2)
      IF (.NOT.LCORNR) RETURN
      CIN=DI(1)*CC(1)
      CIP=DI(2)*CC(2)
      BS=CIN-CIP
      BH=CIN+CIP
      CCP=CP(1)+CP(2)
      BPS=DPS*CCP
      BPH=DPH*CCP
      RETURN
      END
C-----------------------------------------------------------------
      COMPLEX FUNCTION FFCT(X)
C!!!
C!!!  DETERMINES THE TRANSITION FUNCTION RESULT FOR
C!!!  DIFFRACTED FIELD
C!!!
      COMPLEX CJ
      COMMON/PIS/PI,TPI,DPR,CJ
      AX=ABS(X)
      CLIMIT = 1.D5
      IF(AX.LT.CLIMIT) GO TO 10
      FFCT= (1.0,0.0)
      RETURN
10    SOAX=SQRT(AX)
      S=SQRT(2./PI)*SOAX
      CALL FRNELS(CFR,SFR,S)
      FFCT=CMPLX(0.5-CFR,SFR-0.5)
      FFCT=SQRT(TPI)*SOAX*FFCT*CEXP(CMPLX(0.,AX+PI/2.))
      IF(X.LT.0.) FFCT=CONJG(FFCT)
      RETURN
      END
C-----------------------------------------------------------------
      SUBROUTINE FRNELS(C,S,XS)
C!!!
C!!!  THIS IS THE FRESNEL INTEGRAL SUBROUTINE WHERE THE INTEGRAL IS FROM
C!!!  T=0 TO XS. THE INTEGRAND IS EXP(-CJ*PI/2.*T*T), AND THE OUTPUT IS
C!!!  C(XS)-CJ*S(XS).
C!!!
      COMPLEX CJ
      COMMON /PIS/PI,TPI,DPR,CJ
      DIMENSION A(12),B(12),CC(12),D(12)
      DATA A/1.595769140,-0.000001702,-6.808568854,-0.000576361,6.920691
     &902,-0.016898657,-3.050485660,-0.075752419,0.850663781,-0.02563904
     &1,-0.150230960,0.034404779/
      DATA B/-0.000000033,4.255387524,-0.000092810,-7.780020400,-0.00952
     &0895,5.075161298,-0.138341947,-1.363729124,-0.403349276,0.70222201
```

```
     &6,-0.216195929,0.019547031/
      DATA CC/0.,-0.024933975,0.000003936,0.005770956,0.000689892,-0.009
     &497136,0.011948809,-0.006748873,0.000246420,0.002102967,-0.0012179
     &30,0.000233939/
      DATA D/0.199471140,0.000000023,-0.009351341,0.000023006,0.00485146
     &6,0.001903218,-0.017122914,0.029064067,-0.027928955,0.016497308,-0
     &.005598515,0.000838386/
      IF(XS.LE.0.0) GO TO 100
      X=PI*XS*XS/2.0
      FR=0.0
      FI=0.0
      K=13
      IF(X-4.0) 10,40,40
10    Y=X/4.0
20    K=K-1
      FR=(FR+A(K))*Y
      FI=(FI+B(K))*Y
      IF(K-2) 30,30,20
30    FR=FR+A(1)
      FI=FI+B(1)
      C=(FR*COS(X)+FI*SIN(X))*SQRT(Y)
      S=(FR*SIN(X)-FI*COS(X))*SQRT(Y)
      RETURN
40    Y=4.0/X
50    K=K-1
      FR=(FR+CC(K))*Y
      FI=(FI+D(K))*Y
      IF(K-2) 60,60,50
60    FR=FR+CC(1)
      FI=FI+D(1)
      C=0.5+(FR*COS(X)+FI*SIN(X))*SQRT(Y)
      S=0.5+(FR*SIN(X)-FI*COS(X))*SQRT(Y)
      RETURN
100   C=-0.0
      S=-0.0
      RETURN
      END
```

Chapter 5
DIELECTRIC LENS ANTENNAS
C. J. Sletten, W. P. Shillue, F. S. Holt

5.1 INTRODUCTION

There are many books and articles treating dielectric lenses for applications in the optical spectrum. References [1,2,3] are representative of this literature. References to dielectric lenses for microwave and millimeter wave applications are, however, less abundant. Several chapters in antenna books [4,5,6] which are devoted to design approaches that are suitable for lens antennas are recommended. No attempt is made in this chapter to include all published information on lens antenna technology, but rather to develop computer methods that can be used (1) to shape lens contours for control of aperture amplitude distributions, (2) to ray-trace to find the best focal surfaces, and (3) to compute radiation patterns on axially symmetric lenses based on any design technique. Computer design procedures for wide-angle lenses are emphasized.

A primary use of dielectric lens antennas is for multibeam applications such as satellite-borne antennas [7,8] where high-gain and multibeam patterns can be generated by using the wide-angle characteristics of lens antennas. Lenses are also useful as "collimating" caps on corrugated horn feeds or other antennas [9,10]. The Meniscus lens cap on the horn aperture makes the amplitude on the lens nearly uniform, and therefore greatly increases the gain of the corrugated horn.

5.1.1 Basic Advantages and Disadvantages of Dielectric Lenses

Lens antennas can be very wide-angle for beam scanning and multibeam formation. They can be designed to have low losses and wide bandwidths. The two surfaces and choice of dielectric constants (nonhomogeneous lenses are possible) make available lens parameters for independent control of antenna aperture phase and amplitude (and for improving the lens wide-angle scanning characteristics). Shaping of general (aspherical) surfaces is relatively easy at microwave millimeter-wave bands. Lenses can serve, too, as weather protecting caps for antennas. As in optics, dielectric lenses are very versatile, and their advantages become more important in antenna design at the higher microwave bands on up into the infrared spectral region.

Some disadvantages of dielectric lenses must also be recognized. Dielectric lenses tend to be heavier and more costly than metal reflectors. Lens thickness and weight can be reduced by Fresnel stepping. When stepped the bandwidths of the antenna usually decrease. Lens surfaces reflect electromagnetic energy which reduces the antenna aperture efficiency. The reflected energy may reenter the lens feeds causing increases in feed VSWRs. Surface matching, by cutting grooves or using dielectric coatings (quarter wave plates), reduces lens reflections over limited bandwidths. Higher index of refraction dielectric lenses are thinner generally but introduce more surface reflections and cross polarization. Dielectrics used for lenses usually have quite small but significant dielectric propagation losses. Dielectrics are weather sensitive. Humidity and dirt on surfaces cause loss. Lenses can crack under environmental and mechanical stresses. These disadvantages can usually be circumvented by careful design.

5.2 DIELECTRIC (NONSCANNING) LENSES FOR CONTROL OF APERTURE AMPLITUDE AND PHASE

5.2.1 Simple Dielectric Lenses

There are three dielectric lens types that are commonly used for antennas: (1) the plano-convex lens with flat outer and hyperbolic inner surfaces, (2) the plano-convex lens with straight inner face (which is called "straight face lens" in this chapter), and the (3) meniscus lens which has a spherical inner surface and elliptical outer surface. Dielectric lenses are sometimes constructed in two dimensions as cylindrical antennas fed by line sources [4,11]. In this chapter only axially symmetric, three-dimensional models are discussed as developed by rotating the lens cross sections about the axis of symmetry. Figure 5.1(a,b,c,d) shows these common lenses in profile. When these lenses are fed by a point

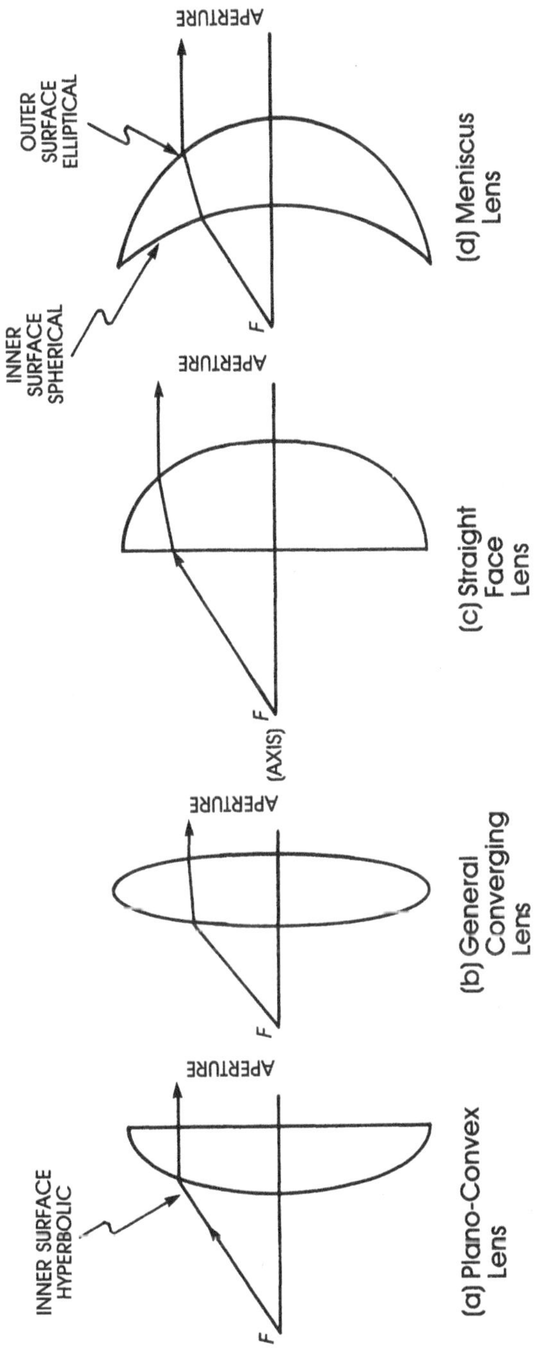

Figure 5.1 Family of Simple Dielectric Lenses

source at the focus, F, a plane wave is generated on the lens apertures (exit faces).

These basic lens types have quite different effects on the amplitude distribution on the lens apertures. For the plano-convex lens Figure 5.1(a), for example, with its inner surface illuminated by a corrugated horn with a rotationally symmetric Gaussian-like radiation pattern, the lens aperture distribution is very highly tapered [9,10]. Generally the lenses of Figure 5.1(b) and (c) have rather small effect on the illuminating feed amplitude distribution, that is, the outer aperture distribution is shaped similarly to the inner horn's angular distribution pattern. The meniscus lens of Figure 5.1(d) tends to flatten the horn pattern, rather than concentrate the energy near the lens axis, as does the plano-convex lens of Figure 5.1(a). The straight face lens of Figure 5.1(c) has good wide-angle characteristic and is suitable for scanning and multibeam applications. The straight face lens configured for wide-angle scanning according to Friedlander's paper [12] is also referred to as the Friedlander lens.

5.2.2 General Equations for Contours of Lenses

The lenses in Figure 5.1 can be considered a family of axially rotational lenses all focusing radiation from a point sourced at F to plane waves on the (outer) aperture but acting as different transducers of the illuminating power pattern $P(\theta,\phi)$. (Power reflected back into the feed at F and cross polarization caused by lens surfaces are somewhat different for each lens geometry.)

Analytically the lenses in this family can be defined by choosing the axis lengths D_1 and D_0 and determining the inner and outer lens contours $f_1(y_1,z_1)$ and $f_2(y_2,z_2)$ such that optical path length from F to the aperture plane (see Figure 5.2) is constant, i.e.,

$$R + ND + t = D_1 + ND_0 \tag{5.1}$$

The contour functions f_1 and f_2 are rotated about the lens axis to generate the axially symmetric lenses. N is the lens index of refraction.

For the lenses in Figure 5.1(a) and Figure 5.1(d), the inner and outer contours are expressible as explicit analytical functions. For the plano-convex lens the inner surface is a hyperboloid of revolution [9]. Using the notation of Figure 5.2 the inner surface, $R(\theta_1)$, is

$$R(\theta_1) = (N - 1)D_1/[N\cos(\theta_1) - 1] \tag{5.2}$$

and the outer surface is a plane given by $z_2 = D_1 + D_0$.

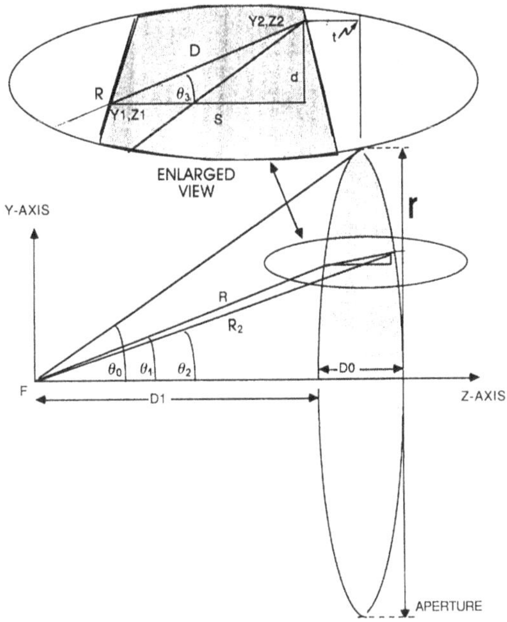

Figure 5.2 Ray Tracing Variables, Dielectric Lenses

For the Meniscus lens in Figure 5.1(d) we write for the spherical inner surface, the inner contour

$$f_1 = (y_1^2 + z_1^2)^{1/2} \tag{5.3a}$$

and for the elliptical outer surface,

$$R_2(\theta_2) = (D_1 + D_0)(N - 1)/(N - \cos(\theta_2)) \tag{5.3b}$$

For the straight face lens [9] of Figure 5.1(c) the inner surface is

$$z_1 = D1$$

and the outer surface f_2 can be expressed parametrically. By applying (5.1) and Snell's law at each surface, employing the variables of Figure 5.2

$$\theta_1 = \sin^{-1}(y_1/R) = \tan^{-1}(y_1/D_1)$$

$$\theta_3 = \sin^{-1}(sin(\theta_1)/N)$$

$$D_3 = ND_0 + D_1$$

$$t = [(D_3 - R)\cos(\theta_3)/N - D_0)]/[\cos(\theta_3)/N - 1]$$

$$s = D_0 - t \tag{5.4}$$

$$\mathrm{d} = s\tan(\theta_3)$$

$$z_2 = s + D_1$$

$$y_2 = y_1 + \mathrm{d}$$

By fixing N, D_1, and D_0 the values of the independent variable y_1 determine $f_2(y_2,z_2)$.

For all the lenses in this family the values of N, D_1 and D_0 also determine the f/D of the lens subject to the analytical design procedures using (5.1) through (5.4). Some thickness near the lens edges is needed for mechanical support and strength so the lenses are usually designed for oversized diameters and truncated leaving some edge thickness.

5.2.3 Computing Lens Aperture Distributions on Axially Symmetric Lenses

For low sidelobe radiation patterns it is necessary to have control of the antenna aperture amplitude and phase distributions, including the edge "taper" levels. The purpose of this section is to provide computer tools for computing the lens aperture power distribution given the feed pattern and the lens contours $f_1(y_1,z_1)$ and $f_2(y_2,z_2)$.

The basic focusing lenses of Figure 5.1 act as power transducers converting the illuminating power pattern $P_p(\theta,\Phi)$ (spherical coordinates) into an aperture power pattern $P_s(r,\phi)$ (cylindrical coordinates). We will assume here the lens feed patterns have no Φ dependence (and therefore no ϕ dependence) which is approximately true for conical corrugated horn feeds.

The power in the feed pattern $P_p(\theta)$ summed from the lens axis to a spherical polar angle, θ, is

$$P_f(\theta) = C_s 2\pi \int_0^{\theta} P_p(\theta)\sin(\theta)\,\mathrm{d}(\theta) \tag{5.5}$$

Power in the feed pattern can be thought of as distributed in cones between cone angles θ_I and θ_0 where $\theta_I < \theta_0$. When $\theta_I = 0$ and $\theta_0 = \theta_{max}$ at the lens edge, then $P_f = P_T$ where P_T is the total feed power incident on the lens inner surface.

The power in the lens aperture from the axis to some radius r is given by,

$$P_r(r) = C_r 2\pi \int_0^r P_s(r) r \, dr \tag{5.6}$$

The aperture power can be considered as distributed in rings of radius R_I and R_0 where $R_I < R_0$. Again $P_r = P_T$ when $r = R_{max}$ at the aperture edge.

The total power inside the feed pattern to θ_0 (edge of lens) and the corresponding power in the lens aperture to $r = R_{max}$ must be equal, neglecting reflection losses and dielectric losses in the lens. We use this fact to evaluate the constants C_s and C_r. The desired radial aperture distribution $P_s(r)$ is usually given by a power series and therefore easily integrated. Horn feed patterns can often be approximated by $P_p(\theta) = \cos^{N_e}(\theta)$ as done in the illustrations to follow. In any case these functions, (5.5) and (5.6), can be integrated numerically.

If we take a simple case where the aperture distribution is constant and the feed pattern given by,

$$P_p(\theta) = \cos^{N_e}(\theta)$$

Then using (5.5) we get

$$P_f(\theta) = C_s(1 - \cos^M(\theta))/M, \quad \text{where } M = N_e + 1 \tag{5.7}$$

and from (5.6) we can write

$$P_r(r) = C_r r^2/2 = C_r y_2^2/2$$

$$C_n = (R_{max})^2/(1 - \cos^M(\theta_0) = 2C_s/(C_r M)$$

$$y_2^2 = C_n(1 - \cos^M(\theta) \tag{5.8}$$

Whenever the outer surface coordinate y_2 is known for given inner surface coordinates y_1, z_1(or $\theta_1 = \tan^{-1}(y_1/z_1)$) the transforming effect of the lens on the feed pattern can be numerically calculated from the computed power in corresponding cones and rings. Equation (5.9) equates the power in cones and rings assuming the feed pattern can be expressed by $\cos^{N_e}(\theta)$,

$$\Delta P = C_n[\cos^M(\theta_1) - \cos^M(\theta_0)]/(R_0^2 - R_1^2) \tag{5.9}$$

These expressions relating the illuminating feed power and aperture power distribution are programmed in LENAMP.BAS. This is not a program to synthesize lens shapes but uses known lens profiles as inputs and calculates the resulting aperture distributions. LENAMP is useful, therefore, for finding the lens aperture distribution when feed pattern and lens contours are known. Also the first step in synthesizing a lens for aperture power control is to select the feed pattern, the semiflare angle of the horn, for example, such that the radial distribution on the lens is close to the desired amplitude function when computed by the program LENAMP.BAS.

Usually the coordinate relations between θ_1 (or y_1, z_1) and y_2 are entered into LENAMP.BAS by data files defining the inner-outer pairs of lens points on the lens contours (see LENAMF.BAS). For this illustration of aperture power distribution, the St. Face (straight face) lens provides the relations between θ_1 and y_2 from (5.4) which is used as a subroutine in LENAMP to relate the inner and outer surface points. By entering $M = N_e + 1 = 2$ at the computer prompt we see that this lens converts a horn pattern with a $\cos(\theta)$ to nearly a uniform aperture distribution for the particular dielectric constant and other lens constants chosen ($R_0 = D_1 = F = 1.8$, $D_0 = 1.25516$, and $N = 1.594$, see Figures 5.1(c) and 5.3). The program LENAMP.BAS can be modified to give the aperture power distributions for lens antennas using the lens types of Figure 5.1 or other axially symmetric lenses with known contours. (The actual feed pattern can be integrated in a subroutine between closely spaced theta intervals based on the lens coordinate pairs instead of using the $\cos^{N_e}(\theta)$ approximation.)

When the lens contours are available in closed form as for the lenses of Figure 5.1(a) and (d), then the differential expression below can be evaluated.

$$P(r)/P(\theta) = \sin(\theta)\, d\theta/(r\, dr)$$

For example for the Meniscus lens (see [9] for derivation of (5.10)) we get, using (5.3b)

$$P(\theta)/P(r) = \frac{(N - 1)^2(D_1 + D_0)^2(N \cos\theta_2 - 1)}{(N - \cos\theta_2)^3} \tag{5.10}$$

When $P(r)$ is constant for uniform aperture power, a feed pattern $P(\theta) = \cos^{N_e}(\theta)$, where $N_e = 7.45$ (or $M = 8.45$) will approximate the function

(5.10). This can be checked by using a Meniscus lens subroutine in LENAMP.BAS with $N = (2.54)^{1/2}$, $D_0 = 1.4265$, $D_1 = 3.62$, for example, and plotting the aperture distribution.

5.3 SYNTHESIS OF LENS CONTOURS FOR CONTROL OF LENS APERTURE DISTRIBUTIONS

When very low sidelobe patterns are required, especially shaped lenses can be used as transducers to convert the feed radiation patterns into lens antenna aperture distributions with plane phase fronts and a radial amplitude function selected from Chapter 1 "ideal" functions. The synthesis procedure must shape both the inner and outer lens contours. It is best to find the feed pattern that is close to the desired function when converted by the lens into the aperture distribution. This design selection can be done using the (rotationally symmetric) feed patterns of Chapter 2 and the program LENAMP.BAS to find the resulting aperture distributions for any lens when the inner and outer contours are known and the y_1, z_1 values can be paired with y_2, z_2 values. The synthesis procedure outlined below can then be used to shape the lens contours f_1 and f_2 to obtain the desired aperture distribution. (In these procedures the effects of lens reflections and of feed spillover are neglected. Absorbing shields can surround the inside volume of lens and feed, keeping the aperture free, to prevent the spillover radiation from altering the aperture radiation pattern.)

To demonstrate the lens shaping procedure we will use the two lenses, straight face and Meniscus, for which we calculated the aperture distributions for known feed illuminations in the last section. We will show that the shape synthesis technique yields approximately the straight face lens and Meniscus lens profiles when illuminated by the radiation patterns that produce (approximately) uniform aperture distributions.

The shaping method requires start-up contours near the lens axis to begin the numerical shaping process. When an antenna with maximum gain is desired, then the Meniscus lens is probably the best model for the axis geometry and the start-up dimensions. For low sidelobe levels with optimum aperture efficiency the straight face lens is preferable.

We will use these two lens configurations as start-up models for the first y_1, z_1 and y_2, z_2 points and slopes near the lens axis in the contour shaping technique. If we illuminated these lenses with feed patterns that produce uniform aperture distributions with LENAMP, then the synthesis process should give us back St. Face lens and Meniscus lens contours. Two numerical lens shaping programs are provided, SYNTH.BAS and

SYNTHS.BAS, based on using the Meniscus lens and the St. Face lens respectively as starting lens contours near the lens axis.

The synthesis procedure is derived using the geometrical relations shown in Figure 5.2. We must find z_1, y_2, and z_2, given y_1 as the independent variable, from the required aperture power function through the relation (5.8). It is assumed also that the feed pattern $P(\theta)$ is given.

The synthesis approach implemented in SYNTH.BAS uses the general pathlength condition for this family of lenses given in (5.1) and Snell's law for refracted rays at the lens inner surface. Equation (5.1) can be written

$$(y_1^2 + z_1^2)^{1/2} + N((y_2 - y_1)^2 + (z_2 - z_1)^2)^{1/2} - z_2 - D_0(N - 1) = 0 \tag{5.11}$$

This is a quadratic relation for finding z_2 when y_1, z_1, and y_2 are given. Referring to SYNTH.BAS the desired quadratic root is found in lines 360–450. The outer lens radius y_2 can be found from the input-output power density relations by using line 340, which is (5.8).

The lines 460–540 are obtained by using Snell's refraction relation,

$$\mathbf{n} \times \mathbf{S}_i = N\mathbf{n} \times \mathbf{S}_t \tag{5.12}$$

where $\mathbf{n}$ is the unit vector inner contour normal, N is the index of refraction, and $\mathbf{S}_i$ and $\mathbf{S}_t$ are incident and transmitted rays (expressed as unit vectors), respectively. The slope on the inner surface of the lens is found to be (line 490) from (5.12)

$$B_8 = (\sin\theta_1 - NM_1)/[(\cos\theta_1 - NN_1)^2 + (\sin\theta_1 - NM_1)^2]^{1/2} \tag{5.13}$$

The Adam's predictor-corrector [13] (line 560) is used to synthesize the inner contour beginning with the "start-up" back derivatives (needed in the numerical calculations) computed for the spherical inner surface near the axis using lines 140–220. Subsequent inner slopes for extrapolation are computed in lines 460–510.

The straight face lens is used for the start-up model in SYNTHS. BAS. This program is written for entering a feed pattern of the $\cos^N(\theta)$ type in line 230. The synthesized lens profiles can be compared with y_2, z_2 values computed by a St. Face lens subroutine as in LENAMP.BAS.

In these lens shaping examples it is assumed that a uniform aperture distribution is desired. This assumption leads to the simple relation between the feed illumination angle θ and the aperture radius, y_2, in Equation (5.8), when the feed pattern is approximated by $P(\theta) = \cos^N\theta$.

Equations (5.5) and (5.6) can be used to convert more general feed patterns into a desired radial amplitude function on the lens aperture, producing a more complicated equation (5.8). Because the contours of f_1 and f_2 of the lens may diverge near the lens edges, it is best to start with simple, known cases. Adjusting the initial parameters, D_1 and D_0, and the edge thickness makes it possible to shape lenses for most f/D ratios. The feed pattern can be varied by changing the semiflare angle of a corrugated horn, for instance. Equation (5.11) makes the aperture phase always constant. When there are known phase errors in the horn patterns these can be included as pathlength errors and added to R in Equation (5.1) and (5.11).

In Figures 5.3 and 5.4 the synthesis programs, SYNTHS.BAS and SYNTH.BAS, were used to compute the profiles of the straight face lens and the meniscus lens using the feed patterns from LENAMP that most nearly produce a constant aperture distribution as starting approximations. As $\cos^N\theta$ functions cannot give exactly the feed illumination on these lenses to produce a uniform aperture distribution, the lens shapes synthesized are not exactly the simple meniscus and straight face profiles, but nearly so. The ends (edges) of the lenses' shapes tend to show that feed errors of approximate $\cos^N\theta$ feed patterns, used to recreate the basic lens contours, are greatest for larger θ. It is seen in Figure 5.3 that the straight face or Friedlander lens is obtained with good approximation when synthesized from a $\cos\theta$ ($M = 1.98 = N + 1$), although the results from LENAMP suggest $M = 2.1$. For the meniscus lens in Figure 5.4 the $\cos^M\theta$ from LENAMP was approximately $M = 8.45$, but $M = 8.35$ produces better lens contours. In both cases the tips of the lenses are not correctly synthesized.

A more useful example of lens shaping to control aperture distribution is programmed into KOUZN.BAS. Starting with a St. Face geometric model for the axis start-up coordinates an "ideal" aperture distribution for low sidelobes and high gain shaped lens contours is computed. The Kouznetsov aperture field distribution discussed in Chapter 1, Section 1.3, is the desired aperture distribution. This aperture field profile assumes no ϕ dependence and is given by

$$E(T) = 1/7 + 6/7(1 - T^2)^2$$

where $T = r/R_0$ is the normalized aperture radius of the lens. We square this equation to get the aperture power distribution and then perform the integration in (5.6). A relationship between the aperture radius, r, and the feed angle, θ, is then found. Again a feed radiation pattern, $P(\theta) = \cos^{M-1}\theta$, is assumed, although a power series representing a corrugated

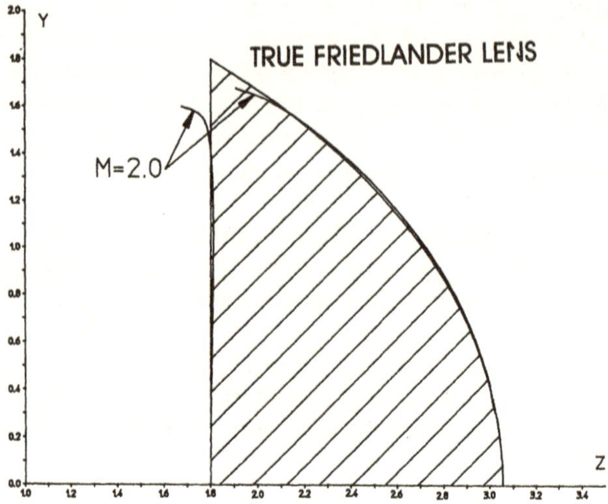

Figure 5.3 Friedlander Lens Shaped from cos θ Feed Pattern

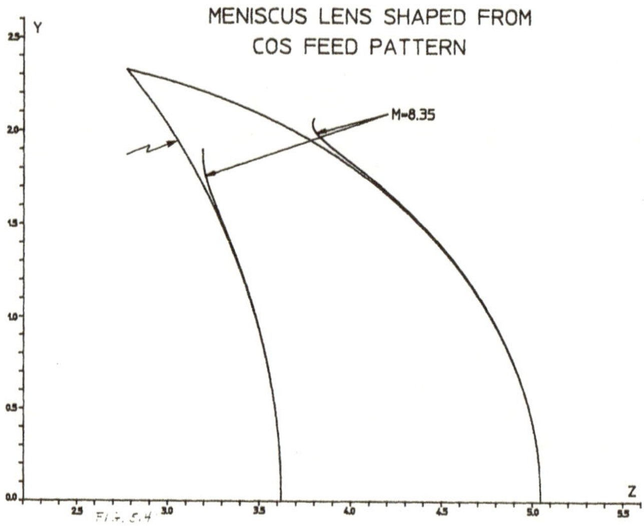

Figure 5.4 Meniscus Lens Shaped from $\cos^{7.35}\theta$ Feed Pattern

horn pattern can be easily integrated in (5.5). The relation programmed in KOUZN.BAS is

$$49T^2 - 84T^4 - 76T^6 + 36T^8 + 7.2T^{10} = C_9(1 - \cos^M\theta)$$

where $C_9 = 12.2/(1 - \cos^M\theta_0)$. A root finder is needed to relate r to θ. When $M = 15.5$ the desired aperture distribution is realized with lens contours not greatly different from the St. Face lens profiles. The synthesized lens profile coordinates are filed by KOUZN.BAS into data files SP1.DAT (inner contour) and SP2.DAT (outer contour). A general program, LENAMF.BAS, that accepts files of lens contours was written, and by running the programs KOUZN.BAS and LENAMF.BAS it is seen that the Kouznetsov aperture power distribution is produced. Using the ray-tracing program HOLT6.BAS, with the data files SP1,SP2 converted to power series representing the inner and outer surfaces of the shaped lens by SLCOD.FOR, it is seen that the lens focuses very well to the point F.

5.4 LENS-SURFACE REFLECTIONS AND CROSS POLARIZATION

We have assumed so far that dielectric lenses were free from losses due to boundary reflections, cross polarization, and dielectric media propagation attenuation. These effects are not always small, and computer methods for evaluating these influences are important in lens design. Reflection losses are often the largest, and lens surface matching techniques, as the lens coatings used with optical lenses, are needed for reducing lens antenna reflections [5, Part I].

Fortunately for most dielectric materials used in microwave and millimeter-wave antennas the dielectric loss tangents are small. For example, Teflon™ and Rexolite 1422™ have loss tangents at 40 GHz of about 0.0006. The dB loss per wavelength due to dissipation along the ray propagation path is given by

$$\text{dB} = 27.3(\varepsilon_r)^{1/2}(\text{loss tangent})/(\text{wavelength})$$

(equation derived from the propagation term in dielectrics with a complex propagation constant). These losses are typically a few tenths of dB and increase with relative dielectric constant, ε_r, and frequency for a given lens thickness.

5.4.1 The Dielectric Prism Antenna

To quantify and program the analysis for evaluating surface reflections and polarization effects in lenses a simple dielectric lens antenna made from flat surfaces in the form of a wedge or prism is described. This antenna is useful for scanning antenna beams to angles 60° or more from the normal to the antenna aperture.

Figure 5.5 shows the prism antenna configuration constructed of a dielectric of index of refraction, N, and a prism angle, A_p. A plane wave source (a paraboloid antenna for example) illuminates the prism lens inner surface with the incident parallel rays, $\mathbf{S}_0$, all coplanar with the normals, $\mathbf{N}_1$ and $\mathbf{N}_2$ to the two prism surfaces. Then the beam of the antenna with prism's outer surface is in the direction $\mathbf{S}_3$. The angle θ_{33} can be made larger than θ_I. Scanning is done by tilting the source in the plane of the prism normals to achieve low angle beams. A beam direction about 70° from the exterior normal to the prism surface can be attained. Scanning in the orthogonal plane is achieved by rotating the prism and plane wave source together. Another method of scanning in one angular plane is to keep the plane wave source fixed, constructing the prism of two wedges of equal angle and rotating them such as to vary the prism angle while keeping the normals coplanar with $\mathbf{S}_0$.

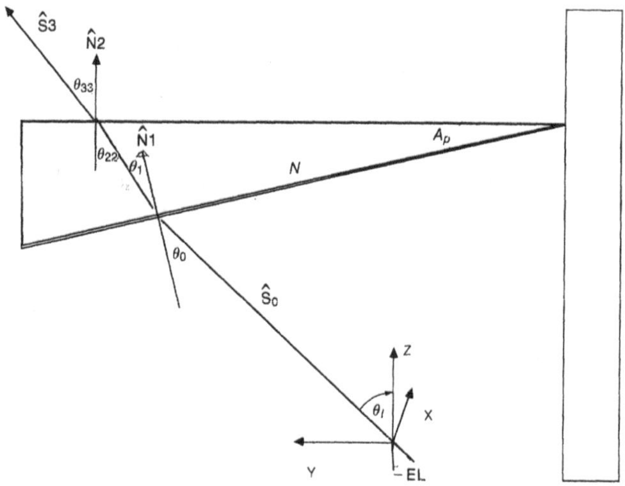

Figure 5.5 Polarization Components for Rays Through Prism Lens

To find the reflection losses for the two surfaces when the incident rays and both prism normals lie in a plane, we can use the Fresnel equations for lossless dielectrics [14] for rays polarized in the plane of incidence and normal to it. The program PRISM.BAS computes the (approximate) transmission losses due to reflections from the prism surfaces and prompts for dielectric constant and prism angle and steps the incidence angle over values of θ_I. The losses from the inner and outer surfaces are combined and printed out for polarization in the plane of incidence, DBI, and normal to the plane of incidence, DBN, along with the incident angle θ_I and the exit ray angle θ_{33}.

Transmission coefficients, based on power flow normal to the prism surfaces, are used in the program at both surfaces. At the inner surface

$$T(\text{perpendicular}) = \sin 2\theta_0 \sin 2\theta_1/\sin^2(\theta_1 + \theta_0)$$

$$T(\text{parallel}) = T(\text{perpendicular})/\cos^2(\theta_0 - \theta_1)$$

This simple program PRISM gives the transmission losses for a wedge antenna, as shown in Figure 5.5, and the refraction bending of rays normal to an incident plane wave passing through it. It is seen that the losses can be high when exit rays are nearly tangent to the prism surface (θ_{33} approaching 90° are required). This two-dimensional program gives the reflection losses for individual rays, although here a bundle of parallel rays represent the plane wave propagation. The program is helpful for tracing individual rays, not necessarily parallel, used in the design of other lenses.

When the depolarization effects are required for polarized fields incident from general elevation and azimuth directions and propagating through a lens of homogeneous (and lossless) dielectric, more complicated expressions are employed. This work is done by the program POLGEN.BAS. This software calculates the polarization components of a general ray passing through a lens when the lens normals at two surfaces and the incident ray direction and polarization components are given. In this program either x-polarized or y-polarized rays can be chosen. The lens surface normals can be entered by data file or subroutine for any lens.

Helpful software for calculating power reflection loss and depolarization through a dielectric object are not generally available. The analytical steps programmed in POLGEN.BAS are outlined here and two procedures are given to solve the polarized ray tracing problem.

We will use the dielectric prism for the example case with the incident rays defined by elevation and azimuth angles as done in later ray tracing on inclined plane waves on lens exit surfaces (receiving mode). The ray directions for incident rays using the coordinates of Figure 5.5 and (5.14) result in the El = elevation angle being minus the θ_I angle in the y-z plane. Alternatively for illuminating the lens or prism with a point source or horn feed the reference polarization can be x- or y-polarized Huygens sources as given in (1.35) and (1.36). We write for the incident ray at any point on the inner surface of the prism lens

$$\mathbf{S}_0 = \cos(\text{EL})\sin(\text{AZ})\mathbf{x} - \sin(\text{EL})\mathbf{y} + \cos(\text{EL})\cos(\text{AZ})\mathbf{z} \qquad (5.14)$$

and

$$\theta_1 = \sin^{-1}[\sin(\theta_0)/N]$$

where N is the index of refraction and the angles are in the first quadrant. The refracted ray $\mathbf{S}_2$ can be found using these unit vectors from

$$\mathbf{S}_2 = \cos\theta_1 \mathbf{N}_1 + \sin\theta_1 \mathbf{T}_0,$$

where in the program $\mathbf{S}_2$ has rectangular vector components A_2, B_2, C_2.

To find the sagittal (normal to plane of incidence) and tangential (in plane of incidence) fields within the dielectric we write

$$\mathbf{T}_A = \mathbf{T}_0 \times \mathbf{S}_0,$$

with components (TAX, TAY, TAZ) and

$$\mathbf{T}_1 = \mathbf{T}_0 \times \mathbf{S}_2,$$

with components (T1X, T1Y, T1Z)

$$\mathbf{E}_{2t} = 2\cos\theta_0 \sin\theta_1/(\sin(\theta_0 + \theta_1)\cos(\theta_0 - \theta_1))(\mathbf{E}_0 \cdot \mathbf{T}_A)\mathbf{T}_1$$

$$\mathbf{E}_{2s} = 2\cos\theta_0 \sin\theta_1/(\sin(\theta_0 + \theta_1))(\mathbf{E}_0 \cdot \mathbf{P}_0)\mathbf{P}_0$$

where $\mathbf{E}_0$ is the polarization of the incident ray.

To trace the polarized ray from region 2, the dielectric region, to region 3, the exit region of relative dielectric constant equal one, we again write the unit vectors for normal and in-plane directions as

$$\mathbf{P}_2 = \mathbf{N}_2 \times \mathbf{S}_2/|\mathbf{N}_2 \times \mathbf{S}_2|$$

with components (S2X, S2Y, S2Z) and

$$\mathbf{T}_2 = \mathbf{P}_2 \times \mathbf{N}_2$$

with components (T2X, T2Y, T2Z) (for in-plane). The angles $\theta_{22} = A_{22}$ and $\theta_{33} = A_{33}$ are computed from

$$\theta_{22} = \cos^{-1}(\mathbf{S}_2 \cdot \mathbf{N}_2)$$

and

$$\theta_{33} = \sin^{-1}(N \sin \theta_{22})$$

in the first quadrant. Also to calculate the exit ray $\mathbf{S}_3$ components we write

$$\mathbf{S}_3 = \cos \theta_{33} n\mathbf{N}_2 + \sin \theta_{33}\mathbf{T}_2$$

with components (A3, B3, C3). We need also the unit vectors

$$\mathbf{T}_3 = \mathbf{T}_2 \times \mathbf{S}_2$$

with components (T3X, T3Y, T3Z), and

$$\mathbf{T}_4 = \mathbf{T}_2 \times \mathbf{S}_3$$

with components (T4X, T4Y, T4Z).

The x-polarized incident ray we write as $\mathbf{E}_{0X} = \mathbf{y} \times \mathbf{S}_0/|\mathbf{E}_{0X}|$ and the y-polarized ray as $\mathbf{E}_{0Y} = \mathbf{x} \times \mathbf{S}_0/|\mathbf{E}_{0Y}|$.

For the dielectric wedge of Figure 5.5, the normals to the inner (lower) and outer (upper) surfaces are everywhere

$$\mathbf{N}_1 = L_1\mathbf{x} + M_1\mathbf{y} + N_1\mathbf{z},$$

for the inner normal where $L_1 = 0$, $M_1 = \sin(\text{AP})$, $N_1 = \cos(\text{AP})$, and AP is the prism angle, and

$$\mathbf{N}_2 = L_2\mathbf{x} + M_2\mathbf{y} + N_2\mathbf{z},$$

for the outer normal where $L_1 = M_1 = 0$ and $N_1 = 1$.

The vectors $\mathbf{S}_0$, $\mathbf{N}_1$, and $\mathbf{N}_2$ can be general unit vectors obtained by ray tracing using methods developed in later sections. The above simple parameters are used in the illustrating examples.

Tracing the polarized wave through the two dielectric surfaces is done by constructing unit vectors in the plane of incidence and normal to the plane of incidence as the incident ray is refracted at the two boundaries. To aid in following the computer code, direction cosines for the unit vectors are indicated with the analytic steps. The first part of the program POLGEN.BAS follows these equations.

A sagittal unit vector normal to $\mathbf{S}_0$ and $\mathbf{N}_1$ is $\mathbf{P}_0 = \mathbf{N}_1 \times \mathbf{S}_0/|\mathbf{N}_1 \times \mathbf{S}_0|$ with direction cosines written in the program as (PX, PY, PZ). Also formed is $\mathbf{T}_0 = \mathbf{P}_0 \times \mathbf{N}_1$ with direction cosines ($T0X$, $T0Y$, $T0Z$) which defines the tangent plane orthogonal to $\mathbf{P}_0$.

To find the angles $\theta_0 = A_1$ and $\theta_1 = A_1$ we can use Snell's law of refraction, that gives

$$\theta_0 = \cos^{-1}(\mathbf{S}_0 \cdot \mathbf{N}_1)$$

The polarized E-fields in the region 3 are

$$\mathbf{E}_{3t} = F_a(\mathbf{E}_2 \cdot \mathbf{T}_3)\mathbf{T}_4$$

$$\mathbf{E}_{3s} = F_a \cos(\theta_{22} - \theta_{33})(\mathbf{E}_2 \cdot \mathbf{P}_2)\mathrm{P}_2, \quad F_a = \frac{2 \cos \theta_{22} \sin \theta_{33}}{\sin(\theta_{22} + \theta_{33}) \cos(\theta_{22} - \theta_{33})}$$

The sagittal and tangential field in region 3 are summed to give the E3X, E3Y, E3Z components displayed by the computer.

The software in the second part of POLGEN.BAS traces the incident ray through the two surfaces of the prism antenna. Transmission coefficients are used as in PRISM. These expressions give the power transmitted normally through the lens boundaries. The rays as vectors are traced using Snell's law as written in (5.13). The refraction angles for general rays are found at the two lens surfaces. The two-dimensional prism lenses solution (PRISM.BAS) can be used as a check on dB power losses, DBN, DBI, on the three-dimensional reflection losses computed by POLGEN.BAS.

To find the polarization components of the rays passing through the two surfaces, unit vectors that lie parallel to the polarization component in the plane of incidence and normal to it are constructed. Because of the refractive bending at the two surfaces when the two surface normals are not coplanar, eight such unit vectors are needed. These vectors are listed here using the computer symbols in POLGEN.BAS for the variables. (Bold type indicates vectors.)

$$\mathbf{V} = \mathrm{VX}\mathbf{x} + \mathrm{VY}\mathbf{y} + \mathrm{VZ}\mathbf{z} = \mathbf{S}_0 \times \mathbf{n}_1/|\mathrm{V}|$$

$$\mathbf{T} = \mathrm{TX}\mathbf{x} + \mathrm{TY}\mathbf{y} + \mathrm{TZ}\mathbf{z} = \mathbf{V} \times \mathbf{S}_0/|\mathrm{T}|$$

$$\mathbf{T2} = \mathrm{T2X}\mathbf{x} + \mathrm{T2Y}\mathbf{y} + \mathrm{T2Z}\mathbf{z} = \mathbf{V} \times \mathbf{S}_2/|\mathrm{T2}|$$

$$\mathbf{V3} = \mathrm{V3X}\mathbf{x} + \mathrm{V3Y}\mathbf{y} + \mathrm{V3Z}\mathbf{z} = \mathbf{S}_2 \times \mathbf{n}_2/|\mathrm{V3}|$$

$$\mathbf{T22} = \mathrm{T22X}\mathbf{x} + \mathrm{T22Y}\mathbf{y} + \mathrm{T22Z}\mathbf{z} = \mathbf{V}_3 \times \mathbf{S}_2/|\mathrm{T22}|$$

$$\mathbf{T3} = \mathrm{T3X}\mathbf{x} + \mathrm{T3Y}\mathbf{y} + \mathrm{T3Z}\mathbf{z} = \mathbf{V}_3 \times \mathbf{S}_3/|\mathrm{T3}|$$

Formed also are:

$DN = \mathbf{EO} \cdot \mathbf{V}$ and $DI = \mathbf{EO} \cdot \mathbf{T}$, where N indicates normal plane and I indicates incident plane.

For the polarized ray within the dielectric we write,

$$\mathbf{E}_{2N} = (PNI)^{1/2} DN(\mathbf{V}) \text{ and } \mathbf{E}_{2I} = (PII)^{1/2} DI(T2)$$

$$\mathbf{E}_2 = \mathbf{E}_{2N} + \mathbf{E}_{2I} \text{ for the ray in the dielectric media}$$

$$DN2 = \mathbf{E2}T \cdot \mathbf{V3} \text{ and } DI2 = \mathbf{E2}T \cdot \mathbf{T22}$$

$$CN2 = (PN2)^{1/2} DN2 \text{ and } CI2 = DI2(PI2)^{1/2}$$

$$\mathbf{E3}N = CN2(\mathbf{V3}) \text{ and } \mathbf{E3}I = CI2(\mathbf{T3})$$

$$\mathbf{E3} = \mathbf{E3}N + \mathbf{E3}I \text{ for the exit ray polarization and amplitude}$$

Part one and part two of POLGEN.BAS give slightly different answers for the fields $\mathbf{E}_2$ in the dielectric and for $\mathbf{E}_3$ the exit fields. The cross-polarization levels calculated from the ratios E_X/E_Z, E_Y/E_Z, E_X/E_Y in regions 1, 2, and 3 are almost identical in the part one and part two computations. The power levels in regions 2 and 3 are different, however. The part two formulation is based on Poynting vector flux, and gives the best approximation to power transmitted into and through the lens. In part one, the E-fields in the dielectric medium are correctly calculated, but they must be squared and multiplied by $\sqrt{\varepsilon_r}$ to find the approximate power levels. For the two parts, the program prints out the values: DBIN = dB level in the dielectric and DBOU = dB level in region 3. Also the variables shown in part two, DBI = dB in region 3 of plane of incidence rays and DBN = dB in region 3 for the normal to plane of incidence, are also useful guides in estimating power losses due to interface reflections.

All of the ray components (but not the field components) and angles are the same in both parts. Part one is preferred for cross-polarization computations, and it can be used as a subroutine for including lens-caused cross polarization in diffraction pattern computations. Part two is preferred for calculating transmission losses through lenses. It is clear that general rays incident with only an x- or y-polarization component have all three components in region 3, the aperture region. The losses due to boundary reflection and cross polarization can be significant.

To "coat" or match the lens surfaces to reduce the reflection losses the quarter-wave matching technique is often used [5]. In this method slots of grooves are formed in the dielectric surfaces of the lens to approximate a quarter-wave plate. The refractive properties of the lens are quite well preserved but the reflection losses are greatly reduced over a restricted bandwidth. The cross polarization components are not significantly changed from the values computed by POLGEN.BAS.

5.5 WIDE-ANGLE, MULTIBEAM DIELECTRIC LENS ANTENNAS

For designing antennas to be scanned by feed motion or fed by cluster feeds (or phased array elements) to generate multiple beams, only the straight face lens or Friedlander lens [12], of the basic types discussed so far has good wide-angle focusing characteristics. We will also develop a lens shaping technique for two-point correction, which term the inventors, Holt and Mayer [15], used to indicate two-angle correction for a two-dimensional (cylindrical) lens [11]. For the rotational lens described here the correction zone is a cone of angles about the lens axis. J.J. Lee [16] provides a lens shaping technique to improve wide-angle coverage analogous to the lens synthesis approach for aperture amplitude control described in Section 5.2. His methods are highly recommended. Lens contours, stepped or unstepped, for any axially symmetric lens can be ray traced and their patterns computed using the programs to be discussed.

5.5.1 Seidel Aberrations of Lens Images

It is useful to borrow from the knowledge and terminology of optical lens designers. For camera lenses, for example, combinations of spherical lenses, generally symmetrically centered on the lens axis, are used to transform faithfully the light source object onto a flat image screen. For antennas working at much longer wavelengths we can use aspherical lens contours, and the focal surface need not be flat. The identification of dielectric lens focusing defects in terms of Seidel aberrations [1,2,3,17], spherical aberration, coma, astigmatism, Petzval curvature of field, and distortion can, however, also be done for microwave dielectric lenses. We can relate to optical language by considering the departure from flatness (linearity) of tilted plane waves on the lens aperture as a feed source is moved laterally from the geometric focus, f, and the lens axis. If we calculate the pathlengths, L, from the feed through the lens to the lens aperture by ray tracing, we can develop a power series for the nonuniform phase ($\Phi = 2\pi L/(\lambda)$) distribution on the aperture of radius r:

$$\Phi = a_0 + a_1 r + a_2 r^2 + a_3 r^3 + a_4 r^4 \ldots$$

In optics the linear term in r is associated with beam tilt, the even terms in r with spherical aberration, and defocusing and the odd terms with coma. The a_n coefficients are functions of the scan angles EL and AZ. This kind of study of focusing aberrations has been done for the paraboloid mirror [5, Part II, p. 57] and for metal-plate lenses in Chapter 6. The aim of these studies is to find the focal positions for locating the feed to minimize the higher terms in the power series which produce loss in antenna gain and rise in sidelobe levels.

It is very difficult to minimize all lens aberrations simultaneously by analytical methods. A numerical computer technique is developed here that makes use of computer graphics to find the best focal surfaces for scanning angular regions. As described in [18] the method is effective in finding the circles of least confusion (Conrady [2], p. 120) where the lens focuses best. Other aberrational effects such as the locations of astigmatic lines are also easily found by this computer ray-tracing plus graphics technique.

5.5.2 Computer-Graphics Method for Finding Best Focal Surfaces

To evaluate the wide-angle scan characteristics of a lens we must know the best focal points and focal surfaces for the highest gain and lowest sidelobe patterns. This task is very time-consuming, either experimentally or by computing diffraction patterns without good information on the correct focal points for a given scan direction. The computer technique developed here (also in Chapter 7 for reflector antennas) first traces a central ray through the lens from the bundle of parallel incident rays on the lens aperture arriving from a given elevation and azimuth angle direction. A flat "computer screen" normal to this central ray is set up and the traces of incident rays, usually on rings over the lens outer surface, are presented on the monitor screen with computer graphics. The focal distance is varied along the central ray by changing the z-coordinate Z_4 until the smallest focal spot is traced on the screen. Three programs, MENISC.BAS, for the Meniscus lens, LENS4.BAS, for the straight face lens, and HOLT6.BAS for the two-point corrected lens are provided in the optional software package with user's manual and example booklet. These three programs are also available with graphics presentation of focal "spot" traces as MENISG.BAS, LENS4G.BAS, and HOLTG.BAS.

These programs can accept different lens parameters and then find the best focal surface over the range of azimuth and elevation values entered. The main work of the programs is in ray tracing the lenses. For lenses where the surfaces cannot be computed from simple closed functions the surfaces and their slopes are obtained from suitable series. To assure good accuracy many power series may be needed to define the lens contours. As all three focal-field finding programs are quite similar to the LENS4.BAS program for the Friedlander lens, we will use LENS4.BAS to illustrate the technique.

To define the incident (receiving mode, see Figure 5.6) ray, $\mathbf{S}_i$, as one of the parallel bundle of unit vectors we use (5.14) again with $\mathbf{S}_i = \mathbf{S}_0$:

$$\mathbf{S}_i = \cos(\mathrm{EL})\sin(\mathrm{AZ})\mathbf{x} - \sin(\mathrm{EL})\mathbf{y} + \cos(\mathrm{EL})\cos(\mathrm{AZ})\mathbf{z}$$

in terms of the ELevation and AZimuth direction coordinates of the incident plane wave. For coordinate definitions in LENS4.BAS see Figure 5.6. The interpolating series for defining the straight face lens outer contour is given in two dimensions only. The general points (and their slopes) on the outer lens surface of revolution are obtained for LENS4.BAS by the substitution $R = (x^2 + y^2)^{1/2}$ for Y_2. In this example 20 sets of power series coefficients are used to define the outer surface contour.

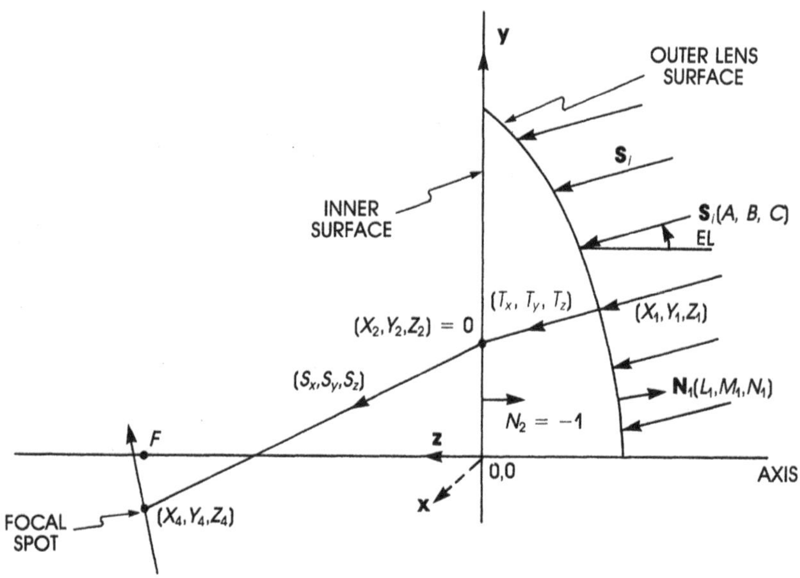

Figure 5.6 Focal-Field Surface Computational Technique for Friedlander Lens

The coefficients of the power series representing the straight face lens outer surface are computed using STCOD.FOR which is a double precision program using the subroutine COOR based on the parametric equations (5.4) along with the Numerical Recipes subroutines POLCOD.FOR and POLIND.FOR. The series coefficients produced by STCOD.FOR are filed in STSET.DAT which is also formatted for the BASIC program LENS4.BAS as STSET.

Snell's law for refraction is used twice in LENS4.BAS for tracing all rays. Initiation of the computer "screen" on the central ray is done first. Focusing is accomplished using Z_4 the z-component of the central ray focal point. Bundles of aperture rays at 30° intervals (on a circle of radius R) in ϕ are traced on the screen for a fixed value of R starting with R near the axis and moving outward. The shape of the trace is usually about the same for all values of R only the focal spots get larger near the lens edge. By varying Z_4 and looking at only a few values of R, the best focal "spot" for a given look (or beam) direction EL and AZ can be found. Also the astigmatic lines are seen to lie at some distance ahead and behind the circle of least confusion "spot." The astigmatic lines can be used to locate the feed source if sharp azimuth plane or elevation plane patterns are desired. Rather low sidelobes in all pattern cuts are usually obtained at the circle of least confusion focal point, X_F, Y_F, Z_F.

By repeating this process for the angular sector of interest the best focal surface can be charted for different elevation and azimuth angles. For the focal length, $F = D_1 = 18$, lens (length units are in wavelengths) with axis thickness, $D_0 = 12.5516$, and index of refraction, $N = 1.594$, the straight face lens has focal spots shown in Figure 5.7. These data were obtained using LENS4.BAS with elevation angle = 8° and azimuth angle = 0°. The best focal point as printed by LENS4 was judged to be at X_F = 0, $Y_F = -3.545499$, $Z_F = Z_4 = 17.4$. This information was used for computing the best diffraction patterns using the programs LENTO4.FOR and PATT.FOR which are described in Section 5.9.

5.6 FOCUSING PROPERTIES OF THE FRIEDLANDER LENS

Friedlander in a paper published [12] in 1946 showed that when the St. Face (inner surface straight) lens is constructed with an index of refraction of $N = 1.618$ then for correct values of D_0 and D_1 the lens closely obeys the Abbe sine condition which greatly reduces coma aberration. He defines the focal length, f, for the straight face lens as the radius of a circle formed by the intersection of the incident and exit rays. He shows that for this focal length, f,

$$f = D_1 + D_0/N$$

The basic design equation to find the lens thickness for a given D_l and maximum aperture angle θ_0 is

$$D_0 = D_1(\sec\theta_0 - 1)/(N - 1) \tag{5.15}$$

and for a lens of total radius H, where $H = D_1 \tan(\theta_0)$, then $(D_1 + D_0)/2H$ is a measure of the lens focal length to diameter. D_1 is labeled F in the program STCOD.FOR which computes accurate series to define the Friedlander lens outer surface.

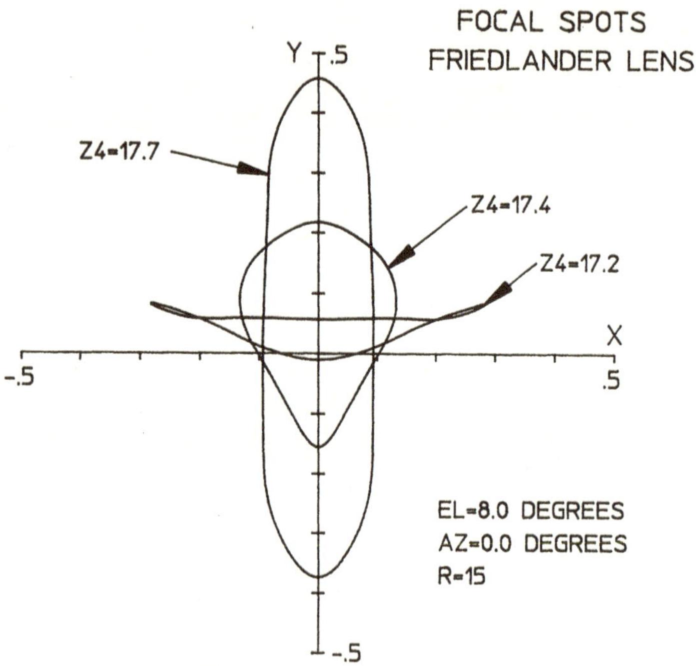

Figure 5.7 Best Focal Spots for Friedlander Lens at EL = 8°, AZ = 0° Scan Angle

These simple relations, which are only approximations to more accurate expressions given in Friedlander's paper, provide a simple design approach for a rather wide-angle antenna made with a flat inner surface. As we have seen, this lens has only a weak effect on the feed horn taper, and this lens type is well suited to the aperture amplitude synthesis technique performed by KOUZN.BAS for "tuning" the aperture distribution for best sidelobe versus gain performance. The common dielectrics for microwave lenses used as example here have index of refraction quite close to 1.618. Departures in index, however, cause some increase in aberrations and the two-point correction method should be used for higher accuracy and for cylindrical lenses, as it is optimized to the value of N used. LENS4.BAS (LENS4S.BAS for stepped beams) can be used to find the straight face lens best focal surfaces. When the lens is "tuned" by amplitude synthesis, the inner surface is no longer flat and must be represented by power series or other interpolating functions as is done in HOLT6.BAS and HOLTG.BAS.

5.7 THE TWO-POINT CORRECTED HOLT-MAYER LENS

The analytical derivation of this lens is given by Holt and Mayer in [15]. The basic two-point correction design concept is illustrated in Figure 5.8. Ray paths are used in the definition of the lens surfaces. Two feed points, F' and F'', located symmetrically above and below the lens axis are identified and pathlengths are defined to a tilted plane wave on the lens aperture. The procedure leads to a finite set of lattice points on the lens surfaces, points P_0 through P_5 in Figure 5.9 are typical, at which both the lens profile points and slopes are calculated. The analysis is incorporated into the program SPL.FOR. The double precision coefficients for the series defining both inner and outer lens surfaces are computed by SLCOD.FOR (and SPLFIT.FOR). The best focal fields are found by the circle of least confusion ray-tracing method described for the Friedlander lens in program HOLTG.FOR. Diffraction patterns are obtained for the Holt lens with program LENP.FOR.

The method for calculating the inner and outer lens contours for either a cylindrical or rotationally symmetric Holt-Mayer lens is reviewed here in some detail correcting typographical errors in [15]. Referring to Figure 5.8 the derivation of the coordinates of the lens contours begins by making the path lengths of the edge rays from F' to the plane bd equal. This leads to the relation,

$$y_f = -(F^2 \tan^2(\theta) + D/2 \sin^2(\theta))^{1/2} \tag{5.16}$$

where y_f is the y coordinate of F', D is the lens diameter, and θ, the correction angle, is the angle of reception relative to the lens axis for which perfect focusing is obtained in the cylindrical design. Based on symmetry of the design the other focus at F'' is Y_f in length units above the z-axis.

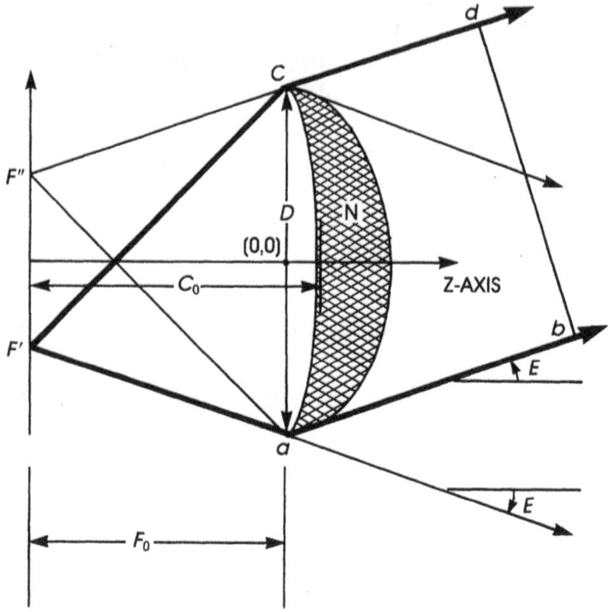

Figure 5.8 Parameters of Two-Point Corrected Holt-Mayer Lens

An iterative ray-tracing procedure beginning with the ray through the center of the lens is then used to obtain lattice point coordinates of both the inner and outer surface. The spacing of these lattice points depends on the correction angle E and values of D and F chosen. The optical path constant for the lens,

$$d_0 = -y_f \csc\theta + (c - f_0)\cos\theta$$

contains a design parameter c that is close to f_0 in value. Lattice point designs are calculated for various values of c and this parameter is considered to be an optimum value c_0 when the associated lattice points tend to converge to the tip points C and A (see Figure 5.8). In the program SPL.FOR the constants of the Holt-Mayer lens have these symbols: $n = N$ = index of refraction of lens, $\theta = E$ = lens correction angle, D = lens diameter, $f_0 = F_0 = F$ the lens focal length as defined in Figure 5.8, and $c_0 = C_0$ = the value obtained by several synthesis trials to make the lens converge to the ideal tip points.

The corrected iterative equations used to compute the lattice points, derived in [15], are listed here. Referring to Figure 5.9 (see also figures in [15]), we have

(a) Outgoing Ray Formulas

$$\alpha_{2k} = \tan^{-1}\left(\frac{Y_{2k} - Y_f}{X_{2k} + c}\right)$$

$$\phi_{2k} = \sin^{-1}\left[\frac{\sin(\alpha_{2k} - \psi_{2k})}{n}\right]$$

$$X_{2k+1} = X_{2k} + \frac{d_0 + X_{2k}\cos\theta + Y_{2k}\sin\theta - \sqrt{(X_{2k} - X_f)^2 + (Y_{2k} - Y_f)^2}}{n\sec(\psi_{2k} + \phi_{2k}) - \cos\theta - \sin\theta\tan(\psi_{2k} + \phi_{2k})}$$

$$Y_{2k+1} = Y_{2k} + (X_{2k+1} - X_{2k})\tan(\psi_{2k} + \phi_{2k})$$

$$\psi_{2k+1} = \tan^{-1}\left[\frac{n\sin(\psi_{2k} + \phi_{2k}) - \sin\theta}{n\cos(\psi_{2k} + \phi_{2k}) - \cos\theta}\right]$$

$$\bar{X}_{2k+1} = X_{2k+1}$$

$$\bar{Y}_{2k+1} = -Y_{2k+1}$$

$$\bar{\psi}_{2k+1} = -\psi_{2k+1}$$

and

(b) Incoming Ray Formulas

$$\bar{\phi}_{2k-1} = \sin^{-1}\left[\frac{\sin(\theta - \bar{\psi}_{2k-1})}{n}\right]$$

$$\bar{X}_{2k} = \bar{X}_{2k-1} + \frac{-B + \sqrt{B^2 - AC}}{A}$$

$$\bar{Y}_{2k} = \bar{Y}_{2k-1} + (\bar{X}_{2k} - \bar{X}_{2k-1})\tan(\bar{\psi}_{2k-1} + \bar{\phi}_{2k-1})$$

$$\bar{\psi}_{2k} = \tan^{-1}\left[\frac{n\sin(\bar{\psi}_{2k-1} + \bar{\phi}_{2k-1})\sqrt{(\bar{X}_{2k} - X_f)^2 + (\bar{Y}_{2k} - Y_f)^2} - (\bar{Y}_{2k} - Y_f)}{n\cos(\bar{\psi}_{2k-1} + \bar{\phi}_{2k-1})\sqrt{(\bar{X}_{2k} - X_f)^2 + (\bar{Y}_{2k} - Y_f)^2} - (\bar{X}_{2k} - X_f)}\right]$$

where

$$A = (n^2 - 1)\sec^2(\bar{\psi}_{2k-1} + \bar{\phi}_{2k-1})$$

$$B = n(d_0 + \bar{X}_{2k-1}\cos\theta + \bar{Y}_{2k-1}\sin\theta)\sec(\bar{\psi}_{2k-1} + \bar{\phi}_{2k-1})$$

$$- (\bar{X}_{2k-1} - X_f)$$

$$- (\bar{Y}_{2k-1} - Y_f)\tan(\bar{\psi}_{2k-1} + \bar{\phi}_{2k-1})$$

$$C = (d_0 + \bar{X}_{2k-1}\cos\theta + \bar{Y}_{2k-1}\sin\theta)^2$$

$$- (\bar{X}_{2k-1} - X_f)^2 - (\bar{Y}_{2k-1} - Y_f)^2$$

$$X_{2k} = \bar{X}_{2k}$$

$$Y_{2k} = -\bar{Y}_{2k}$$

$$\psi_{2k} = -\bar{\psi}_{2k}.$$

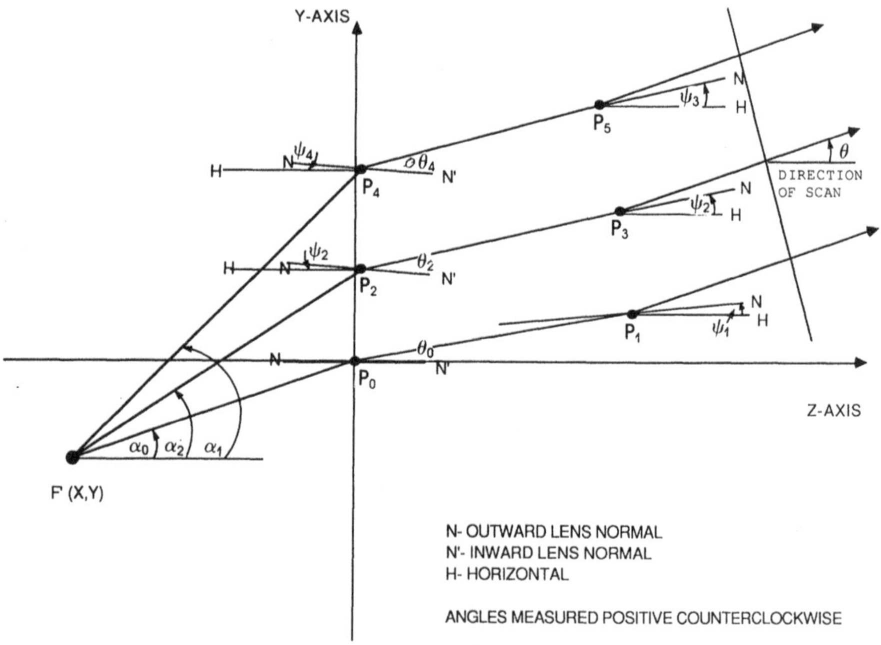

SYNTHESIS OF HOLT-MAYER LENS

Figure 5.9 Synthesis of Lattice Points on Holt-Mayer Lens

The synthesis procedure is programmed in SPL.FOR. The letters SPL refer to stepped prism lens which was the name used for the cylindrical lens designed for wide-angle collection of solar energy [11]. The input lens parameters are entered by the data file SPL.IN. The lens contours are calculated with the inner and outer surface coordinates and slopes given only at the lattice points. The coordinates and slopes for the two lens surfaces are stored in the files SP1.DAT (for the inner surface) and SP2.DAT (for the outer surface). For large correction angles the lattice points are further apart, making it useful to use both the slopes and contour points for computing the precision series needed to define the lens surfaces accurately using SPLIT.FOR. However, using a simpler and quite accurate procedure, double precision coefficients for power series representing the two lens surfaces are computed by SLCOD.FOR and stored in data files SP1FIT.DAT (inner surface) and SP2FIT.DAT (outer surface). SLCOD.FOR must be used LENP.FOR in the computation of radiation patterns of the Holt-Mayer lens. We rotate the two-dimensional curves to create a three-dimensional rotationally symmetric lens which is ray traced for best focal surfaces using HOLTG.BAS.

The focal trace data shown in Figure 5.10 was obtained using HOLTG.BAS for a Holt lens with $N = 1.594$, EL = 8.5°, AZ = 0.0°, $F_0 = 60$, $C_0 = 59.958$, $D = 40$, and $E = 8.5$. Unlike the cylindrical model of the two-point corrected lens, this rotationally symmetric antenna is not free of aberrations at the two correction angles. The plots were made for $R = 3.56$ units from the lens axis. There is a circle of least confusion curve lying between values of Z_4 that produce nearly horizontal and vertical astigmatic lines. The diameter of the circle of least confusion and the lengths of the astigmatic lines depend on aperture radius, R, and scan angle generally growing with angle and R as shown in Figures 5.11(a) and (b). The actual focal surface is rotationally symmetric and is plotted in Figure 5.12 for $D = 60$ and $F_0 = 60$, $F_0 = 40$. The lens thickness actually becomes thinner for larger correction angles. The lens thickness can be reduced by stepping or cutting λ pathlength difference steps in the surface. Figure 5.13 shows such steps. Their vertical heights, H_s, are

$$H_s = \lambda_0/(N - 1)$$

where λ_0 is the wavelength in free space.

It can be easily shown that such stepping reduces the bandwidth of the lens antenna as the above equation introduces the wavelength, λ. The percent bandwidth [4,7] due to stepping is approximately,

Bandwidth = **25/*n*** where n is the number of steps.

There is also some increase in phase errors and sidelobe levels due to shadowing of the steps. For the model shown in Figure 5.13, these

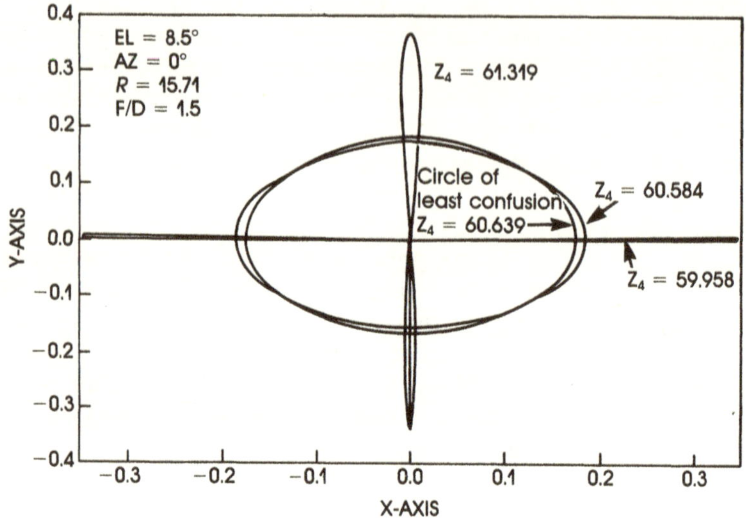

Figure 5.10 Focal Spots or Circles of Least Confusion and Astigmatic Lines for Holt-Mayer Lens

errors were found to be small using ray tracing with HOLT6.BAS. Stepping can also be done on the inner surface of the lens, which may be preferable in conditions of dust or snow.

The Holt-Mayer lens and the Friedlander lens are quite thick, and stepping is recommended to reduce weight when only limited bandwidth is needed. A stepping procedure for the outer surface of the Friedlander lens is available in program STEPF.BAS (STCODS.FOR) that uses different Friedlander lens segments in each step to reduce stepping aberrations such that axis-parallel rays are perfectly focused. LENS4S.BAS is provided for stepping the inner surface of the Friedlander lens, but now even the axis-parallel rays are poorly focused. The outer surface stepping procedure for the Friedlander lens is described further in the optional software manual.

The Holt-Mayer lens designs can lead, as in these examples, to a nearly flat inner surface antenna. Both the Friedlander and Holt-Mayer lenses are quite wide-angle. The Holt-Mayer lens allows for optimization at a variety of index of refraction values, whereas Friedlander's are better near to $N = 1.618$. The Holt-Mayer lens also permits a choice of correction angles which, for cylindrical lenses, gives perfect correction at two symmetric angles with very good performance for all angles between the correction angles. The Friedlander lens is easier to construct with its truly flat inner surface.

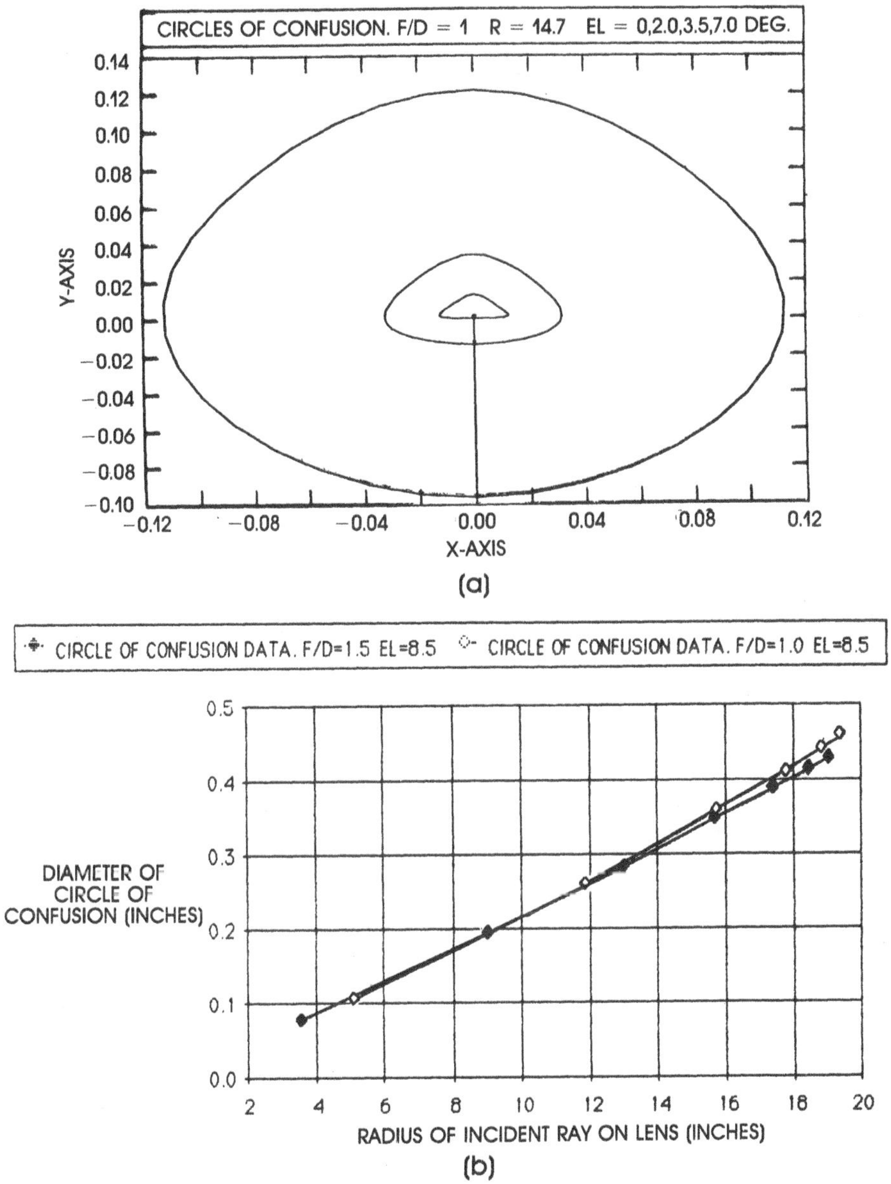

Figure 5.11 Focal Spot Size Depending on (a) Scan Angle and (b) Lens Radius for Holt-Mayer Lens

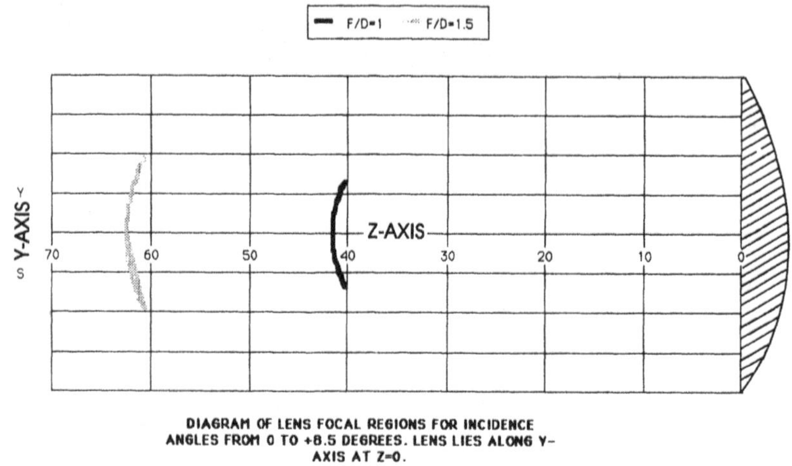

Figure 5.12 Contour of Focal Surface for Holt-Mayer Lens

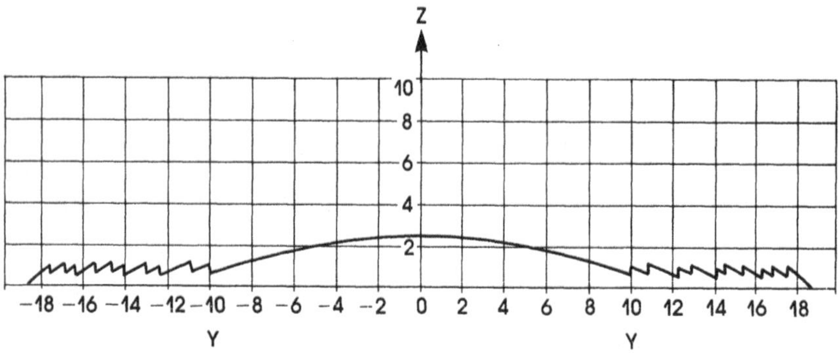

Figure 5.13 Profile of Holt-Mayer Lens Stepped to Reduce Thickness

5.8 COMPUTING LENS RADIATION PATTERNS

For computer-aided design of dielectric lenses the lens surfaces must be very accurately computed. In fabrication some roughness in the machine tolerances is allowed (larger surface errors than on reflector antennas are permitted for a given gain-sidelobe level performance). Because many lens synthesis procedures are iterative, accumulative accuracy is important in computer design. It is lens surface slopes or normals that determine Snell's Law in ray-path calculations. The surface contours must be accurately known to be used for computing surface slopes. On surfaces that cannot be described by closed-form functions the lens surfaces are generally defined by a system of power series or splines. These interpolation series must define both the surface coordinates and slopes accurately.

The electrical phase (from path lengths) and amplitude aperture distributions on the lens can be computed by ray-tracing methods. Usually three-coordinate root-finder programs are needed to find intersection points of rays on the lens surfaces defined by power series. Fourier transforms of phase and amplitude aperture distributions for different scan angles are used to compute diffraction patterns of the lens antenna. The program PATT.FOR described in Chapter 1 is used.

The program LENTO4.FOR will be used to illustrate the techniques used to compute radiation patterns on lenses using the Friedlander lens as the example. LENTO4.FOR computes the phase and amplitude distribution on the lens aperture plane and files this data for points in r, ϕ aperture coordinates for use by PATT.FOR which performs a Fourier transform on the data to give the radiation pattern.

The same precision series coefficients used in LENS4.BAS for finding the best focal surfaces for the Friedlander lens are used again in LENTO4.FOR to define the outer surface of the lens accurately. To compute patterns on scanned beams remote from the lens axis direction, it is necessary to tilt the aperture plane on which the phase (by path-lengths from the best focal point) and amplitude are computed. This means that the gain computed by PATT.FOR will be nearly correct as the approximate phase front on the aperture lies parallel to the tilted aperture plane.

An input data file, LEN.DAT, is used to enter the lens parameters which for the example used are $A = 17$ wavelengths (A is radius of lens aperture), LAM $= \lambda = 1$, $NN = 1.595$ the index of refraction, $D_0 = 12.5516$ the axis thickness of the lens, THM $= 9$ is the largest scan angle anticipated and enlarges the aperture to accept the tilted plane wave. On the second line of the file LEN.DAT are the best focal coordinates for an 8° elevation angle scan or $F_X = 0.0$, $F_y = -3.5455$, $F_Z = 17.4$. Also EL = 8.0 and AZ = 0.0 in the example. In line three $F = 18.0$, $Z_{11} = -14$ (This center of tilted aperture plane is moved away from the lens to prevent the tilted plane from intersecting the lens during beam scan), NRAD = 51, NANG = 101 are the radial and angular increments for the r, ϕ cuts on the tilted aperture.

Referring to Figure 5.14 the incident bundle of rays have direction cosines defined by Equation (5.14) as AA,BB,CC. To find where a ray from a given aperture point will strike the lens we use a root finding technique with *Numerical Recipes* programs ZBRAK and ZRENT. This

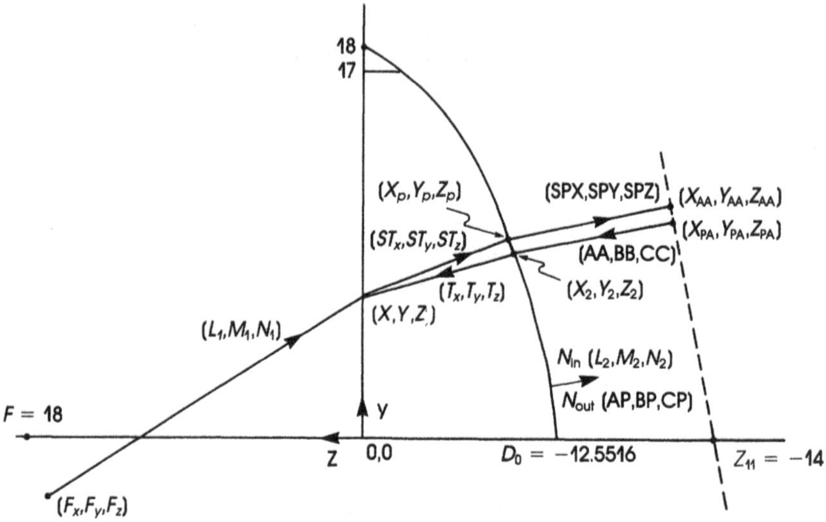

Figure 5.14 Ray-Tracing Technique for Pattern Computations on Friedlander Lens

routine solves where the incident ray through the aperture point intersects the lens using the particular power series expression for the lens surface that is most accurate for that sector of the lens. This work is performed in the subroutine PHSAMP of LENTO4.FOR. The normal ray to the lens is computed at the ray intersection points using the lens surface power series and its partial derivatives. Using Snell's law the transmitted (or refracted) ray with direction cosines T_X, T_Y, T_Z is determined and traced to its intersection point on the flat inner surface $Z = Z_1 = 0$. Once the ray intersection point with the inner surface is found, another ray from the best focal point, F_X, F_Y, F_Z is constructed and its component L_1, M_1, N_1 written down from the geometry. This ray is refracted by the flat surface with normal unit vector $\mathbf{N} = -\mathbf{z}$. Application of Snell's law defines the components STX,STY,STZ of the returning vector. Again root finding must be used to find where this ray penetrates the outer surface of the lens and Snell's law is again used to find the direction of the refracted ray headed for the tilted aperture plane. Its intersection point on that plane is labeled XAA, YAA, ZAA. This point is the same as the launching aperture point for EL = AZ = 0, when all rays focus to F. But for general tilted incidence there is a small difference in the aperture points for the incoming and returning rays. The phase calculated from the pathlengths is correct, however, for the return point XAA, YAA, ZAA. The phase errors computed exist on the aperture for points slightly displaced from the regular grid of aperture point and therefore there is a small error in patterns when the values of the error parameters, ERX, ERY, are large compared to the grid increments.

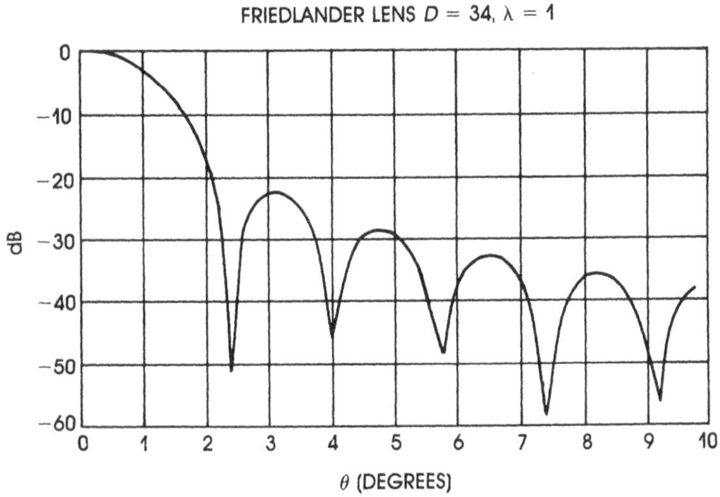

Figure 5.15 Radiation Pattern for Friedlander Lens at EL = 0°, AZ = 0°

The diffraction pattern for an 8° elevation scan is shown in Figure 5.16(b). The pattern for $\lambda = 1$ unit length has about 0.1 dB of loss in gain due to path errors when compared with the EL = AZ = 0.0 pattern shown in Figure 5.16(a). The pattern peaks slightly toward the axis on the ϕ = PHI = 90 cut and the sidelobes are not quite the same in the up and down directions. The good wide-angle characteristics of the Friedlander lens, however, are evident. The circle of confusion data on the Friedlander and Holt-Mayer lenses being very similar suggests they have rather equivalent scanning capabilities.

In Figure 5.17(a) and (b) the on-axis and EL = 8.5°, AZ = 0° patterns for the Holt lens are shown as computed by LENP.FOR.

The pattern program LENP.FOR is able to ray-trace curved lens profiles on both inner and outer surfaces of lenses. The program SLCOD .FOR computes interpolating power series to represent rather general lens contours. The programs HOLT6.BAS and HOLTG.BAS use these same power series to find the best focal surfaces for general lens contours. This set of software can be used to compute the radiation patterns of most axially symmetric lenses.

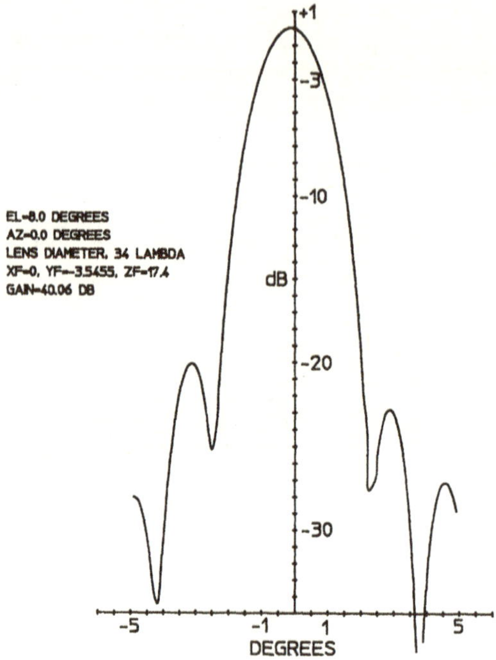

Figure 5.16 Radiation Pattern for Friedlander Lens at EL = 8°, AZ = 0°

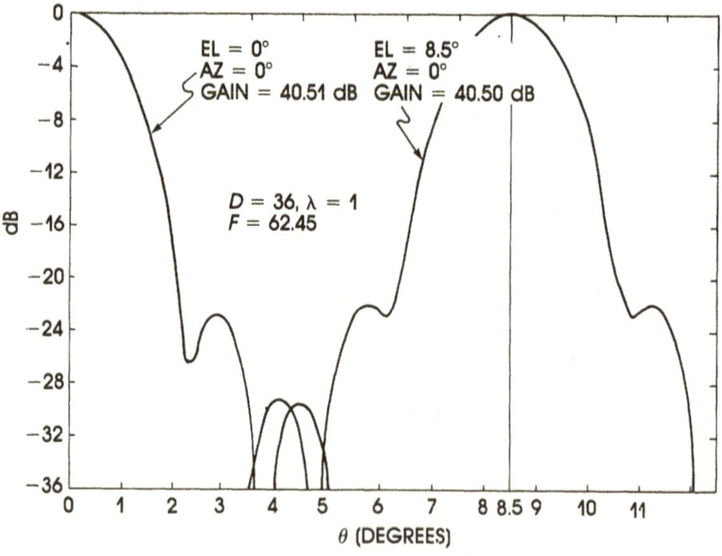

Figure 5.17 (a) Radiation Pattern for Holt-Mayer Lens at EL = 0°, AZ = 0°; (b) Radiation Pattern for Holt-Mayer Lens at EL = 8.5°, AZ = 0°

References

[1] Born, M., and E. Wolf, *Principles of Optics,* New York: Pergamon Press, 1964.

[2] Conrady, A.E., *Applied Optics and Optical Design,* New York: Dover, 1957, p. 120, pp. 270–271.

[3] Schwarzschild, K., *Astr. Mitt. Konigl.,* Göttingen: Sternwarte, 1905.

[4] Jasik, H., *Antenna Engineering Handbook,* Chapter 14, S.B. Cohn, New York: McGraw-Hill, 1961.

[5] Collin, R.E., and F.J. Zucker, *Antenna Theory,* Part II, New York: McGraw-Hill, 1969, Chapter 18, J. Brown, pp. 55–57, Part I, pp. 650–653.

[6] Rudge, A.W., *et al., The Handbook of Antenna Design,* Vols. 1 & 2, London: Peter Peregrimus, 1986, pp. 293–315, A.R. Dion.

[7] Rotman, W., "EHF Dielectric Lens Antenna for Satellite Communication Systems," Lincoln Lab. Tech. Rept. 620, ESD-TR-82-102, 3 Jan 1983.

[8] Peng, S.Y., "Computer-aided bifocal dielectric lens antenna design for MM wave applications," Proc. of 1984 Ant. Appl. Symposium, RADC, Hanscom AFB, RADC-Tech Report 85-14, Jan. 1985.

[9] Silver, S., *Microwave Antenna Theory and Design,* New York: McGraw-Hill, 1949, pp. 388–402.

[10] Clarricoat, P.J.B., and P.K. Saha, "Radiation Pattern of Lens-Corrected Conical Scalar Horn," Electronic Letters, Vol. 5, pp. 592–593, Nov. 1969.

[11] Sletten, C.J., F.S. Holt, and S.B. Herskovitz, "Wide-angle Lenses and Image Collapsing Subreflectors for Nontracking Solar Collectors," Applied Optics, Vol. 19, No. 9, 1 May 1980.

[12] Friedlander, F.G., "A Dielectric-lens Aerial for Wide-angle Beam Scanning," Proc. IEE, 93, Pt. 3A, pp. 658–662, 1946.

[13] Press, W.H., *et al., Numerical Recipes,* "The Art of Scientific Computing," Cambridge Univ. Press, 1986, pp. 569–572.

[14] Stratton, J., *Electromagnetic Theory,* New York: McGraw-Hill, 1941, pp. 494–500.

[15] Holt, F.S., and A. Mayer, "A Design Procedure for Dielectric Microwave Lenses of Large Aperture Ratio and Large Scanning Angle," Trans. IRE, Vol. AP-5, pp. 20–25, Jan. 1957.

[16] Lee, J.J., "Dielectric Lens Shaping and Coma Correcting," Part I, "A Coma-Corrected Multibeam Shaped Lens Antenna," Part II—Experiments, Trans. IEEE, Vol. AP-31, Jan. 1983, pp. 211–216, pp. 216–223.

[17] Ruze, J., "Axial Astigmatic Fields in the Focal Region," Trans. IEEE, Vol. AP-23, pp. 734–735, Sept. 1975.
[18] Sletten, C.J., and R.A. Shore, "Focal Surfaces of Offset Dual-Reflector Antennas," Proc. IEE, Vol. 129, Pt. H, No. 3, June 1982.

Appendix 5A
List of Chapter 5 Computer Programs and Source Code Listings of Selected Programs

In this appendix the names of computer programs developed especially for Chapter 5 are listed. The source code for some of the simpler programs is listed here. Source code on 5¼ inch diskettes is available for all programs in the optional software package with manual giving operating instructions. All programs are written in Microsoft BASIC and FORTRAN-77 for IBM-PC type computers. The program names are written in the first few lines of the source code. The suffix .FOR indicates a FORTRAN program and .BAS a BASIC program. Data files have the suffix .DAT.

SOFTWARE LIST, CHAPTER 5

PROGRAM NAME:	ASSOCIATED PROGRAMS, FILES AND SUBROUTINES:

*LENAMP.BAS
*SYNTH.BAS
SYNTHS.BAS
KOUZN.BAS; SP1.DAT, SP2.DAT
LENAMF.BAS; SP1.DAT, SP2.DAT
*PRISM.BAS
POLGEN.BAS
MENISC.BAS
MENISG.BAS
*LENS4.BAS; STSET
LENS4G.BAS; STSET
LENS4S.BAS; STSET, STEP.DAT

STEPF.BAS
HOLTG.BAS; SP1F6, SP2F6
HOLT6.BAS; SP1F6, SP2F6
HOLTT.BAS; SP1FIT.DAT,SP2FIT.DAT

STCOD.FOR; FRIED.DAT, POLCOD, POLIND, STSET.DAT
STCODS.FOR; FRIED.DAT, STEP1.DAT, POLCOD, POLIND, STSET.DAT
SPL.FOR; SPL.IN, SP1.DAT, SP2.DAT
SLCOD.FOR; SP1.DAT, SP2.DAT, POLCOD, POLIND, SP1FIT.DAT, SP2FIT.DAT
SPLFIT.FOR; SP1.DAT, SP2.DAT, SP1FIT.DAT, SP2FIT.DAT, GAUSSS
LENT04.FOR; LEN.DAT, SOLV.FOR, ZBRAK, ZBRENT, STSET.DAT, PATT.FOR
LENP.FOR, LEN2.DAT, SOLV1.FOR, SOLV2.FOR, SP1FIT.DAT, SP2FIT.DAT, ALAM.DAT, PHAMP.DAT, ZBRAK, ZBRENT, PATT.FOR

* Source code for these programs is printed in this appendix.

The FORTRAN programs GAUSSS, POLCOD, POLIND, ZBRAK, and ZRENT are Numerical Recipes subroutines.

```
10 REM Computes the radial aperture distribution of len given
20 REM illumination of inner surface P=cos(theta)^N and lens
30 REM coordinates.  CJS 6 DEC 1986
40 REM Example here is St. Inner Face Lens.
50 PRINT "LENAMP.BAS"
60 REM M=2.1 gives almost uniform power on aperture
70 PI=3.14159265359#
80 A0=45*PI/180 : D=.8
90 INPUT "M= ";M
100 S1=0 : R1=0 : I=0 : F=20 : EP=.00001
110 R0=20  : N=1.594
120 REM INPUT " ENTER F0,R0,D,N, "; F,R0,D,N
130 A0=ATN(R0/F)
140 C3=R0^2/(1-COS(A0)^M)
150 D0=F*(1/COS(A0)-1)/(N-1)
160 REM Y1 is height or inner radius of Straight face of lens.
170 FOR Y1=D  TO R0 STEP D
180 S2=ATN(Y1/F)
190 GOSUB 290
200 R2=Y2
210 P=C3*(COS(S1)^M-COS(S2)^M)/(R2^2 -R1^2)
220 IF I=0 THEN P0=P : I=1
230 PRINT "Y1=";Y1;"R=";(R1+R2)/2;"Pn="P/P0
240 R1=R2 :S1=S2
250 IF A2>A0 THEN GOTO 270
260 NEXT Y1
270 END
280 REM Subroutine calculating outer surface of Friedlander lens.
290 D3=N*D0+F
300 T2=SIN(S2)/N
310 A2=ATN(T2/SQR(1-T2*T2))
320 R=Y1/SIN(S2)
330 T=((D3-R)*COS(A2)/N-D0)/(COS(A2)/N-1)
340 S=D0-T
350 D1=S*TAN(A2)
360 Z2=S
370 Y2=Y1+D1
380 BN=SIN(A2)*N/SQR(N*N-2*N*COS(A2)+1)
390 CN=SQR(1-BN*BN) : PHI=ATN(BN/CN)
400 REM PRINT "A1=";S2*180/PI;"A2=";A2*180/PI;"PHI=";PHI*180/PI;"SLOPE=";BN/CN
410 RETURN
420 END
```

```
10 REM Shaping inner and outer surface for UNIFORM aperture AMPLITUDE.
20 REM The meniscus lens with parameters in next lines
30 REM is the start-up model for synthesis.
40 PRINT "SYNTH.BAS"
50 REM DEFDBL A-Z
60 N=1.5937# : M=8.350001      : R0=3.62# : R1=5.0465#
70 Y0=2.32689 : A0=40
80 INPUT "M=N+1 of Cos(A)^N horn pattern ",M
90 PI=3.14159265359#
100 A0R=A0*PI/180
110 C0=Y0*Y0/(1-COS(A0R)^M)
120 D0=R1-R0
130 DIM S(4)
140 REM Start up slopes for inner surface of Meniscus Lens.
150 Y1=0# : Z1=R0
160 D=.05# : K=1
170 FOR J=1 TO 4
180 Z1=SQR(R0^2-Y1^2)
190 S5=-Y1/Z1
200 S(J)=S5
210 Y1=Y1+D
220 NEXT J
230 Y1=3*D :Z1=SQR(R0^2-Y1^2)
240 K7=S(1) : K6=S(2) : K5=S(3)
250 REM PRINT "Y1=";Y1;"Z1=";Z1
260 REM Getting horn pattern angle, A.
270 RI=SQR(Y1^2+Z1^2)
280 W=Y1/RI
290 A=ATN(W/SQR(1-W*W))
300 REM R2 equation of meniscus lens, no shaping.
310 R2=R1*(N-1)/(N-COS(A))
320 Y3=R2*SIN(A)
330 REM Y2 level for horn power to give uniform ap. illum.
340  Y2=SQR(C0*(1-COS(A)^M))
350 Z3=R2*COS(A)
360 REM Computing Z2 for lens in-phase condition on aperture.
370 C1=R0-D0*(N-1)
380 A5=N*N-1
390 B5=2*(C1-N*N*Z1)
400 C5=N*N*(Y2-Y1)^2+(Z1*N)^2-C1^2
410 D5=B5^2-4*A5*C5
420 PRINT
430 REM PRINT "Radical, D5=";D5
440 IF D5<.00001# THEN GOTO 680
450 Z2=(-B5+SQR(D5))/(2*A5)
455 REM Snell's law on inner surface to find slope.
460 R6=SQR((Y2-Y1)^2+(Z2-Z1)^2)
470 M1=(Y2-Y1)/R6
480 N1=(Z2-Z1)/R6
490 B8=(SIN(A)-N*M1)/SQR((COS(A)-N*N1)^2+(SIN(A)-N*M1)^2)
500 C8=SQR(1-B8^2)
```

```
510 K8=B8/C8
520 REM PRINT "K8i=";K8
530   KC8=-Y1/Z1
540 REM PRINT "K8c=";KC8
545 REM Adam's Projector for inner surface.
550 Y1=Y1+D
560 Z1=Z1+D/24*(55*K8-59*K5+37*K6-9*K7)
570 K7=K6 : K6=K5 : K5=K8
580  PRINT "A=";A*180/PI
590  PRINT "Y2=";Y2;"Z2=";Z2
600  PRINT "Y2c=";Y3;"Z2c=";Z3
610 Y5=R0*SIN(A)+D
620 Z5=R0*COS(A)
630       PRINT "Y1=";Y1;"Z1=";Z1
640       PRINT "Y1c=";Y5;"Z1c=";Z5
650 IF A*180/PI>=A0*3/4  THEN GOTO 700
660 REM PRINT
670 GOTO 270
680 PRINT "Radical negative." : D5=.000001# : GOTO 450
690 GOTO 80
700 END
```

```
10 REM  Prism Lens Antenna, incident rays coplanar with lens normals.
15 REM Reflection losses are computed for range of incident angles.
20 PI=3.141592654#
30 CO=.4342944819#
40 PRINT "PRISM.BAS"
50 INPUT "DIELECTRIC CONSTANT and PRISM ANGLE,degrees: ",E,ANG
60 PRINT "DIEL. CONST.=";E;"PRISM ANGLE=";ANG
70 PRINT
80 N=SQR(E) : AP=ANG*PI/180 + .00001
90 FOR A=-40 TO 55 STEP 5
100 AI=A*PI/180
110 AO=AI-AP
120 T=SIN(AO)/N
130 A1=ATN(T/SQR(1-T*T))
140 AE=A1+AP
150 TE=N*SIN(AE)
160 A2=ATN(TE/SQR(1-TE^2))
170 PI1=SIN(2*AO)*SIN(2*A1)/(SIN(AO+A1)^2*COS(AO-A1)^2)
180 DB1=10*CO*LOG(PI1)
190 PI2=SIN(2*AE)*SIN(2*A2)/(SIN(AE+A2)^2*COS(AE-A2)^2)
200 DB2=10*CO*LOG(PI2)
210 DBI=DB1+DB2
220 PN1=PI1*COS(AO-A1)^2
230 PN2=PI2*COS(AE-A2)^2
240 DBN=10*CO*(LOG(PN1)+LOG(PN2))
250  PRINT "AI=";AI*180/PI;"A2=";A2*180/PI;"DBI=";DBI;"DBN=";DBN
260  REM PRINT
270 NEXT
280 END
```

```
DIEL. CONST.= 2.8 PRISM ANGLE= 10

AI=-40 A2=-29.74003 DBI=-.2290703 DBN=-1.141094
AI=-35 A2=-25.65843 DBI=-.2908028 DBN=-.982492
AI=-30 A2=-21.39151 DBI=-.351807 DBN=-.862014
AI=-25 A2=-16.9711 DBI=-.407757 DBN=-.7706211
AI=-20 A2=-12.42154 DBI=-.4561693 DBN=-.70199
AI=-15 A2=-7.761024 DBI=-.4956143 DBN=-.6516761
AI=-10 A2=-3.002663 DBI=-.5252946 DBN=-.6166207
AI=-5 A2= 1.844789 DBI=-.5447322 DBN=-.5948186
AI= 0 A2= 6.776511 DBI=-.5536125 DBN=-.5851453
AI= 5 A2= 11.79143 DBI=-.5516625 DBN=-.5872544
AI= 10 A2= 16.89242 DBI=-.538605 DBN=-.6015993
AI= 15 A2= 22.08692 DBI=-.5141323 DBN=-.6295403
AI= 20 A2= 27.38819 DBI=-.4779588 DBN=-.673638
AI= 25 A2= 32.81755 DBI=-.4299256 DBN=-.7381571
AI= 30 A2= 38.40859 DBI=-.3703168 DBN=-.830106
AI= 35 A2= 44.21506 DBI=-.3005731 DBN=-.9613103
AI= 40 A2= 50.32738 DBI=-.2251634 DBN=-1.15307
AI= 45 A2= 56.91219 DBI=-.1572273 DBN=-1.448508
AI= 50 A2= 64.33008 DBI=-.1409307 DBN=-1.955259
AI= 55 A2= 73.68302 DBI=-.4075325 DBN=-3.099477
```

```
10 REM A PROGRAM FOR RAY TRACING ON ST. FACE LENS
20 PRINT "LENS4.BAS"
30      DEFDBL A
40 PI=3.141592654#
50 J=1 : K=1
60 OPEN "I",#2,"STSET"
70 GOTO  1070
80 REM For example Z4=18 for El=AZ=0
90 A5=0# : N=1.594#
100 INPUT "ELEVATION ANGLE: ";EL
110 INPUT "AZIMUTH ANGLE: ";AZ
120 INPUT "Z4= ";Z4
130 J=1 : K=1
140 X1=.00001 : Y1=0 : GOTO 180
150 INPUT "R,increasing "; R
160 PHI=0
170 X1=R*COS(PHI) : Y1=R*SIN(PHI)
180 R=SQR(X1*X1+Y1*Y1)
190 IF R>=Y22 THEN GOTO  1070
195 REM Finding normal to outer surface.
200 Z1=-(A0+A1*R+A2*R^2+A3*R^3+A4*R^4+A5*R^5)
210 N1X=-X1*(A1/R+2*A2+3*A3*R+4*A4*R^2+5*A5*R^3)
220 N1Y=-Y1*(A1/R+2*A2+3*A3*R+4*A4*R^2+5*A5*R^3)
230 R1=SQR(N1X^2+N1Y^2+1)
240 L1=   N1X/R1
250 M1=    N1Y/R1
260 N1=-1/R1
270  REM PRINT "X1=";X1;"Y1=";Y1;"PHI=";PHI*180/PI;"R=";R
280 ELR=EL*PI/180:AZR=AZ*PI/180
290 A=COS(ELR)*SIN(AZR):B=-SIN(ELR):C=COS(ELR)*COS(AZR)
300 REM PRINT"L1=";L1;"M1=";M1;"N1=";N1
310 REM PRINT "A=";A;"B=";B;"C=";C
320 ANG=ATN(N1Y)
330 REM PRINT "Z1=";Z1;"dz/dx=";N1X;"dz/dy=";N1Y
340 REM PRINT "ANG=";ANG
345 REM Snell refraction at outer lens surface.
350 AL=M1*C-N1*B
360 BL=-(L1*C-N1*A)
370 CL=L1*B-M1*A
380 REM LPRINT "AL=";AL;"BL=";BL;"CL=";CL
390 AA=N*N*(N1^2+M1^2+L1^2)
400 BB=2*N*(L1*BL-M1*AL)
410 CC=AL^2+BL^2-(N*N1)^2
420 TZ=(-BB+SQR(BB^2-4*AA*CC))/(2*AA)
430 TY=(M1*N*TZ-AL)/N1/N
440 TX=(BL+L1*N*TZ)/N1/N
450 REM PRINT "TZ=";TZ;"TY=";TY;"TX=";TX
460 Z2=0
470 X2=TX/TZ*(Z2-Z1)+X1
480 Y2=TY/TZ*(Z2-Z1)+Y1
490 REM PRINT "X2=";X2;"Y2=";Y2;"Z2=";Z2
495 REM When inner surface not flat enter its
497 REM series coeficients here.
500 GOTO 640
570 RR=SQR(X2*X2+Y2*Y2)
580 Z2=-(B0+B1*RR+B2*RR^2+B3*RR^3+B4*RR^4+B5*RR^5)
590 N2X=-X2*(B1/RR+2*B2+3*B3*RR+4*B4*RR^2+5*B5*RR^3)
600 N2Y=-Y2*(B1/RR+2*B2+3*B3*RR+4*B4*RR^2+5*B5*RR^3)
```

```
610 R2=SQR(N2X^2+N2Y^2+1)
620 L2=   N2X/R2
630 M2=   N2Y/R2
640 L2=0 : M2=0 :N2=-1
650 REM PRINT "L2=";L2;"M2=";M2;"N2=";N2
660 REM PRINT "Z2=";Z2;"dz/dx=";N2X;"dz/dy=";N2Y
670 ANG2=ATN(N2Y)
680 REM PRINT "ANG2=";ANG2
685 REM Snell refraction at inner lens surface.
690 AL2=M2*N*TZ-N2*N*TY
700 BL2=-(L2*N*TZ-N2*N*TX)
710 CL2=L2*N*TY-M2*N*TX
720 AA2=N2^2+L2^2+M2^2
730 BB2=2*(BL2*L2-M2*AL2)
740 CC2=AL2^2+BL2^2-N2^2
750 SZ=(-BB2+SQR(BB2^2-4*AA2*CC2))/(2*AA2)
760 SY=(M2*SZ-AL2)/N2
770 SX=(BL2+L2*SZ)/N2
780 REM PRINT "SZ=";SZ;"SY=";SY;"SX=";SX
790 REM PRINT
800 IF J=1 GOTO 950
805 REM Sets up image screen perpendicular to
807 REM to initial central ray.
810 Z0=(L3*SX/SZ*Z2+Z2*M3*SY/SZ-(D3+L3*X2+M3*Y2))/(N3+L3*SX/SZ+M3*SY/SZ)
820 X0=SX/SZ*(Z0-Z2)+X2
830 Y0=SY/SZ*(Z0-Z2)+Y2
840 X5=(X0-X4)/COS(A7)
850 Y5=(Y0-Y4)/COS(A6)
860  REM PRINT "X0=";X0;"Y0=";Y0
870 PRINT "PHI=";PHI*180/PI;"R=";R;"      ";"X5=";X5;"Y5=";Y5
880 REM LPRINT
890 IF K=1 THEN K=0 : GOTO 150
900 IF PHI>359*PI/180 THEN GOTO 930
910 PHI=PHI + 30*PI/180
920 GOTO 170
930 CLOSE #2
940 END
950 L3=SX:M3=SY:N3=SZ
960 T3=M3/N3 : A6=ATN(T3)
970 T4=L3/N3 : A7=ATN(T4)
980 X4=L3/N3*(Z4-Z2) + X2
990 Y4=M3/N3*(Z4-Z2) + Y2
1000 D3=-(L3*X4+M3*Y4+N3*Z4)
1010 LPRINT
1020 LPRINT
1030 LPRINT "CENTRAL RAY INITIATION"
1040 LPRINT "X4=";X4;"Y4=";Y4;"Z4=";Z4;"AZ=";AZ;"EL=";EL
1045 LPRINT
1050 J=0
1060 GOTO 810
1065 REM  Recalls outer surface coefficients.
1070 INPUT#2,Y11,Y22,Z22
1080  REM PRINT "Y11=";Y11;"Y22=";Y22;"Z22=";Z22
1100 INPUT#2,A0,A1,A2
1110 INPUT#2,A3,A4
1120 REM PRINT "A0=";A0;"A1=";A1;"A2=";A2;"A3=";A3;"A4=";A4
1125 REM PRINT
1130 IF EOF(2) THEN END
1140 IF K=1 THEN GOTO 90
1150 GOTO 190
```

Chapter 6

CONSTRAINED LENSES

Daniel T. McGrath

6.1 INTRODUCTION

Constrained lens antennas can serve the same functions as dielectric lens and reflector antennas. They are a separate class of beamformers, and, like normal lenses or reflectors, they collimate energy from a transmitting feed into a narrow beam, or focus energy incident on a large aperture into a small receiving feed. Yet, unlike those other devices, the constrained lenses do their collimating or focusing discretely, using arrays of small antennas to collect energy from one direction and reradiate it in another.

The most general type of constrained lens is illustrated in Figure 6.1. The lens has two arrays facing in opposite directions with pairs of elements connected by transmission lines. There are a variety of antennas that can serve as the array elements, such as dipoles, microstrip patches, and open-ended waveguide. Similarly, a variety of transmission line types could be used to interconnect the array elements. One or both of the arrays is usually curved and the transmission lines are normally of varying length. The combination of line lengths and array geometries acts as a phase transformer, converting the spherical wave emitted by the feed into a plane wave.

This general class of beam-forming devices is called "constrained" because a wave striking one face does not necessarily obey Snell's law in traveling to the other face and back into free space. Instead, the wave is constrained to follow the transmission line paths. There are several geometric "degrees of freedom" available, such as front and back face curvature and line lengths, which can be combined to produce a beamformer

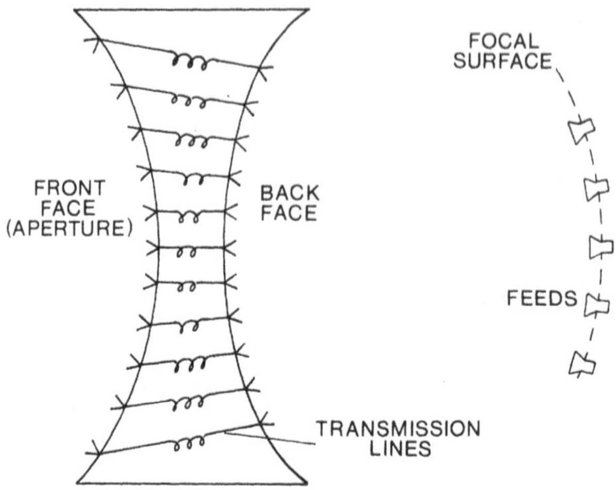

Figure 6.1 Schematic of a General Constrained Lens

with a wide field of view, which is critical to the most important application of constrained lenses as electronic scanning antennas. Each of the feeds in Figure 6.1 produces a beam directed at a different angle in space. Switching the feeds on and off in sequence accomplishes a beam-steering function. This approach is sometimes preferred over a phase array, particularly for scanning over a limited angular region because it avoids the need to populate a large aperture with phase shifters. As we will show in this chapter, the angular region that a lens can cover depends on how many degrees of freedom it uses and how effectively it uses them.

The development of constrained lenses will be traced historically from the early metal plate lenses of Kock and Ruze through the multifocal lenses of recent development. Design procedures for various well-known lens types such as "egg crate," Rotman, and R2R will be presented. The advantages and possible applications of some of these will be shown, as will their disadvantages. Last, recent work on three-dimensional wide angle constrained lenses will be summarized. There are five computer programs that illustrate the key concepts and lens types. They are mentioned briefly in this chapter and described in detail in the optional software package with user's manual and example booklet.

6.2 METAL PLATE LENSES

The best starting point for a discussion of constrained lenses is the metal plate lens, both because it was the earliest example historically and because it represents a transition in thinking from continuous media, such as dielectrics and mirrors, to discrete media. In fact, the original intent for the metal plate lens, made up of a stack of parallel-plate waveguides, was to imitate a dielectric; the perception of the device as an array structure came much later. For example, John Brown's 1953 monograph, *Microwave Lenses,* refers to them as "metal plate dielectrics" [1]. Since the phase velocity of waveguide is faster than that of free space, a stack of waveguides open at both ends appears electrically as a dielectric with an index of refraction less than unity.

Kock noted that a stack of thin metal plates spaced distance a apart, as shown in Figure 6.2, has an index of refraction of [2]

$$\eta = \sqrt{1 - (\lambda_0/2a)^2} \tag{6.1}$$

for a wave polarized parallel to the plates. The incident wave polarized in that direction will excite the TE10 mode in the waveguides. A structure like the one in Figure 6.3(a) imitates the focusing properties of the natural dielectric lens in Figure 6.3(b).

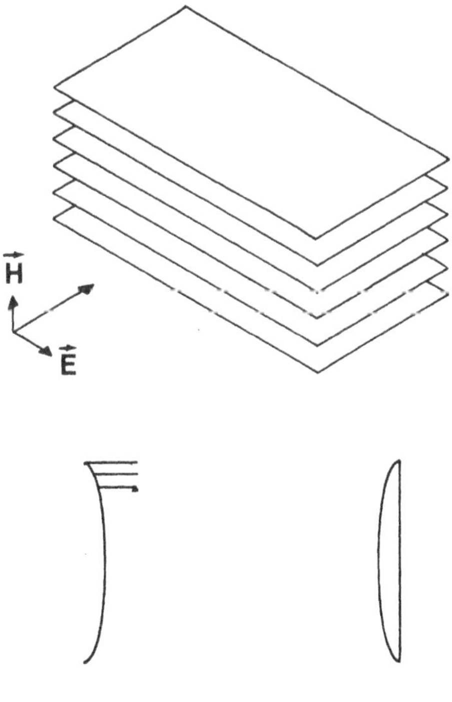

Figure 6.2 **A "Metal Plate Dielectric" Structure**

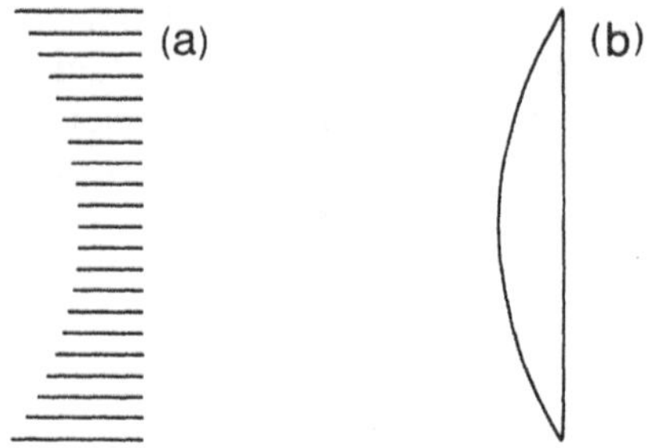

Figure 6.3 Equivalent (a) Metal Plate and (b) Dielectric Lenses (Cross-Sectional Views)

There are many variations on the metal plate lens, but all use parallel plates of varying shape and height. We can choose to make both lens faces curved, or one a curved face and one a flat face. We can also use waveguides of varying height to obtain an index of refraction that is different at each point on the lens. By shaping the plates in their horizontal dimension as well, we can obtain continuous (unconstrained) focusing in one dimension and discrete, constrained focusing in the other, to focus a single point source feed. This section will retrace the derivation of the design equations for those types of parallel plate lenses. A later section will extend those principles to treat the "egg crate" lenses that are made up of a lattice of square, rectangular, or round waveguides. The program METALENS.BAS simulates most of the structures discussed in this section.

6.2.1 Metal Plate Lens with Flat Front Face: Constrained Focusing in the H-Plane

The simplest form of metal plate lens is one with a single curved face and with waveguides that have the same height, like the one illustrated in Figure 6.3(a). We would like to know how thick to make the lens as a function of height, y, to focus a feeding antenna located some distance F behind the lens. Suppose a plane wave is incident on the flat front face from the broadside ($+z$) direction. For the lens to focus perfectly, it must transform that plane wave into a cylindrical wave converging toward the focal point. This demands also that the electrical path length from the focus to any point on the back face of the lens and through the lens be constant. This "path length equality" condition is better illustrated in Figure 6.4,

$$F = R + \eta w \tag{6.2}$$

where R is the distance from the focus to an arbitrary point on the lens back face and w is the physical length of waveguide in the z-direction. The electrical length is ηw with η given by (6.1). The total electrical path length from the focus to the point at height y in the aperture is $R + \eta w$.

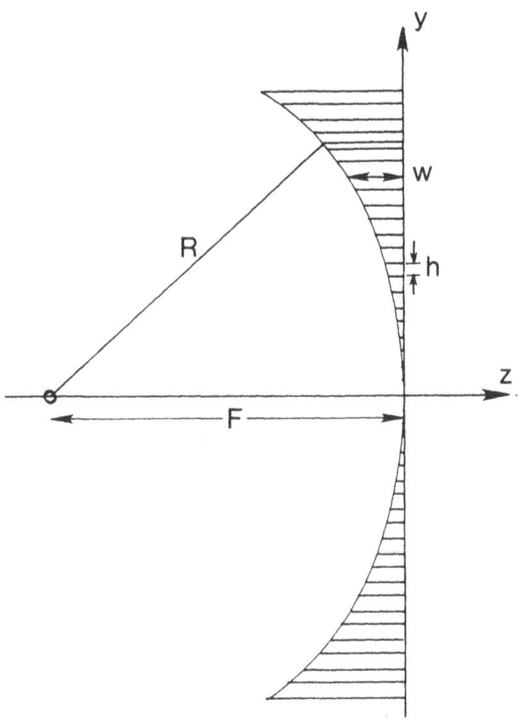

Figure 6.4 Lens Geometry and Equal Path Length Rays

That path length must be equal to that of a ray through the centerpoint, and the resulting path length equality condition is

$$\sqrt{y^2 + [F - w]^2} + \eta w = F \tag{6.3}$$

which, solved for $w(y)$, gives the lens thickness as a function of height:

$$\frac{w}{F} = \frac{1}{1 + \eta}\left[1 - \sqrt{1 - \frac{(y/F)^2(1 + \eta)}{1 - \eta}}\right] \tag{6.4}$$

There are a few restrictions inherent in (6.1) and (6.4). First, the waveguide height must be greater than 0.5λ (for the TE_{10} mode to propagate). Second, the maximum value of y is given by

$$(y/F)^2 \le (1 - \eta)/(1 + \eta) \tag{6.5}$$

which restricts the allowable range of F/D, the ratio of focal length to aperture diameter. This may be important if a compact antenna is required. For example, with $\eta = 0.5$ ($h = 0.577\lambda$) the smallest focal length possible is 0.866 times the aperture diameter (or aperture height in this case).

Figure 6.5 illustrates another trade-off governed by the choice of the index of refraction. Using a large h is desirable because fewer plates are needed to build the lens. On the other hand, there is a greater variation in the lens width with large h, and that is of concern because of dispersion in the waveguide. A change in frequency, and hence in λ, causes a change in η, and there will be a path length error proportional to $w(y)$. Changing frequency will cause a progressively greater phase error in the aperture distribution. Ultimately it will become severe enough that the gain is reduced to 3 dB below that at the design frequency, and the low and high frequencies at which this occurs limit the antenna's bandwidth. Since a wider plate spacing dictates a greater variation in $w(y)$, that lens will have less bandwidth than one with narrow plate spacing.

Another limit on the plate spacing, discussed in detail by Cohn [3], is apparent when we recognize that the open waveguides form an array, which can be susceptible to grating lobes. If the lens is to be able to scan an angular sector of $\pm\gamma$ then the maximum plate spacing is

$$h_{\max} = \frac{1}{\sin(\gamma) + 1} \tag{6.6}$$

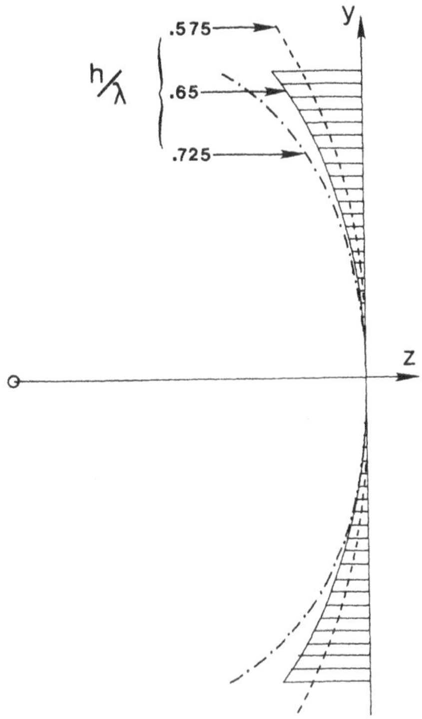

Figure 6.5 Effect of Index of Refraction Variation on Inner Lens Contour

Figures 6.4 and 6.5 show only a two-dimensional cross section of a lens (in the H-plane). Figure 6.6 illustrates two three-dimensional antenna structures that can be formed from such a cross section. In Figure 6.6(a), all of the plates are rectangular. That cylindrical structure only focuses in the H-plane (elevation) and requires a line source feed to obtain gain in the E-plane. Figure 6.6(b) shows a lens that would focus in both planes, and thus needs only a single point source feed. The lens plates have the back edge curved to cause refractive, unconstrained focusing in the E-plane, as described in the following section.

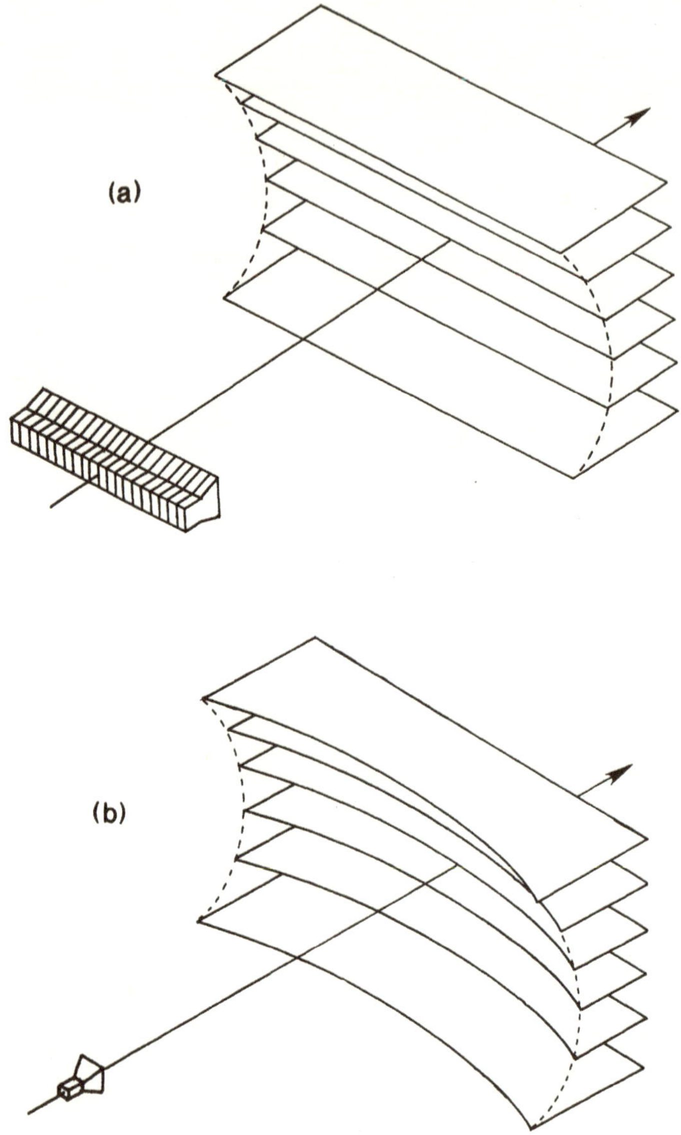

Figure 6.6 Three-Dimensional Antenna Structures Using Metal Plate Lenses: (a) Rectangular Plates for Constrained H-Plane Focusing; (b) Curved Plates Add Refractive E-Plane Focusing

6.2.2 Metal Plate Lens with Flat Front Face: Unconstrained Focusing in the E-Plane

Its refractive index less than one allows the parallel plate to bend rays at its interface with free space, and thus to focus in the plane parallel to the plates. In fact, that was the essence of Kock's metal plate lens [2]; the constrained H-plane lens was a later development by Ruze [4]. Figure 6.7 shows the geometry used to derive the back face contour, assuming again that the front face is flat. Snell's law must be enforced at the back face:

$$\sin \theta_1 = \eta \sin \theta_2 \tag{6.7}$$

The angle θ_2 is also the angle the surface normally makes with the lens axis (z-axis). It is not immediately obvious, but nonetheless true, that the back face contour derived for the constrained lens (6.4) also satisfies this condition. A ray incident normally on the front face will not be refracted since that face is flat, but its intersection with the back face given by (6.4) at $z = -w(y)$ will be at such an angle as to depart refracted in exactly the right direction to pass through the focus. Thus we see an equivalence between the constrained and unconstrained lenses (in the special case of a flat front face) and see how the use of the path length equality condition can simplify derivation of contours for an unconstrained lens.

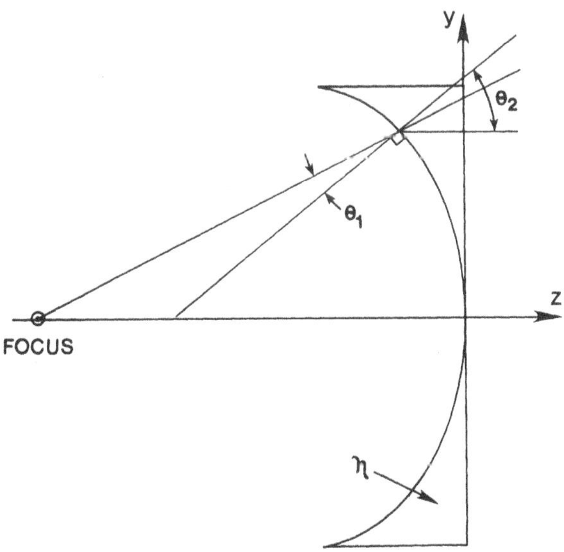

Figure 6.7 Geometry for Refractive (Unconstrained) Focusing

6.2.3 The Bifocal Metal Plate Lens with Two Curved Surfaces

The constrained and unconstrained lenses described in the previous section are "unifocal," that is, they only have one perfect focal point. Supposing an array of feeds were set behind the lens, only one of them could be perfectly focused. The lens cannot collimate a wave from an off-axis feed very well, and the off-axis beams are "defocused." Defocusing effects, which will be discussed in greater detail in Section 6.3, may include broadening of the main beam, an increase in sidelobe level, and decreased gain. Hence unifocal lenses are poorly suited for electronic scanning, or even for those mechanical scanning systems that scan by moving the feed over some "focal arc" behind the lens.

A bifocal lens, like the one depicted in Figure 6.8, is an improvement because it has two perfect focal points. More importantly, it also focuses quite well nearby and in between those points. Obtaining two perfect foci, however, requires employing two geometric degrees of freedom. This section will describe a design that uses front face curvature and back face curvature as those independent degrees of freedom. The next section will discuss a design with only one curved face, but in which the plate spacing and index of refraction vary in the y direction.

A general rule of thumb in designing constrained lenses with more than one perfect focus is this: for each focal point there is a path length equality condition, and that set of conditions must be solved simultaneously to derive the lens equations. The lower feed shown in Figure 6.8 must produce a plane wave directed at the angle α. Path length equality implies that the distance R_1 from the focus to an arbitrary point on the back face plus $\eta(\omega + w)$, the electrical thickness, must equal the length of a ray from the focus through the origin to the same wavefront:

$$F + s = R_1 + \eta(\omega + w)$$

$$s = y \sin \alpha + \omega \cos \alpha \tag{6.8}$$

$$R_1^2 = (F \cos \alpha - w)^2 + (F \sin \alpha + y)^2$$

A feed at the other focus must produce a wave directed at $-\alpha$:

$$F - t = R_2 + \eta(\omega + w)$$

$$t = y \sin \alpha - \omega \cos \alpha \tag{6.9}$$

$$R_2^2 = (F \cos \alpha - w)^2 + (y - F \sin \alpha)^2$$

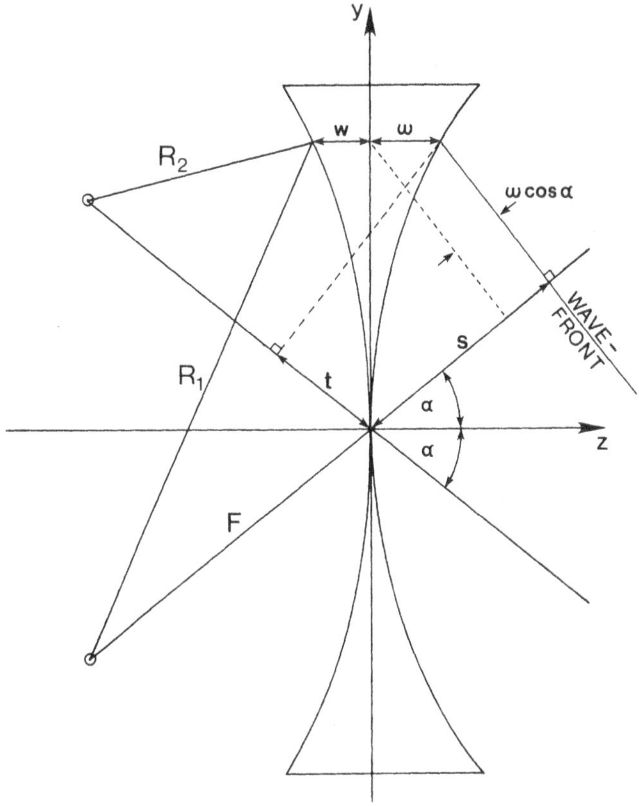

Figure 6.8 **Bifocal Metal Plate Lens Geometry**

Subtracting (6.9) from (6.8) yields $s + t = R_1 - R_2$, or $2y \sin \alpha = R_1 - R_2$. Squaring both sides, then rearranging and squaring again leads to

$$y^2 \sin^2 \alpha = R_1^2 + R_2^2 - 2F^2.$$

Substituting R_1 and R_2 from (6.8) and (6.9) leaves a quadratic equation in w which can be solved as

$$w = F \cos \alpha(1 - \sqrt{1 - (y/F)^2} \qquad (6.10)$$

Substituting (6.10) into the expressions for R_1 and R_2 we find that

$$R_1 = F + y \sin \alpha \text{ and } R_2 = F - y \sin \alpha \qquad (6.11)$$

Since $R_1 + R_2 = 2F$, a constant, we conclude, as Ruze did, that the back face is an ellipse whose foci are the two focal points. Its minor axis is $2F \cos \alpha$ and its major axis is $2F$. Finally, we can derive an expression for the front face contour by adding (6.8) and (6.9) and substituting (6.11):

$$2F + s - t = R_1 + R_2 + 2\eta(\omega + w)$$

$$\omega \cos \alpha = \eta(\omega + w) \tag{6.12}$$

$$\omega = \eta w/(\cos \alpha - \eta)$$

Since ω is proportional to w, the front face must also be elliptical.

Equations (6.10) and (6.12) describe the inner and outer contours of a lens in one plane only. If the plates making up the lens were rectangular, we would need to feed it with two line sources. Alternatively, we could shape the plates to obtain unconstrained focusing in the E-plane. The resulting structure would appear similar to the one shown in Figure 6.6(b), but it would have two perfect focal points.

6.2.4 The Bifocal Metal Plate Lens with One Curved Surface

The final case of metal plate lens that we will consider is one with a curved back face, a flat front face, and variable waveguide height. This combination of varying index of refraction and back face curvature gives us the two degrees of freedom needed to obtain two perfect foci.

The path length equality conditions are found from Figure 6.9:

$$\begin{aligned} F + y \sin \alpha + \eta_0 w_0 &= R_1 + \eta(w + w_0) \\ F - y \sin \alpha + \eta_0 w_0 &= R_2 + \eta(w + w_0) \end{aligned} \tag{6.13}$$

Note that the lens has a constant "offset" thickness w_0, and η_0 is the index of refraction at the lens center. Subtracting the two conditions, we get $2y \sin a = R_1 - R_2$, the same as in the previous case for the two-curved-surface lens. Thus, the back face is again elliptical with the contour given by (6.10). Since $R_1 + R_2 = 2F$, adding the conditions in (6.13) gives the following expression for η as a function of y (the y dependence is contained in w):

$$\eta = \frac{\eta_0 w_0}{(w + w_0)} \tag{6.14}$$

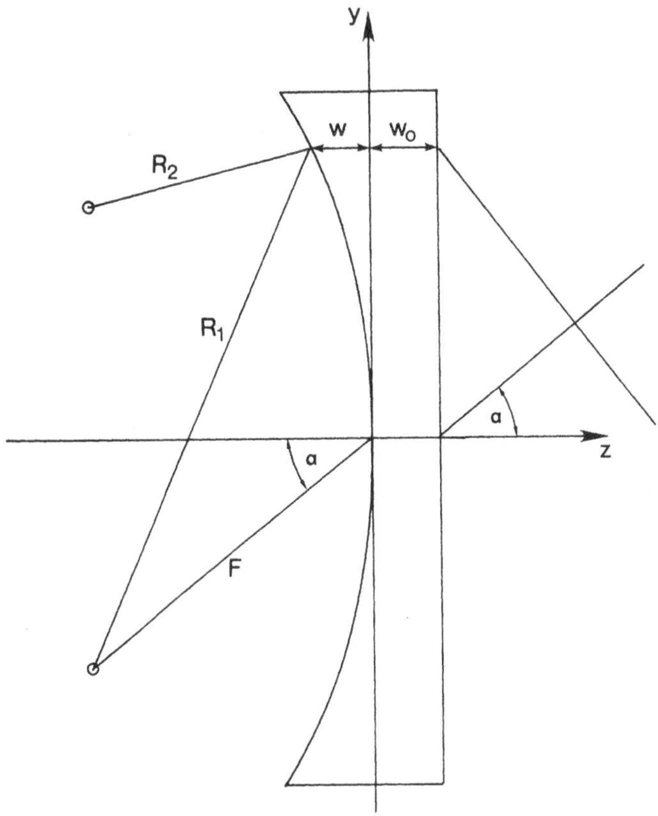

Figure 6.9 Bifocal Lens Geometry with Variable Refractive Index

The performance of this bifocal design will be compared later with that of the one with two curved surfaces. Program METALENS.BAS can be used to calculate contours for either bifocal type, as well as for the unifocal lens discussed above, and an additional unifocal that has constant thickness and a variable index of refraction.

6.3 SCANNING ABERRATIONS AND REFOCUSING

Before moving on to discuss more complicated lenses, it will be worthwhile to address the quality of beams that would be produced by feeds located somewhere other than at the point or points of perfect focus. This is a vital consideration for antennas (like the one depicted in

Figure 6.1) that scan electronically by switching between elements of an array feed, or that produce shaped beams by weighting several elements of the array feed. In the case of bifocal lenses, for instance, only two feeds can be perfectly focused. All the others will be defocused to varying extents, and will therefore produce distorted beams. In some cases, that distortion may be no more than a slight filling of the pattern nulls, while in others it may broaden the main beam (with a consequent loss in gain), raise the sidelobe levels, or cause the main beam to "squint" away from the chosen scan angle.

This section will explain the concepts of path length error and "scanning aberrations" and show how those errors depend on the lens design. It will also show how the errors can be minimized by proper choice of a *focal arc,* or *scanning locus.* The numerical results presented in this section were generated by program METALENS.

6.3.1 Path Length Error

Figure 6.10 depicts a feed located at an arbitrary point behind a lens. Its distance from the origin is G, and it is at an angle β from the axis. For simplicity, assume the lens has constant η and a flat front face.

The path length equality condition that would make the point a perfect focus is

$$G = R + \eta w + y \sin \beta \tag{6.15}$$

where

$$R = [(y - G \sin \beta)^2 + (G \cos \beta - w)^2]^{1/2}$$

Suppose, however, the lens was already designed to have an on-axis focus. The path through the lens at $+y$ will not equal the path through $y = 0$ and there will be a path length error E:

$$E = G - R - \eta w - y \sin \beta \tag{6.16}$$

We will denote $\varepsilon = E/F$, the path length error normalized to the design focal length. Figure 6.11 shows how ε varies with y for various choices of angle β with $G = F$. As expected, the error becomes progressively worse the farther the feed moves from the perfect focus. The *shape* of the path length error curve is also important. It contains a strong third-order term, which is known as *coma aberration,* or *coma phase error.* This is generally considered the most severe kind of error because it causes beam squint as well as a pronounced increase in the sidelobe level on one side of

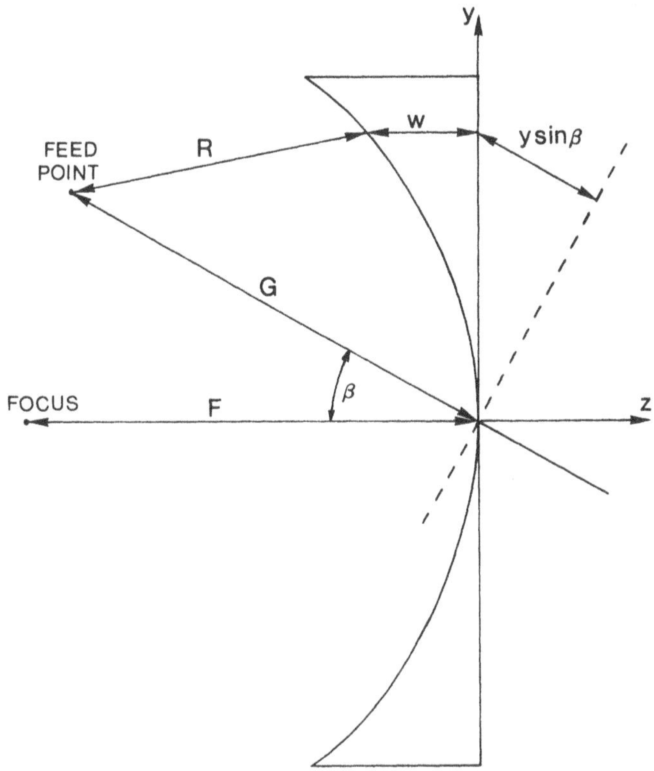

Figure 6.10 Reference Geometry for Path Length Error Analysis

the main beam. Figure 6.12 shows the radiation patterns corresponding to the phase distributions of Figure 6.11.

The sample parameters used for Figure 6.12 were $F/D = 1$ and $D = 10\lambda$. With $F/D = 1$ the maximum y coordinate in the aperture is $y/F = 0.5$. If the F/D ratio were larger, the maximum value of y would be smaller, and so would the maximum path length error. That is why lenses with larger F/D ratios are generally less susceptible to scanning aberrations, but this comes at the expense of higher loss due to spillover.

The path length error for the most general metal plate lens is

$$E = G - R - \eta(w + \omega) - w_0(\eta - \eta_0) - (y \sin \beta - \omega \cos \beta) \tag{6.17}$$

which allows for variation in η as well as front and back face curvatures w and ω, respectively. For the case of the bifocal lens with one curved surface, (6.14) reduces to

$$E = G - R - y \sin \beta - \eta(w + w_0) + \eta_0 w_0. \tag{6.18}$$

R is the same as given in (6.15) and w and η are the same as in (6.10) and (6.14). Figure 6.13 shows ε for three feed positions: 0°, 10°, and 15°, all with $G = F$, for a lens designed with $\alpha = 5°$. These errors are more than an order of magnitude smaller than those in Figure 10 from the single-focus lens (note that the scales on the two graphs are different). This illustrates the advantage of the bifocal, or two-point-corrected, lens. Even at 15° scan, there is no substantial degradation in the radiation pattern. The phase error is mostly second-order "focus error." Its primary effect is to fill the pattern nulls near the main beam. We can, however, suppress that error even further using a technique called "refocusing."

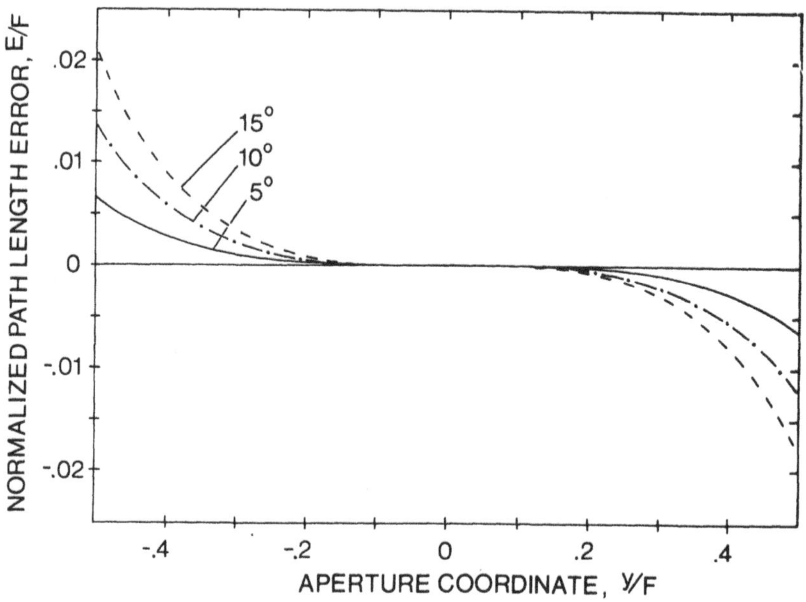

Figure 6.11 Path Length Error *versus* Aperture Coordinate (for the single focus lens of Figure 6.10), $F/D = 1$, $\eta = 0.45$

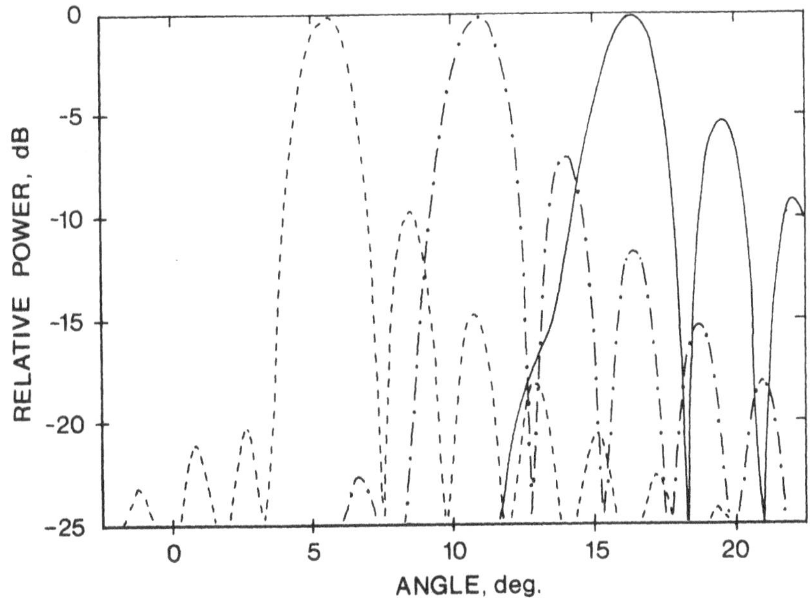

Figure 6.12 Radiation Patterns from a Unifocal Lens (line source aperture with uniform amplitude and the path length errors from Figure 6.11)

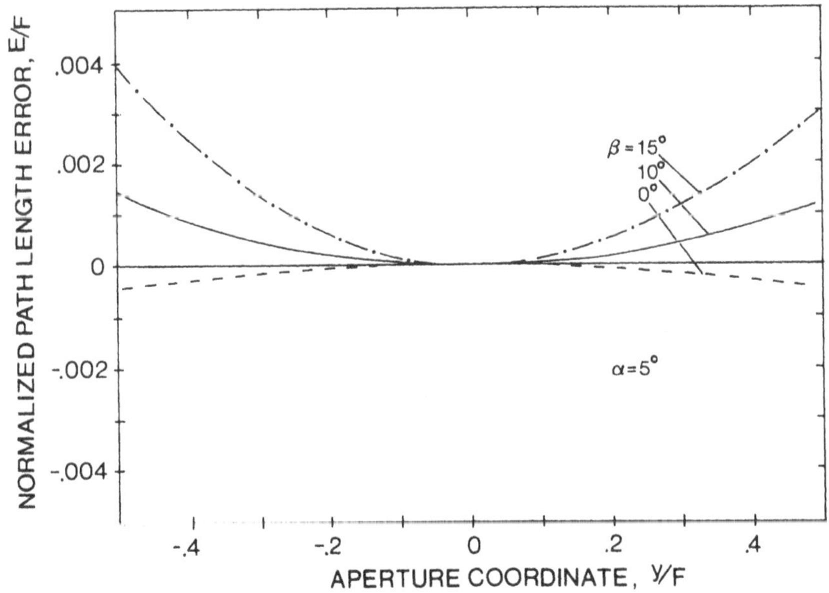

Figure 6.13 Path Length Error *versus* Aperture Coordinate for a Bifocal Lens (one curved surface, variable η), $F/D = 1$, $\eta_0 = 0.45$, $\alpha = 5°$

6.3.2 Refocusing

For angles away from the focal point or points, the best focal length is usually something other than F. Rays arriving from an arbitrary angle $\beta \neq \alpha$ do not converge to a single point, but there is some point a distance G behind the lens where they come closest. If we could find the distance G for every angle, we would know the proper shape of the "focal surface," or "scanning locus." This is an important part of any lens design, since it tells us where to put the feeds.

There are two strategies we could use in trying to find the best focal surface: we could minimize the peak error, which usually appears at the aperture edge, or we could minimize the root-mean-square (rms) error over the entire aperture. The former can often be found analytically, while the latter ensures the highest gain for the given lens and scan angle. The two conditions are sometimes equivalent, since the error distribution with the smallest peak value tends also to have the lowest rms value. Figure 6.14 shows the normalized error for various values of G/F, for the 15° scan case for the bifocal lens with varying η. Figure 6.15 shows similar curves for the two-curved surface design. Judging these qualitatively, we would choose $G = 0.975F$ and $G = 0.999F$ as the best focal lengths. A more rigorous strategy would be to calculate the root-mean-square (rms) value of E/F for each value of G/F and choose the one that gives the smallest ε_{rms}. Repeating that procedure for each scan angle, β, would result in a curve of G versus β. Figure 6.16 shows the shape of those minimum error focal arcs for both types of bifocal lenses with $\alpha = 15°$. The best focal arc for the lens with two curved surfaces is very nearly circular, while the one for the variable-η lens is quite eccentric. These are consistent with focal arcs derived by Ruze for various lens types [4].

Finally, we have reached a point where we can make a quantitative comparison of the lens designs we have discussed. It is obvious that the bifocal lens is far superior to the unifocal lens in terms of scanning aberrations and, hence, beam quality. One might presume, on the other hand, that one bifocal lens is much the same as another in those terms, but that is not the case. Figure 6.17 shows the rms path length error *versus* scan angle for the two bifocal lens designs discussed, both with $F/D = 1$, and $\alpha = 5°$ and $\alpha = 15°$, and both with $\eta_0 = 0.45$. The dashed curves are for the straight-front-face lens and show that design to be superior to the one with two curved surfaces. Hence, even though two lenses have the same number of focal points, their scanning properties may be quite different. The remaining curve in Figure 6.17 is for a single focus lens (again with $\eta = 0.45$, $F/D = 1$, and optimum focal arc), and serves to illustrate how much improvement can be gained with two-point correction.

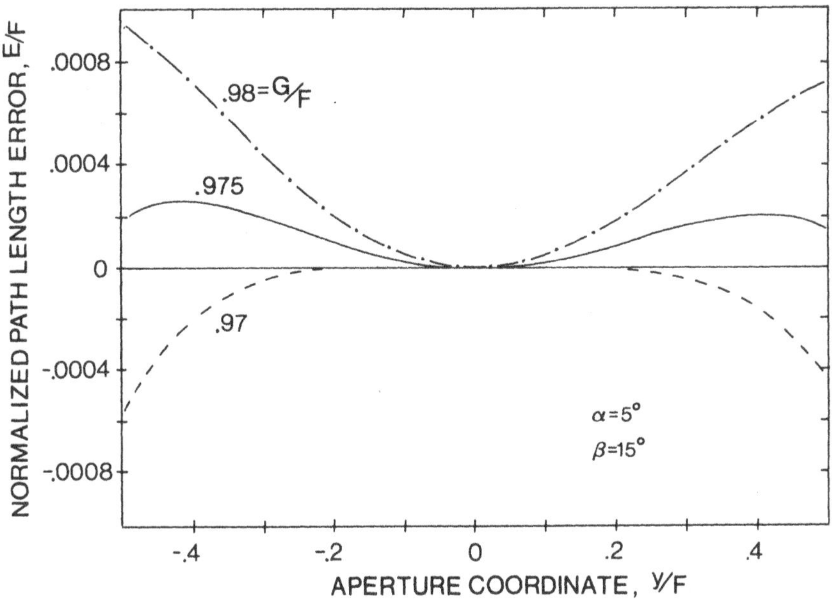

Figure 6.14 Path Length Error for a Bifocal Lens with Refocusing (one curved surface, variable η), $F/D = 1$, $\eta_0 = 0.45$, $\alpha = 5°$, $\beta = 15°$

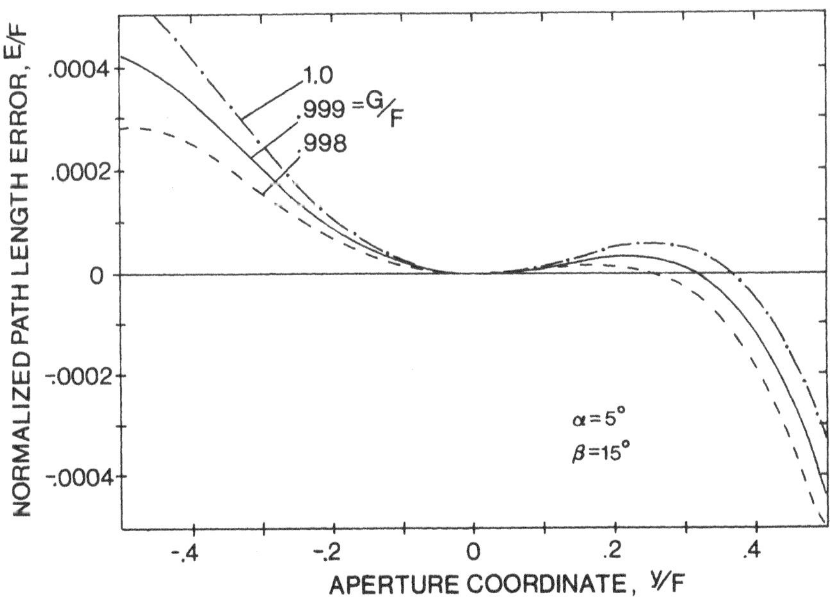

Figure 6.15 Path Length Error for a Bifocal Lens with Refocusing (two curved surfaces, constant η), $F/D = 1$, $\eta = 0.45$, $\alpha = 5°$, $\beta = 15°$

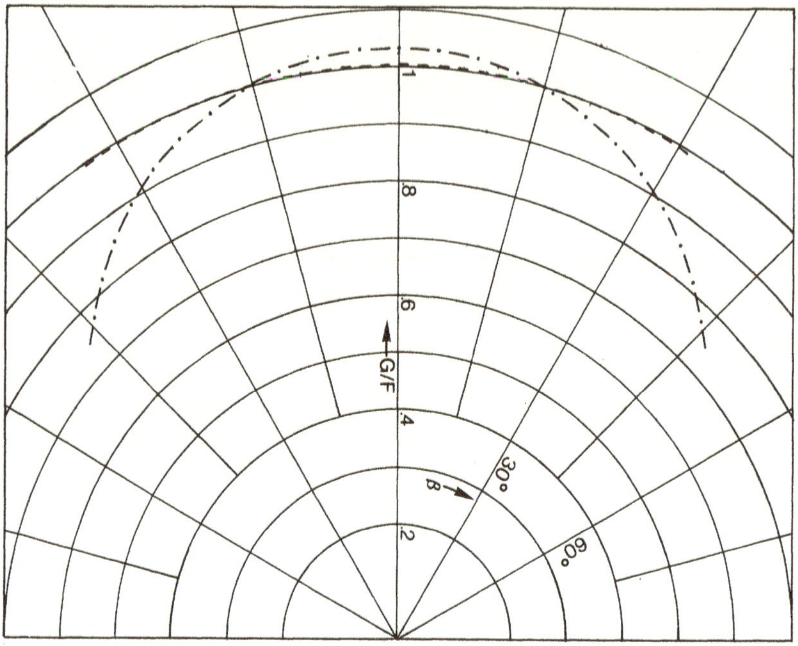

Figure 6.16 Minimum-Error Focal Arc Contours for Bifocal Metal Plate Lenses (*solid*—two curved surfaces with constant η; *dashed*—straight front face with variable η), $F/D = 1$, $\eta = 0.45$, $\alpha = 15°$

6.3.3 Phase Compensation for Refocusing

In the case of array-fed lenses, where scanning is done electronically, we must often combine two or more adjacent beams to "synthesize" a radiation pattern. For example, to obtain a low sidelobe beam, one might combine the power from three adjacent feeds, with amplitude weighting to reduce the power accepted from the outside two. This beam combining must be done coherently, adding the beams in phase. The refocused feeds are different distances from the center of the lens, however, and the beams they produce will be slightly out of phase when they combine in the far field. To compensate, we usually add an electrical length of transmission line to the inputs to all the feeds. The electrical lengths of those lines will equal the feed's displacement from the circle of radius F, or $G - F$.

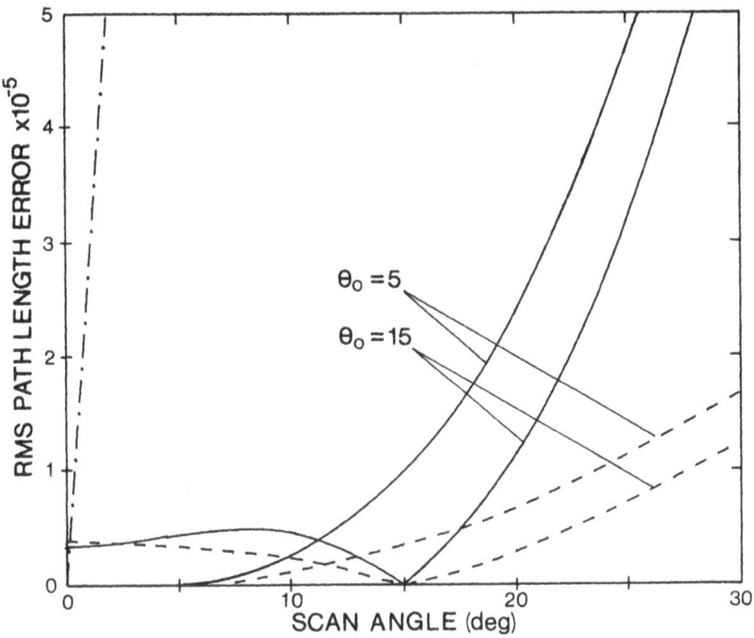

Figure 6.17 RMS Path Length Error *versus* Scan Angle for Two Bifocal Designs (*solid*—two curved surfaces with constant η; *dashed*—straight front face with variable η) and a One Degree of Freedom Unifocal (*dash-dot*)

6.4 WAVEGUIDE LENSES FOR SCANNING IN TWO PLANES

The parallel plate lenses we have discussed thus far have limited application because they only focus in one plane. Furthermore, they can only focus one polarization because there is no phase delay for a wave polarized perpendicular to the plates. The natural trend, to try to obtain dual polarization, is to combine a row of vertical parallel plates with a stack of horizontal plates. The result is a lattice of rectangular waveguides. The phase velocity in a rectangular waveguide is always faster than in free space, so we end up with a structure similar to Figure 6.18, which is thinnest at the center, growing progressively thicker toward the edges. This "egg crate" structure has an added advantage over the parallel plate lenses in that the rows and columns support each other mechanically.

Figure 6.18 **Rectangular Waveguide "Egg Crate" Lens**

It appears at first glance that the two dimensions of the lattice afford more degrees of freedom, i.e., curvature of both faces and varying refractive index along the x-axis, and another set of three variations along the y-axis. Should we not be allowed six perfect focal points? Unfortunately, no. The problem is the three degrees of freedom in x are not *independent* of the three in y. This is easiest to visualize by considering the simplest "constant thickness" lens, in which we can only vary the index of refraction: variation of either $\eta(x)$ or $\eta(y)$ would cause focusing onto a line parallel to the y or x-axis, respectively. We must specify both simultaneously to obtain focusing to a point. Similarly, where three degrees of freedom are available in the front and back face curvature as well as the index of refraction, we must specify the contours as functions of x and y (or as functions of ρ and ϕ).

6.4.1 Design Procedure for Three Dimensions

The general design procedure for three dimensions is the same as for two dimensions: write a set of path length equality conditions, solve for the lens contours and index of refraction, and find an optimum scanning locus.

A common practice is to design three-dimensional beamformers by calculating element locations and transmission line lengths for a two-dimensional case first, then simply replacing y with ρ, creating a lens with rotational symmetry. It will not have any perfect foci except in the special case where the two-dimensional design only had a single on-axis focus. This section will show the derivation for an alternative three-dimensional lens that does have two perfect foci, with the intent of illustrating the design principles and comparing symmetric and asymmetric lenses.

The path length equality conditions must now specify both θ and ϕ, where before they only specified θ (α) for the focal point or points. The angles of the focal points in spherical coordinates will be denoted θ_0 and ϕ_0. The most general geometry is illustrated in Figure 6.19. The lens front face radial coordinates are expressed in cylindrical coordinates as (ρ,ϕ_l,ω) and the back face coordinate as $(\rho,\phi_l,-w)$. Restricting our attention to a two-point (bifocal) lens, the focal point coordinates are $(\rho,\phi,z) = (\pm F \sin\theta_0, \phi_0, \pm F\cos\theta_0)$. Define, by convention, the $\phi = 0$ plane as that plane containing the lens axis and the two foci ($\phi_0 = 0$). The path length equality conditions are similar to (6.8) and (6.9):

$$F + s = R_1 + \eta(\omega + w)$$

$$s = \rho \sin\theta_0 \cos\phi_l + \omega\cos\theta_0 \tag{6.19}$$

$$R_1^2 = (F\cos\theta_0 - w)^2 + (\rho\sin\phi_l)^2 + (F\sin\theta_0 + \rho\cos\phi_l)^2$$

$$F - t = R_2 + \eta(\omega + w)$$

$$t = \rho\sin\theta_0\cos\phi_l - \omega\cos\theta_0 \tag{6.20}$$

$$R_2^2 = (F\cos\theta_0 - w)^2 + (\rho\sin\phi_l)^2 + (F\sin\theta_0 - \rho\cos\phi_l)^2$$

Subtracting (6.19) and (6.20) results in the following expression for w:

$$w = F\cos\theta_0\left[1 - \sqrt{1 - \frac{\rho^2(1 - \sin^2\theta_0\cos^2\phi_l)}{F^2\cos^2\theta_0}}\right] \tag{6.21}$$

Note that when $\phi_l = 0$ this reduces to (6.10). In other words, in the plane containing the foci, the inner lens contour is the same as it was for the bifocal parallel plate lens. Substituting w back into the expressions for R_1 and R_2:

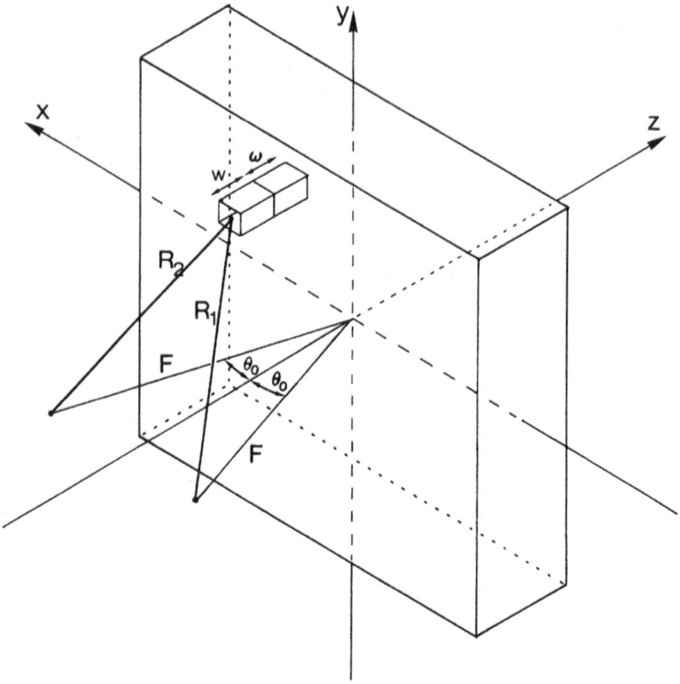

Figure 6.19 General Geometry for a Three-Dimensional Waveguide Lens

$$R_1 = F + \rho \sin \theta_0 \cos \phi_l$$
$$R_2 = F - \rho \sin \theta_0 \cos \phi_l \tag{6.22}$$

and once again $R_1 + R_2 = 2F$, a constant, implying that the inner lens face is an ellipsoid. Finally, adding (6.19) and (6.20) results in the following expression for the front face depth:

$$\omega = \eta_0 w/(\cos \theta_0 - \eta_0) \tag{6.23}$$

(In the above, η_0 denotes the index of refraction at a selected center frequency. This distinction will become apparent in a later discussion of bandwidth.) Equation (6.23) is identically the same relationship between w and ω as for the bifocal parallel plate lens, so we can conclude that the 3D lens will have an ellipsoidal front face as well.

The design procedure is fairly straightforward now, because (6.22) and (6.23) are completely general. The waveguide spacing should be chosen to prevent grating lobes at the maximum feed scan angle, and this spacing determines the allowable waveguide size and, hence, the allowable index of refraction. If the guides cannot be packed closely enough and still allow the TE_{10} mode to propagate, some dielectric loading can be used to reduce the waveguide dimensions for the desired frequency. Once the lattice spacing is determined, (6.22) is used to find how far each guide will extend behind the aperture plane (w), then (6.23) is used to find how far it will extend in front of the aperture (ω). Alternatively, (6.10) and (6.12) (with ρ substituted for y and θ_0 substituted for α) could be used to generate a ϕ-symmetric design. The scanning properties of those two alternatives will be compared in the following section.

6.4.2 Comparison of Scanning Regions

As pointed out earlier, a ϕ-symmetric lens does not, in general, have any perfect focal points. It may be "two point corrected," but it is not "bifocal." In the special case where $\theta_0 = 0$ it will have a single focus on the lens axis, and for that condition the ϕ-symmetric and bifocal designs would be identical.

For illustration, an example parameter of $F/D = 1$ was chosen, and the rms error was found for angles out to $\theta = 30°$, varying G/F to find the minimum error feed point. Figure 6.20(a) shows ε_{rms} for various scan planes for the bifocal lens with $\theta_0 = 10°$. As we would expect, the error goes to zero at 10° in the $\phi = 0$ plane, because that is where one of the foci is, and the error is highest when the feed is scanned in the plane orthogonal ($\phi = 90°$) to that containing the foci. Figure 6.20(b) is the equivalent result for a ϕ-symmetric lens, both with $\theta_0 = 0$ and $\theta_0 = 10°$. There is only one curve for each, because the results would be the same for all ϕ planes. Observe that the error in both cases is less than for the bifocal lens in planes where $\phi < 45°$. If the available scan region is defined as that range of feed scan angles for which the rms error is smaller than some threshold value, then the axially symmetric design will have a circular region centered on the axis and the bifocal lens will have two regions centered around the focal angles. Figure 6.21 contrasts the scan regions for the bifocal lens with $\theta_0 = 10°$ and the ϕ-symmetric lens with $\theta_0 = 0$. In spite of the fact that the latter only has one perfect focus, its total scan region is actually slightly larger. It is very important to note, however, that even though it only has one focus, it still uses two degrees of freedom.

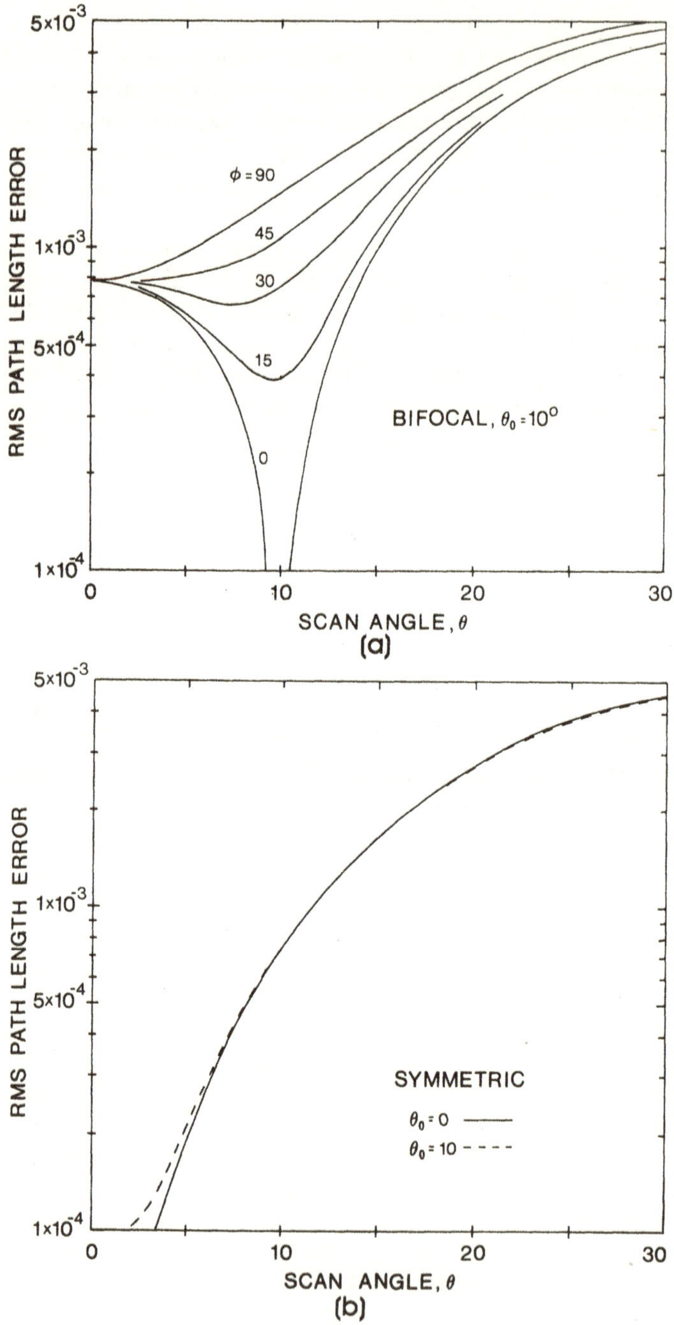

Figure 6.20 RMS Path Length Error *versus* Scan Angle for (a) a Bifocal 3D Waveguide Lens with $\theta_0 = 10°$ and (b) 3D ϕ-Symmetric Lens with $\theta_0 = 0, 10°$

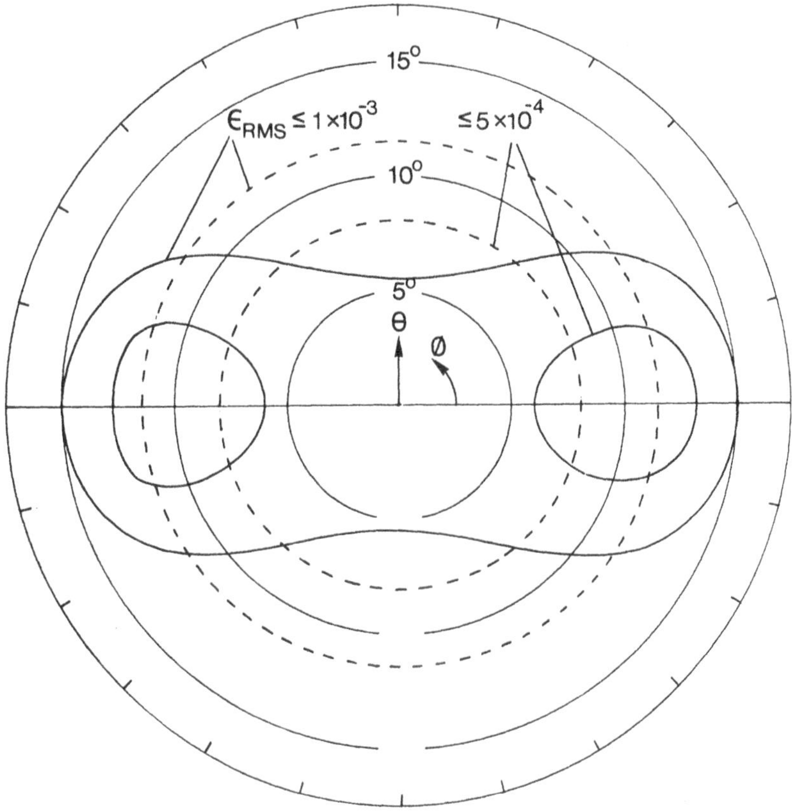

Figure 6.21 Comparison of Effective Scan Regions for Two Degree of Freedom Waveguide Lenses (*solid*—bifocal; *dashed*—ϕ-symmetric)

A common misperception is that more focal points will result in a larger scan region. Yet that is only a coincidence stemming from the requirement that one must use more degrees of freedom to get more focal points. The above comparison illustrates that we can get wide angle scanning with a single-focus lens. Whether one uses the available degrees of freedom to get two or more perfect foci depends on the application since, as Figure 6.21 showed, the improvement near the focal points comes at the expense of other directions.

The program EGGCRATE.FOR was used to generate Figures 6.20 and 6.21. It can calculate waveguide lengths for both symmetric and bifocal lenses. Once those are found for a particular choice of F and α, it can find the path length error for an arbitrary feed location, or find the optimum focal length for a given scan angle.

6.4.3 Bandwidth

One disadvantage of egg crate lenses is that they tend to have limited bandwidth, due to dispersion in the waveguides. The phase velocity of the waveguides varies with frequency, which causes a corresponding change in the index of refraction. To study the effects of frequency variation, we will derive an expression for path length error versus frequency, which is known in optics as *chromatic aberration.*

From (6.19), we find the error at one of the foci as

$$E = F + s - R_1 - \eta(\omega + w)$$

where it is understood that η is now a function of frequency, with η_0 denoting the index of refraction at the center frequency, f_0. Substituting $R_1 = F + \rho \sin \theta_0 \cos \phi_l$ and (6.23):

$$E = \eta_0 w \frac{\cos \theta_0 - \eta}{\cos \theta_0 - \eta_0} - \eta w \tag{6.24}$$

The index of refraction in any waveguide is related to its cut-off frequency by

$$\begin{aligned} \eta &= \sqrt{1 - (f_c/f)^2} \\ &= \sqrt{1 - (f_0/f)^2(1 - \eta_0^2)}. \end{aligned} \tag{6.25}$$

From (6.24) it is apparent that given any η_0 and θ_0, the error is directly proportional to the lens thickness, w. The maximum error depends only on the difference in the lens thickness between its center and its edge. This error causes a corresponding loss of gain as f moves away from f_0. At some low and high frequencies f_L and f_H, the gain will have decreased by 3 dB, and those are the bandwidth limits. Figure 6.22 shows the percentage bandwidth *versus* η_0 with $F = D = 20\lambda$, assuming a uniform aperture amplitude distribution. As pointed out earlier, smaller values of η_0 result in less bandwidth because they cause a larger variation in the lens' thickness between its center and its edges.

Note that other authors define the bandwidth limits as those frequencies where the path length error exceeds $\lambda/8$ [2,5]. This will yield lower estimates of bandwidth than those shown in Figure 6.22.

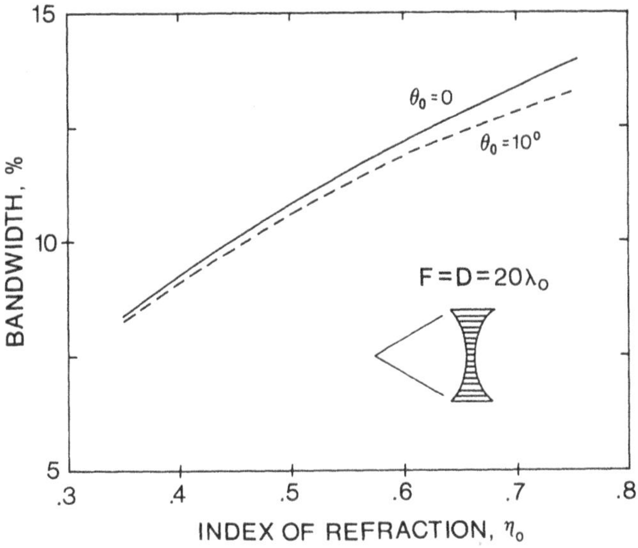

Figure 6.22 Percentage Bandwidth *versus* Index of Refraction

6.4.4 Zoning

In the forms examined thus far, metal plate lenses would tend to be quite heavy. To make them practical, it is usually necessary to reduce their weight by *zoning*, or *stepping*. This procedure involves cutting the thickness down everywhere it exceeds one wavelength in electrical thickness. This section will describe how the zoned lens surfaces are derived and discuss some of the consequences.

It is easiest to understand zoning for a simple lens form, so we shall examine a single-focus, one degree of freedom lens. The equation for the back face contour is given by (6.4) for a 2D structure. A similar three-dimensional lens would be described by the same equation, but with ρ substituted for y. Figure 6.23(a) shows back face contours for a family of lenses that have the same focus, but have focal lengths of $F + \delta$ and $F + 2\delta$. The electrical length between the various contours must be stepped in increments of one wavelength in the refractive medium, therefore [2]:

$$\delta = \lambda/(1 - \eta)$$

and the resulting equation for the zoned back face contour is

$$w_n = \frac{F + n\delta}{1 + \eta}\left[1 - \sqrt{1 - \frac{[y/(F + n\delta)]^2(1 + \eta)}{1 - \eta}}\right] \tag{6.26}$$

Figure 6.23(a) shows the lens design that uses this formula, while Figure 6.23(b) shows an alternative zoned lens that uses the back face for $n = 0$ with front face stepping in increments of the same δ.

There are two undesirable effects of zoning: increased sidelobes due to shadowing and reduced bandwidth. The latter is not really of concern for waveguide lenses, because it usually turns out that the zoning *increases* the bandwidth. It reduces the maximum difference in lens thickness between the center and the edges, so that the variation in waveguide phase velocity with frequency is less significant. Zoning, however, will in general limit the percentage bandwidth to ([5], p. 16-11):

$$\text{Bandwidth} = \frac{25\eta_0}{1 + (N + 1)/\eta_0} \text{ (percent)} \tag{6.27}$$

where N is the number of zones.

Shadowing produces a more severe effect, because the waveguides near the steps will not receive as much power from the feed as their neighbors. That causes dropouts in the aperture amplitude distribution, resulting in higher sidelobes [3].

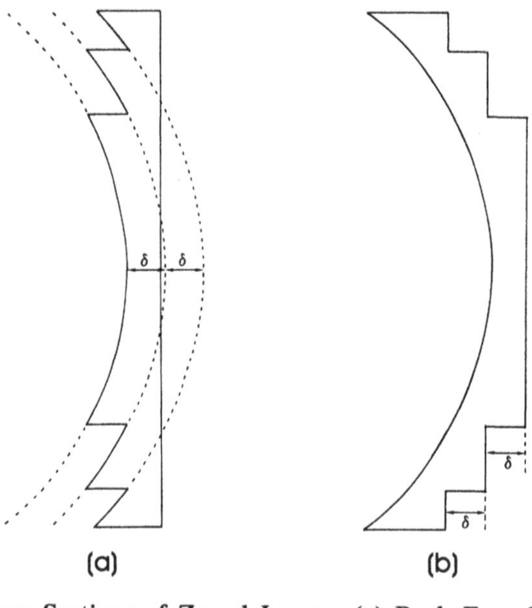

Figure 6.23 Cross Sections of Zoned Lenses: (a) Back Face Stepped; (b) Front Face Stepped

6.5 THE BOOTLACE AERIAL

Up to now, this chapter has treated constrained lenses as though they were merely a convenient refractive medium. The perception of these devices as array antennas was not widespread until Gent introduced the "Bootlace Aerial" in 1957 [6]. His experimental lens was made of pairs of folded dipole radiators separated by a ground screen and interconnected by parallel wire cables, as illustrated in Figure 6.24. The transmission lines were simply television cable, and the folded dipoles were short pieces of the same cable stock with short wires soldered across their ends. Since the cables were longest near the lens center and progressively shorter going out toward its edge, an incident wave from a feed suspended behind the lens would be slowed down near the center relative to the edges, or "collimated."

The importance of the bootlace aerial was in showing that a constrained lens need not be made of waveguides. Many types of transmission line connecting pairs of any of several types of radiator could result in a lens. Possibly the most important capability of the new lens was for very broadband antennas, since it could use nondispersive transmission lines like coaxial and parallel-wire cable. Gent went on to show that there are four geometrical degrees of freedom available in the general design, and further to suggest several new antenna architectures based on constrained lenses that encompass most present-day variations.

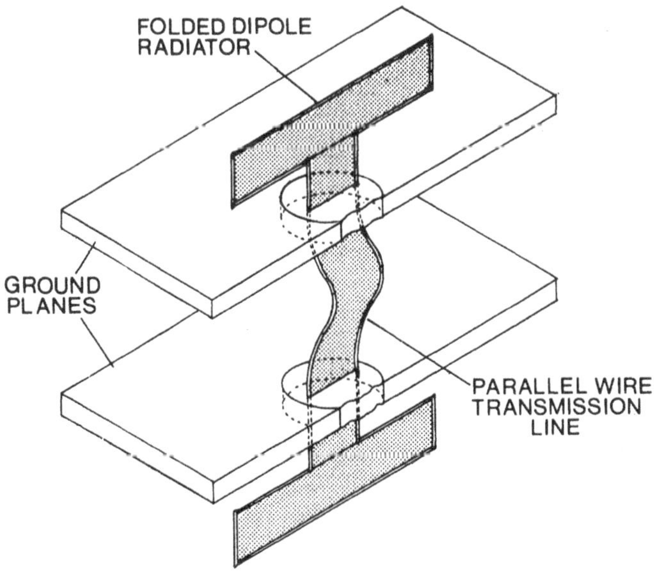

Figure 6.24 A Pair of Elements in Gent's Experimental Bootlace Aerial

6.5.1 Degrees of Freedom

Gent's lens only embodied a single degree of freedom—the length of the parallel wire cables, which varied as a quadratic function of radius measured in the plane of either lens face. He recognized, however, that there are four features available:

(1) the shape of the input surface (that nearest to the feed),
(2) the shape of the output surface,
(3) the relative position of corresponding elements in the two surfaces, and
(4) the length of the transmission line joining corresponding elements in the two surfaces.

The first two features are incorporated in a bifocal waveguide lens, as derived in the previous section. Feature (4), however, was not independently variable, since the length of the waveguides is dictated by the shape of the lens faces. We could achieve this feature only by filling the waveguides with different dielectrics or varying their dimensions to obtain a variation of η with radius. Also, feature (3) would be difficult or impossible to achieve with waveguides. This feature in particular has led to some very useful lens designs that will be discussed later in this chapter.

6.5.2 Novel Antenna Architectures

Gent, Jones, and Browne listed several antenna designs based on bootlace lenses, which collectively describe most practical constrained lens antennas, including reflectarrays, phased array lenses, and active lenses [7]. The six embodiments they described are illustrated in Figure 6.25 and discussed below:

1. *Parasitic Array Mirror*. Better known as a "reflectarray," this antenna uses only one surface, the elements of which are short-circuited at the ends of their transmission lines. The combination of surface shape and trimmed line lengths can be used to control scanning aberrations. Reflectarrays can also use phase shifters imbedded in the transmission lines for beam steering. Furthermore, adding the line length term to the beam steering phases removes the need for phase-trimmed transmission lines.

2. *Variable Line-Length Scanning Lens*. Gent suggested inserting rotary phase shifters between the lens faces. Electronically variable phase shifters have long since replaced mechanical rotary phase shifters, but the basic idea of the *phased array lens* was established. Like reflectarrays,

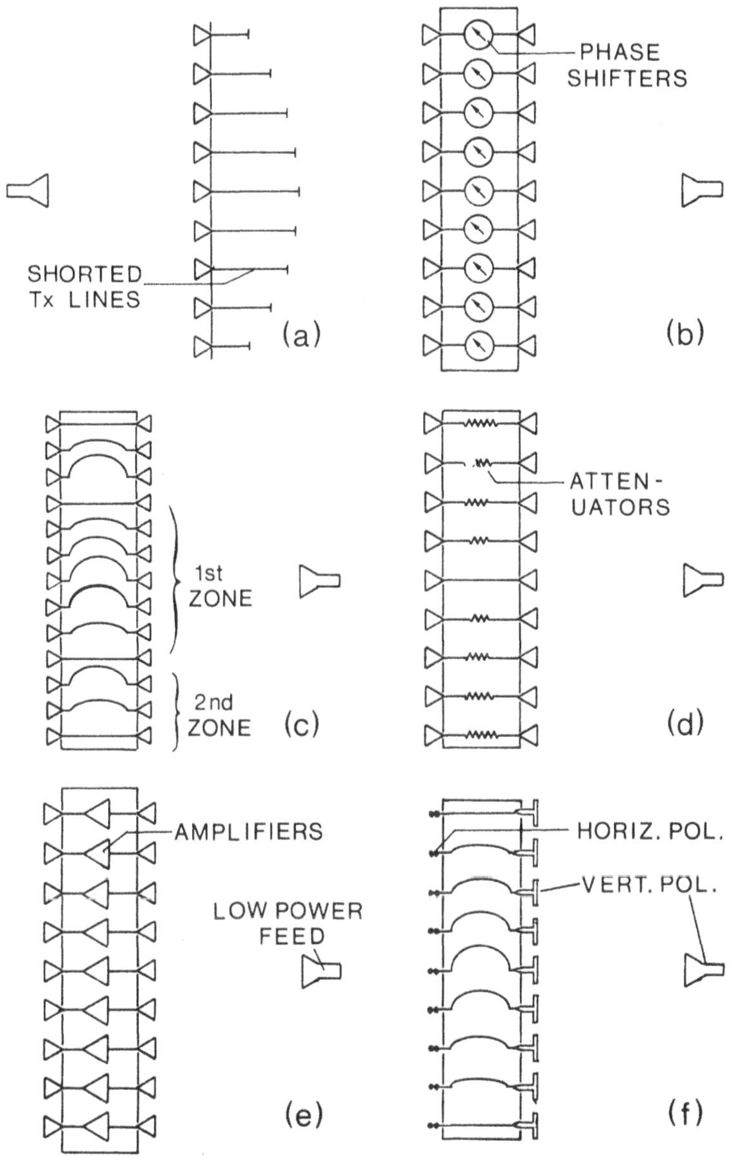

Figure 6.25 Bootlace Structures [7]: (a) "Parasitic Array Mirror" (reflect-array); (b) "Variable Line-Length Scanning Lens" (phased array lens); (c) Stepped (zoned) Parasitic Array Lens; (d) Attenuation Control for Low Side-lobes; (e) Active Aerial; and (f) Polarization Changing Aerial

phased array lenses use phase shifts for collimation as well as beam steering. They typically use only one degree of freedom (the line length phase shifts). They do not have to be wide-angle lenses, since scanning is not performed by feed switching or feed movement but by the phase shifters in the lens. This type of antenna is popularly regarded as a *space-fed phased array* instead of a true lens.

3. *Stepped Parasitic Array Lens.* The transmission line lengths may be zoned in increments of a wavelength. This reduces the total length of line required, but also reduces the bandwidth by creating a zoning effect. Phased array lenses also fall into this category, since their phase shifters are only variable within the range from 0° to 360°.

4. *Attenuation Control of Polar Diagram.* (Polar diagram refers to the polar plot of the antenna's radiation pattern.) Attenuators inserted in the transmission lines can be used to adjust the aperture amplitude distribution for low sidelobes or beam shaping. This is useful in cases where it is impossible or impractical to arrange and weight an array of feeds to obtain a desired aperture distribution. This technique allows pattern shaping using relatively simple feed elements.

5. *Active Aerial.* Amplifiers in the transmission lines will allow high transmit power without using a high power feed. This concept has evolved into using transmit-receive modules that contain high-power transmitting amplifiers, low-noise receiving amplifiers, and phase shifters, all in a single package to make an active phased array lens.

6. *Polarization Changing Aerial.* The front and back faces may have orthogonal polarizations. This technique is often used to reduce spillover. An extension of this concept is using the lens as a polarization transformer by making the aperture side elements circularly polarized, while the feed side elements and the feeds have the same linear polarization. This would obviate the need for circularly polarized feed elements.

The importance of Gent's work cannot be overemphasized. It represents a transition from the perception of constrained lenses as metal dielectrics to one of arrays plus transmission lines. It offered an alternative to metal plate lenses with their inherent weight and bandwidth disadvantages, making possible a variety of new antennas.

6.6 THE ROTMAN LENS

Rotman's lens is possibly the best known of all constrained lenses. First reported in 1962 [8], it was one of the first exploitations of Gent's theory, and undoubtedly the most successful. It has been widely used in

ECM antennas, which require wide bandwidth and wide angle scanning. This section will discuss its design and design trade-offs and will review some measured data to illustrate a few practical considerations for lens design in general as well as for Rotman lenses in particular.

6.6.1 Trifocal Design Equations

Although Ruze and Gent both alluded to the possibility of "three-point corrected" or trifocal lenses, Rotman was apparently the first to show a solution of the path length equality conditions to find lens parameters that would produce three perfect focal points. His solution results in expressions for the back face element locations and transmission line lengths for the two-dimensional geometry shown in Figure 6.26.

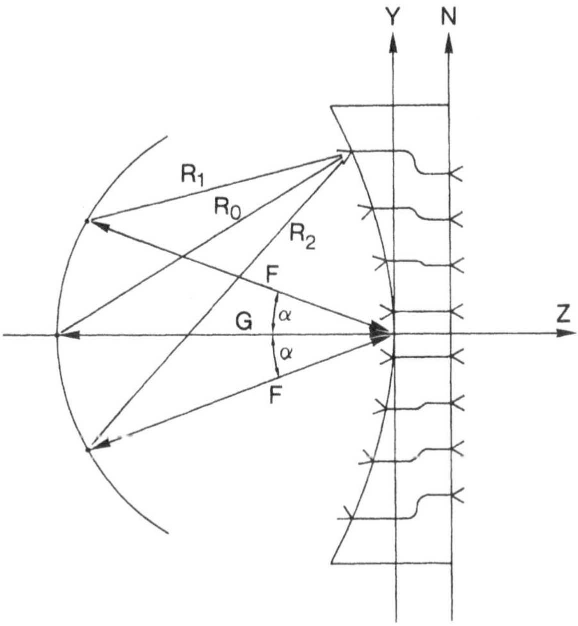

Figure 6.26 Rotman Lens Geometry

The three focal points include one located on the axis at a distance G from the back face and two at symmetric points $(y,z) = (\pm F \sin \alpha, -F \cos \alpha)$. The off-axis foci will produce plane wavefronts directed at angles $\pm\alpha$ from the axis. The path length equality conditions for the lower, center, and upper foci are, respectively

$$R_1 + W = F + W_0 + N \sin \alpha \tag{6.28a}$$

$$R_0 + W = G + W_0 \tag{6.28b}$$

$$R_2 + W = F + W_0 - N \sin \alpha \tag{6.28c}$$

where

$$R_1^2 = (F \cos \alpha - Z)^2 + (Y + F \sin \alpha)^2$$

$$R_2^2 = (F \cos \alpha - Z)^2 + (Y - F \sin \alpha)^2$$

$$R_0^2 = (G - Z)^2 + Y^2.$$

N is the y-coordinate of the aperture element, which is not the same as that of the feed side element (Y). This lateral displacement between front and back face elements is one of the three degrees of freedom. The other two are the back face shape, defined by Z, and the line lengths, W. W_0 is an arbitrary offset line length. The details of the derivation, given in Rotman and Turner, result in the following coefficients of the quadratic equation $a\omega^2 + b\omega + c = 0$ where $\omega = (W - W_0)/F$:

$$a = 1 - \eta^2 - \left(\frac{g - 1}{g - \cos \alpha}\right)^2 \tag{6.29a}$$

$$b = 2g \frac{g - 1}{g - \cos \alpha} - \eta^2 \sin^2\alpha \frac{g - 1}{(g - \cos \alpha)^2} + 2\eta^2 - 2g \tag{6.29b}$$

$$c = g \frac{\eta^2 \sin^2\alpha}{g - \cos \alpha} - \frac{\eta^4 \sin^4\alpha}{4(g - \cos \alpha)^2} - \eta^2 \tag{6.29c}$$

(*Note:* In reference [8], the second equation was missing the last term, $-2g$.) In the above, $g = G/F$ and $\eta = N/F$. Once ω is obtained from (6.29), Y and Z are found from

$$Y/F = \eta(1 - \omega) \tag{6.30}$$

$$z^2 + 2gz + (y^2 - \omega^2 + 2g\omega) = 0; \; y = Y/F, \; z = Z/F \tag{6.31}$$

The element locations, N, are those of a linear array, usually chosen to be one-half wavelength apart, or slightly greater. The solutions for W are the electrical lengths of cables joining the front and back face elements. And, finally, y and z give the back face element coordinates.

The resulting lens will have three perfect focal points, but we are also interested in how well the lens will focus in between and outside those points. Thus, the remaining task is to find a good focal arc. Rotman's approach was to assume a circular focal arc. Then, varying G will move that arc's center of curvature along the z-axis. Since G is contained in the equations for W, Y, and Z, this procedure changes the lens as it changes the focal arc. It is usually desirable to vary the two independently, but, nevertheless, Rotman and Turner found that when $G = 1.137F$, the path length error is less than $0.0005F$ over all angles between 0° and ±40° for $F/D = 0.83$ and $\alpha = 30°$.

Quite often, Rotman lenses are used as multiple-beam feeds. Several feeds on the focal arc will transmit or receive simultaneously, and their beams are added in a power combiner. They may be amplitude weighted to synthesize a shaped beam or a low sidelobe beam, but they must be added in phase. Unless $G = F$, however, the feeds are not equidistant from the lens center, and the beams they produce are not in phase. To compensate, fixed lengths of cable can be added to the feeds. The correct electrical lengths are $l_e = l_o - D$, where l_o is an arbitrary constant and D is a solution of

$$
\begin{aligned}
&D^2 + D(-2a \cos \beta) + (2aG - G^2) = 0 \\
&a = (F^2 - G^2)/2(F \cos \alpha - G)
\end{aligned}
\qquad (6.32)
$$

and β is the angle of the feed from the lens axis.

The program ROTMAN.BAS implements Equations (6.29) through (6.32) to solve for port coordinates and line lengths for any choices of F, G, and α. It also calculates path length error and beam patterns.

6.6.2 Parallel Plate Rotman Lens

A parallel plate Rotman lens is a pillbox structure in which the space feed region, between the feeds and back face elements, is a parallel plate waveguide. In contrast to the "metal plate lenses" discussed earlier, the parallel plates do not do the focusing, but serve only to confine the propagation between the feeds and the back face. Figure 6.27 depicts the original Rotman-Turner lens, which used parallel plate as the medium between

H-plane sectoral horn feeds and the back face array and which was made up of coaxial monopoles. Those are *N*-type connectors with the center conductors extending into the waveguide. The aperture array was also made up of coaxial probes in a short waveguide section, which transitioned to free space through an E-plane flare section. (The waveguide's TEM mode has its E-field perpendicular to the plates.) It is the back face probe array, the front face probe array, and the cables joining the two that collectively perform the focusing, so it is somewhat of a misnomer to refer to this as a "parallel-plate lens," but that terminology is widely accepted, nonetheless. The feeds and back face elements are usually referred to, respectively, as "beam ports" and "antenna ports" because they are analogous to the inputs and outputs of beamformers such as the Butler matrix and the Blass matrix.

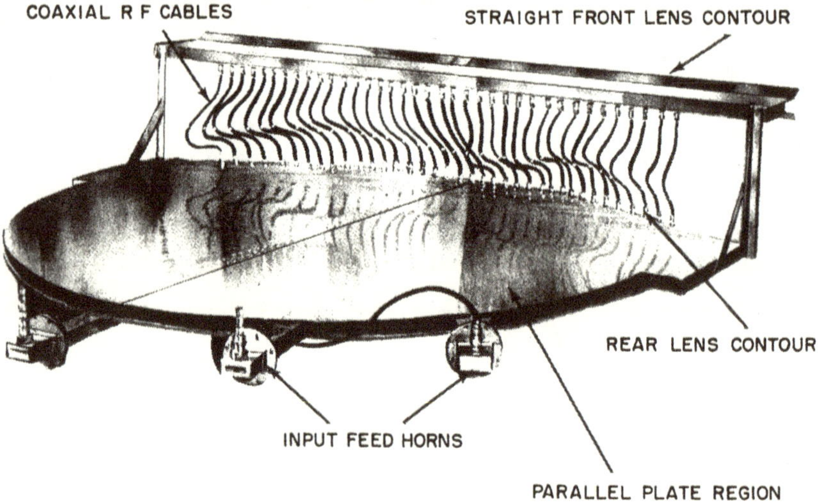

Figure 6.27 Experimental Rotman-Turner Lens [8]

6.6.3 Printed Circuit Rotman Lens

The parallel plate space-feed region can be replaced with printed circuit board, in either a microstrip or a stripline configuration [9]. The bottom layer of the microstrip will be solid copper, while the top layer will have a pattern similar to Figure 6.28. The center tabs of RF connectors will be soldered to the narrow microstrip lines at the board edge. The flares are the actual beam port (feeds) and antenna port (back face) elements. They provide a match between the high-impedance microstrip

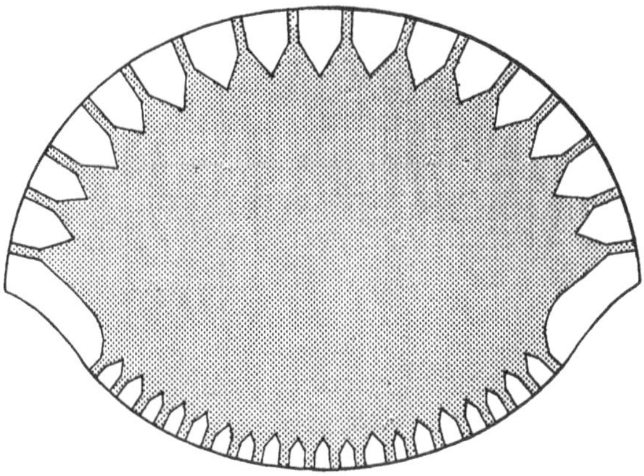

Figure 6.28 Printed Circuit Rotman Lens

lines (usually 50Ω) and the low-impedance, dielectric-loaded, parallel-plate region formed by the circuit board. A stripline lens would be similar, except the pattern of Figure 6.28 would be sandwiched between two copper-clad dielectric layers. Either type has an advantage over parallel plate lenses because the flared launchers are more broadband than coaxial probes. Furthermore, they are smaller, because their size is proportional to $\sqrt{\varepsilon_r}$, where ε_r is the relative dielectric constant of the circuit board material. The lens line lengths, W, are sometimes etched on the circuit board, but it is more common to use phase-trimmed coaxial cables because microstrip transmission line is dispersive. Archer discusses several variations of printed Rotman lenses that use a variety of radiator types for front face elements [9], and these are usually connected to the antenna ports with coaxial cables.

6.6.4 Practical Design Considerations

Throughout this chapter, we have compared lens designs in terms of path length error alone. Yet there are several hardware design considerations that, taken together, determine how closely the lens will approach that theoretical performance. Among them are:

1. array impedance matching,
2. element pattern,
3. element phase center, and
4. internal reflections.

This section will discuss how these factors influence the lens design and will show experimental measurements that illustrate their importance.

1. *Array Impedance Matching.* The feeds, the back face, and the front face elements are antenna arrays, and must be impedance matched to the array environment. It is not sufficient to use an element that is impedance matched (with low S_{00}) when isolated and line it up amongst many identical elements. The array's reflection coefficient is the sum of that isolated element reflection, S_{00}, and mutual coupling from all other array elements, $S_{0,n}$[10]:

$$\Gamma_a(\theta) = \sum_{n=-N/2}^{N/2-1} S_{0,n} e^{jnkd\sin\theta} \tag{6.33}$$

(k is the wavenumber and d is the interelement spacing). Thus, designing the array elements requires determining the coupling coefficients $S_{0,n}$, by theory or measurement. Measurement is usually an expensive cut and try process, best reserved for verifying theoretical predictions. Most theoretical models use moment methods to solve for the currents on the elements in the presence of the array. The input impedance and element pattern can be found from those currents.

An example of such a model is Tomasic and Hessel's solution for coaxial monopole arrays [11]. Figure 6.29 shows their prediction with experimental measurements of $|\Gamma_a(0)|$ for an array of coaxial probe elements. Also shown is $|S_{00}|$, to illustrate the pronounced difference between the two. If the desired center frequency was 9 GHz, the result of using this element design in an array would be substantial surface reflection. An amount of power from the feed equal to $|\Gamma_a|^2$ would reflect from the back face. Assuming the front face array was similar, roughly the same amount would reflect off the front face, back into the transmission lines, and the resulting loss would be approximately 20 $\log(1 - |\Gamma_a|^2)^2$. Note also that Γ_a is a function of angle, and in a Rotman lens it must be made small for all angles of θ over which the beam ports are distributed.

2. *Element Pattern.* The beam ports should illuminate the antenna ports with nearly equal amplitude, and *vice versa*. But mutual coupling effects may cause the arrayed element's pattern to become narrowed. Interelement spacings of $d > 0.5\lambda$ will typically cause an element pattern roll-off inside the angle

$$\phi = \sin^{-1}(\lambda/d - 1), \tag{6.34}$$

which is apparent in Figure 6.30, the calculated patterns of a coaxial probe array with $d = 0.6\lambda$ and $d = 0.7\lambda$, from [11].

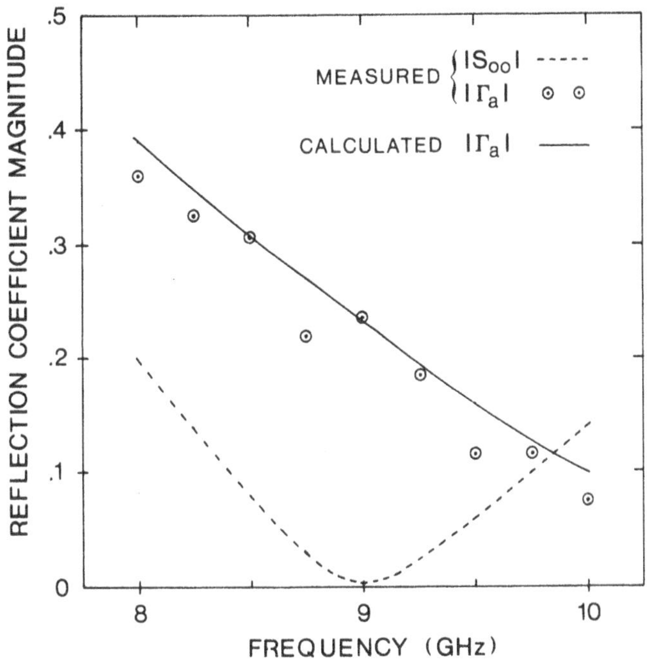

Figure 6.29 Active Reflection Coefficient of an Array of Coaxial Monopoles in a Parallel Plate Waveguide

Rotman lens beam ports are usually more than 0.6λ apart when the beams are intended to overlap at −3-dB points (orthogonal beams). Unfortunately this will cause them to illuminate only part of the back face, resulting in a broader beam and lower gain. A common remedy is to "pack" the beam port array with "dummy" (resistively terminated) elements to reduce the effective beam port spacing. Figure 6.31 (Southall and McGrath [12]) shows the relative amplitude received at the antenna ports of a small Rotman lens when one beam port is transmitting, before and after addition of dummy elements to reduce the beam port spacing from 0.7λ to 0.35λ. The densely spaced beam ports have broader element patterns, and any one of them will illuminate the lens back face with nearly constant amplitude.

3. *Element Phase Center.* Another problem with wide element spacings is that they cause irregularities in the element's phase pattern, which, in turn, cause a shift in the phase center. "Phase center" is defined as the center of a circle of constant phase. It is vital that the true phase center coincide with the desired beam port or antenna port location, or else a

quadratic path length error will be introduced. Figure 6.30(c) shows that the phase pattern has a notch at the angle given by (6.34), and that causes the element's apparent location to differ from its physical location. Situations like Figure 6.30(c) pose a dilemma because the phase variation is not quadratic, which is a condition for having a true phase center. It may nevertheless be possible to find an average location over the angular region of interest. In the case of a beam port, for instance, that region is the arc subtended by the antenna ports. Packing the beam ports (and antenna ports) closer than 0.5λ is also desirable because it ensures a smooth element phase pattern, without the "notches" apparent in Figure 6.30(c).

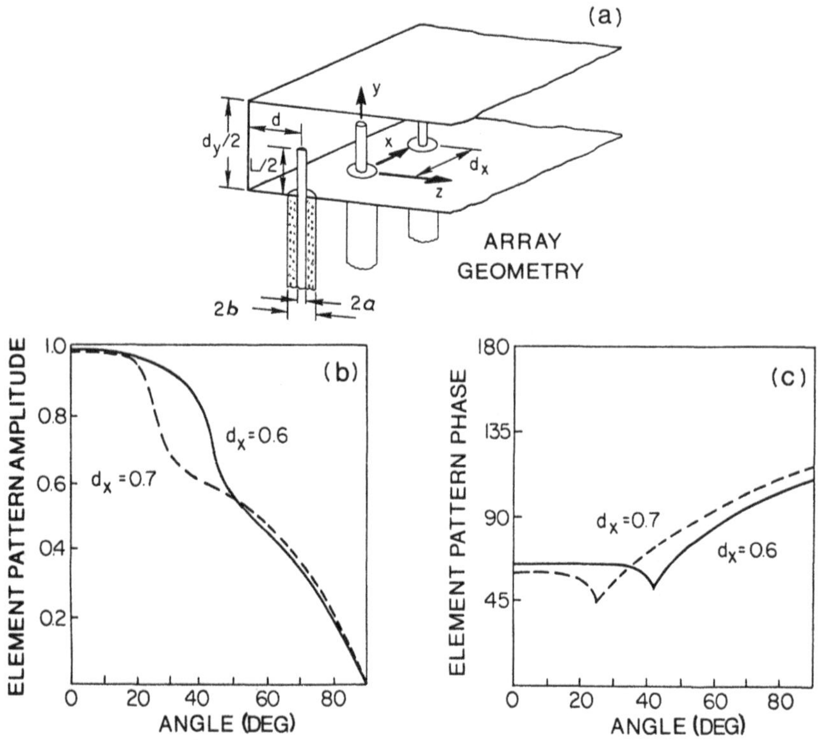

Figure 6.30 Calculated Element Patterns for Coaxial Monopole Arrays: (a) Array Geometry; (b) Amplitude; (c) Phase [11]

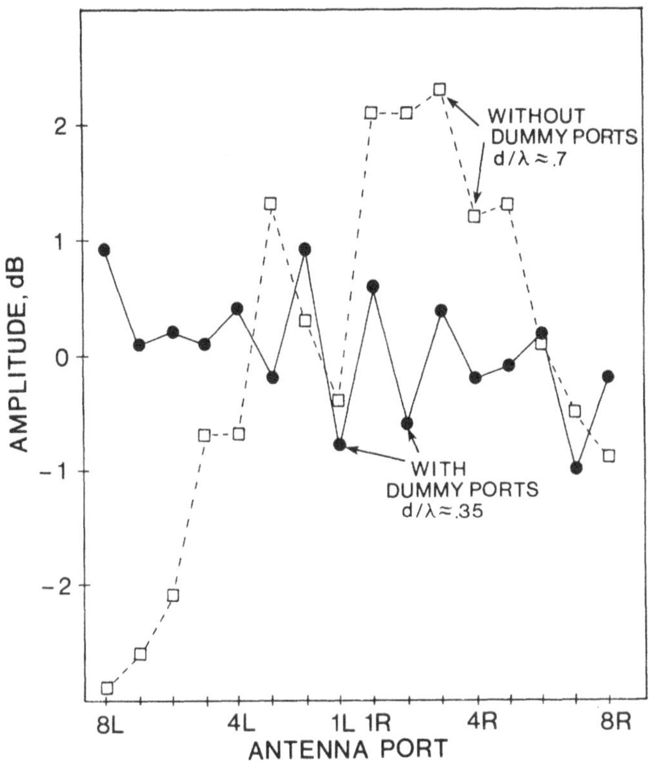

Figure 6.31 Effect of Dummy Beam Ports on the Aperture Amplitude Distribution in a Rotman Lens

4. *Internal Reflections.* Wide angle lenses must have feeds distributed over a wide focal arc. Especially in Rotman lenses, that arc may be nearly 180° in extent relative to its center of curvature. As illustrated in Figure 6.32(a), an edge beam port radiates less toward the antenna ports than toward the other beam ports. Some of that energy will reflect off the beam port arc toward the antenna ports, resulting in a sidelobe like the one on the right side of Figure 6.32(b). Wide angle impedance matching the beam ports will help to reduce that effect, since they will then reflect less. But impedance matching does nothing for the fact that in Figure 6.32(a) (and similar lenses) the antenna ports are on the edge of the beam port's radiation pattern, and they will be illuminated with unequal amplitude. It is often possible to choose lens surfaces that avoid the problem altogether.

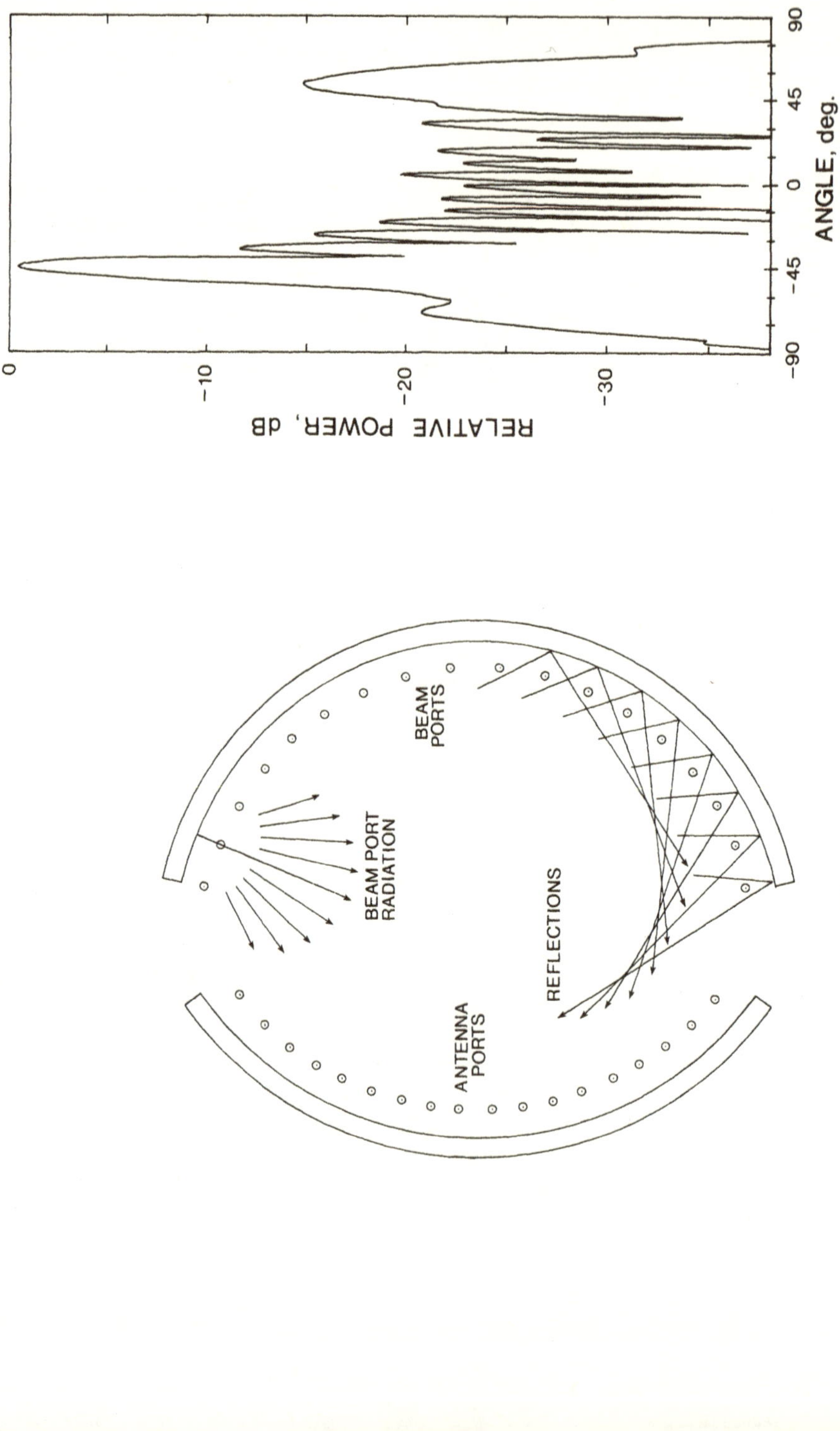

Figure 6.32 Internal Reflections in a Rotman Lens: (a) Ray Paths; (b) Resulting Radiation Pattern

The lens shown in Figure 6.32(a) was designed for optimum *theoretical* performance, but it could have been designed better in a *practical* sense. Figure 6.33 shows the port locations for Rotman lenses with different G/F ratios. Only the one with $G = 1.137F$ is optimum in terms of minimizing path length errors [8]. But the one with $G = 1.05F$ is a more forgiving geometry, allowing less internal reflection. All four designs shown use $F/D = 0.85$ and $\alpha = 30°$. The maximum path length error for a beam port at 45° is $0.00059F$ for the $G = 1.05F$ case. With an eight-wavelength wide aperture, that equates to less than 1.7° phase error, which is probably insignificant next to the error caused by internal reflections in the $G = 1.137F$ case.

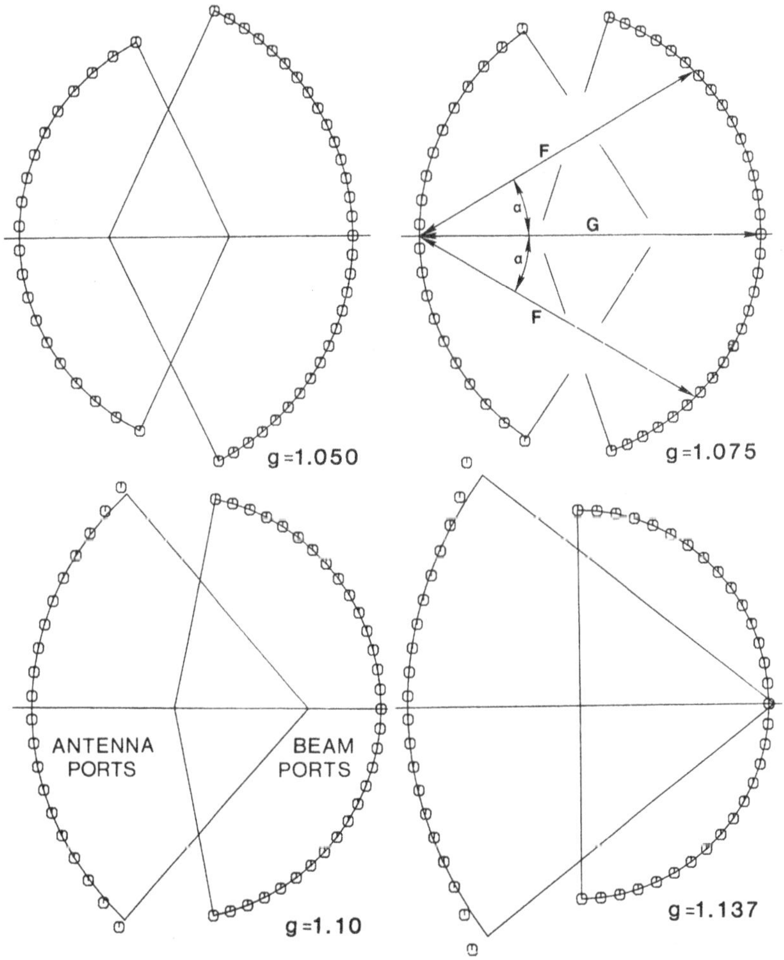

Figure 6.33 Rotman Lens Port Locations for Various Values of *G/F*

This serves to illustrate that any lens design is only as good as its implementation. Extremely low path length error is a good quality for rating a theoretical design, but it is not useful unless it can be realized in practice. The practical considerations reviewed above, array impedance matching, element pattern, element phase center, and internal reflections are important, but often overlooked.

6.7 CONSTRAINED LENSES FOR CIRCULAR APERTURES

There are three types of constrained lens popularly used to feed convex circular apertures. The R2R and Luneburg designs have the interesting property that every point on the feed arc is a perfect focus, and the same is very nearly true for the third, the RKR lens. This section will show the design procedures for those lenses and discuss some of their advantages and weaknesses.

6.7.1 The R2R Lens

Figure 6.34 shows a convex circular array that covers 180°. Its elements are connected with equal line lengths to ports lining the edge of a cylindrical cavity. If the outer surface is to produce a plane wave directed at an angle α, its elements must have a phase difference (expressed in path length d) of

$$d = A[1 - \cos(\delta - \alpha)] \tag{6.35}$$

where δ is the angle to the outer surface element and A is the outer radius. Let γ be the angle of the inner surface element. A feed at the angle (β + 180°) will illuminate the opposing interior elements with a path length difference equal to

$$R = 2B \cos\left(\frac{\gamma - \beta}{2}\right) \tag{6.36}$$

where B is the radius of the interior surface. If all transmission lines are the same length, then the path length equality condition is $R + d = 2B$, or

$$2B[1 - \cos(\gamma/2 - \beta/2) = A[1 - \cos(\delta - \alpha)] \tag{6.37}$$

It is easy to see that in the special case where $A = 2B$, $\gamma = 2\delta$, and $\beta = 2\alpha$, the above condition is always satisfied. The inner radius is one-half the

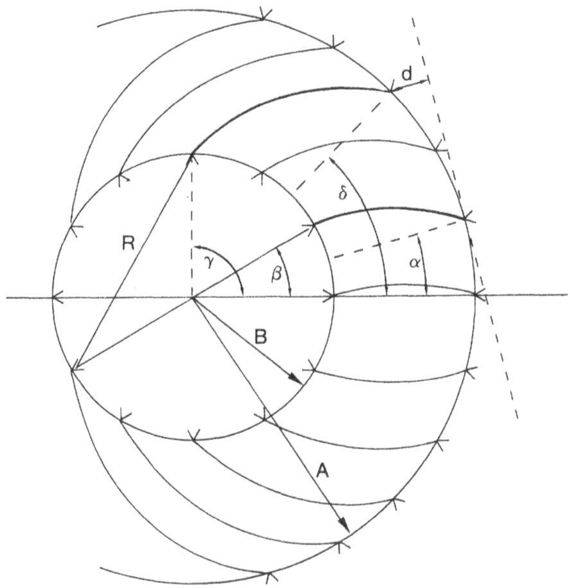

Figure 6.34 **Convex Circular Aperture Array Fed from a Concave Circular Cavity (R2R Lens)**

outer radius, and the elements on the inner surface are spaced the same distance apart (but twice the angular spacing) as those on the outer surface. This design is known as the R2R lens because of the 2 : 1 ratio of the radii of its two surfaces [5].

The fact that (6.37) is always satisfied implies that every feed on the inner surface is a perfect focus. Hence, the R2R lens is capable of aberration-free scanning. It is limited to angular regions of less than 180°, however, because the 360° of the inside surface maps into only 180° of the outer surface. The abrupt truncation of the aperture at ±90° does not allow good beams to be formed much past ±60°. Clapp has devised a means to use four or six R2R lenses to feed a full 360° aperture, but the hardware design is further complicated by several hybrids for each beam [13].

6.7.2 The RKR Lens

The RKR lens avoids the inherent scanning limitation of the R2R lens by making a direct mapping from the inner to the outer surface as shown in Figure 6.35. A circulator is placed in each transmission line so that the interior elements may be used simultaneously as feeds (beam ports) and pickup elements (antenna ports). Since interior and exterior elements are at the same angles relative to each other, (6.37) becomes

$$2B[1 - \cos(\delta/2 - \alpha/2)] = A[1 - \cos(\delta - \alpha)] \tag{6.38}$$

There is no choice of B/A that will make this true for all angles $(\delta - \alpha)$. $B = 1.92A$, however, comes closest ([5], p. 16–22). This is evident from Figure 6.36, which shows the path length error

$$\Delta l/A = 2K[1 - \cos(\delta/2 - \alpha/2)] - [1 - \cos(\delta - \alpha)] \tag{6.39}$$

for five choices of K ($B = KA$). Figure 6.35 is somewhat misleading because with $B = 1.92A$ the feed circle would be nearly twice the aperture (cylindrical array) radius. RKR feeds are commonly filled with dielectric to reduce the lens size enough to fit inside the aperture array.

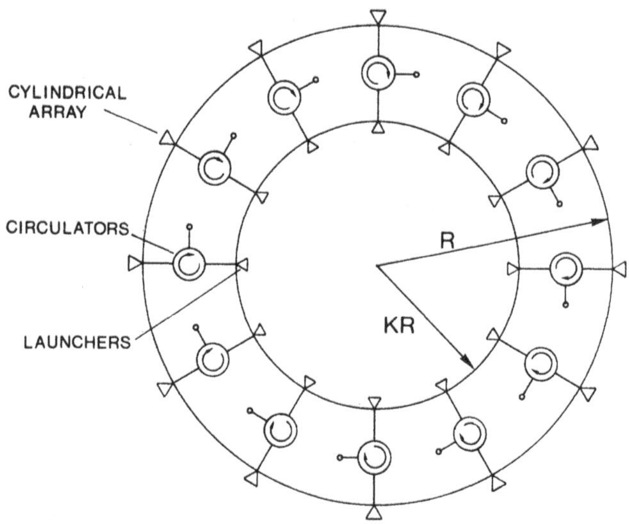

Figure 6.35 Schematic of an RKR Lens

6.7.3 The Luneburg Lens

The Luneburg lens is a third device in which the beam ports double as antenna ports. Like the R2R lens, it has the property that any point on the feed circle is perfectly focused. It is unlike any other lens discussed in

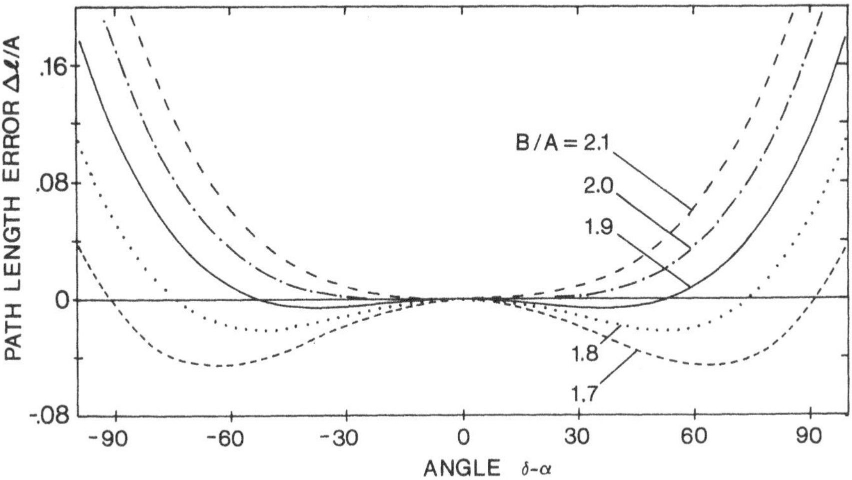

Figure 6.36 Path Length Error *versus* Aperture Angle for an RKR Lens with Various Values of $K = B/A$

this chapter because focusing is achieved by a varying index of refraction material in the space feed region.

Figure 6.37 depicts a dielectric Luneberg lens. The spherical region has an index of refraction of [14]:

$$\eta = \sqrt{(2 - r^2/a^2)} \tag{6.40}$$

A two-dimensional bootlace lens based on this device would be similar to the RKR lens shown in Figure 6.35, except that the inner and outer surfaces would have the same radius.

In practice it is very difficult to construct a dielectric with a varying index of refraction. Even approximating a Luneburg lens by using layers of successively lower η to approximate (6.40) is impractical. An alternative 2D lens that uses a uniform dielectric was demonstrated by Peeler and Archer [15]. As illustrated in Figure 6.38, the cylindrical parallel plate region is deformed so that its height is a function of radius:

$$h/\lambda = [2\sqrt{(\varepsilon_r - 2 - r^2/a^2)}]^{-1} \tag{6.41}$$

where ε_r is the dielectric constant of the filler material. The lens was used in the TE_{10} mode (E-field parallel to the bottom plate).

As with an RKR lens, the 2D Luneburg lens requires circulators between the interior ports and the exterior ports if it is to cover more than about ±60°. Archer proposes an alternative, shown in Figure 6.39(b) that uses two lenses (either RKR or Luneburg) with 90° hybrids [9]. The advantage to that approach is that hybrids are more compact and broadband than circulators.

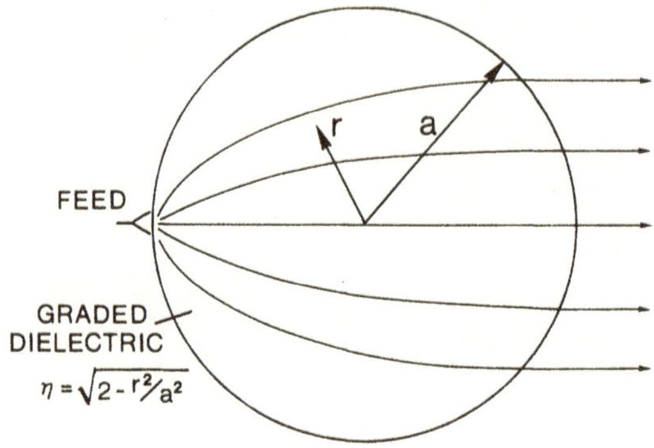

Figure 6.37 Ray Paths in a Dielectric Luneburg Lens

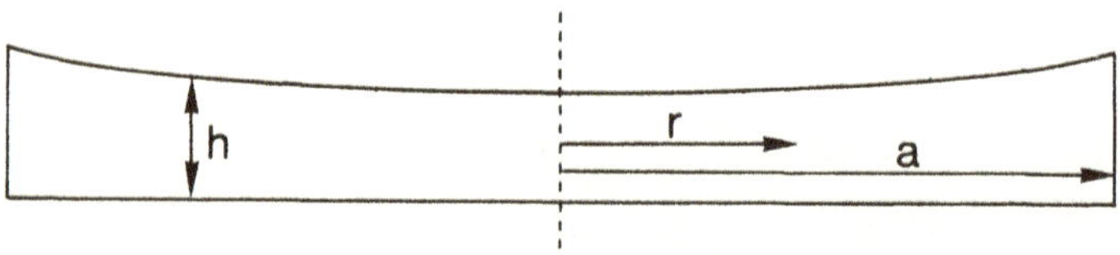

Figure 6.38 Cross Section of a Parallel Plate Luneburg Lens (circular cavity)

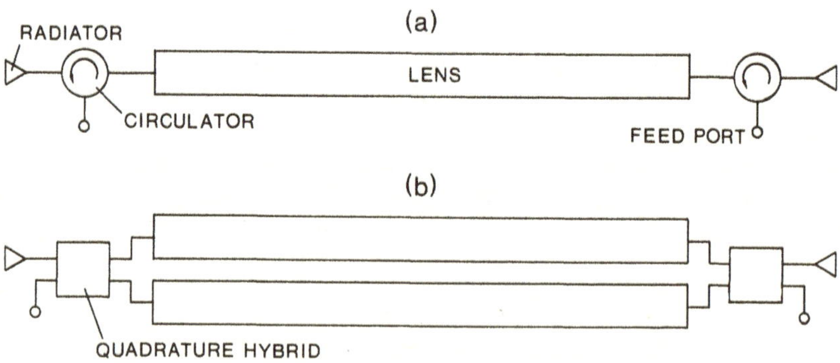

Figure 6.39 Alternative Designs for RKR and Luneburg Lenses with 360° Coverage: (a) Single Lens with Circulators; (b) Dual Lens with Quadrature Hybrids

6.7.4 Effect of Internal Reflections

Section 6.6 discussed a difficulty with internal reflections in Rotman lenses, the beam port surfaces of which had large angular extents. Since the feed surfaces of R2R, RKR, and Luneburg lenses extend over 360°, they suffer from the same problem to an even greater extent. Figure 6.40 shows the paths of incident rays from a transmitting beam port and reflections from the receiving antenna ports. Assuming that each port radiates and receives with an ideal cosine element pattern (with the angle measured from the surface normal at the port), the reflected power may be approximated by

$$P_2 = \cos^2\theta_1 |\Gamma_a(\theta_2)|^2 \tag{6.42}$$

where Γ_a is the active array reflection coefficient. Neglecting higher order reflections, the excitation of any antenna port will be the sum of the direct ray from the beam port plus two reflections from points on opposite sides of the cavity. Figure 6.41 shows simulated radiation patterns for three sizes of RKR lenses. In those analyses, the array reflection coefficient was assumed to be $\Gamma_a(\theta) = 1 - \cos(\theta)$. The build-up of sidelobe power in the $\pm 90°$ directions is the result of those internal reflections. Unlike the Rotman lens, we do not have the option of altering the lens geometry to reduce reflections.

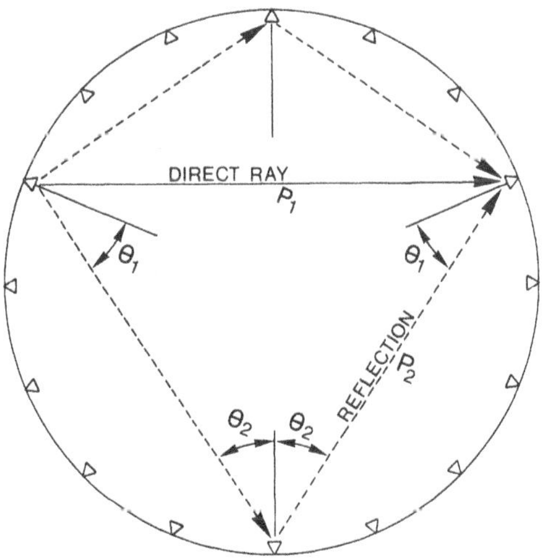

Figure 6.40 Internal Reflections in an RKR Lens

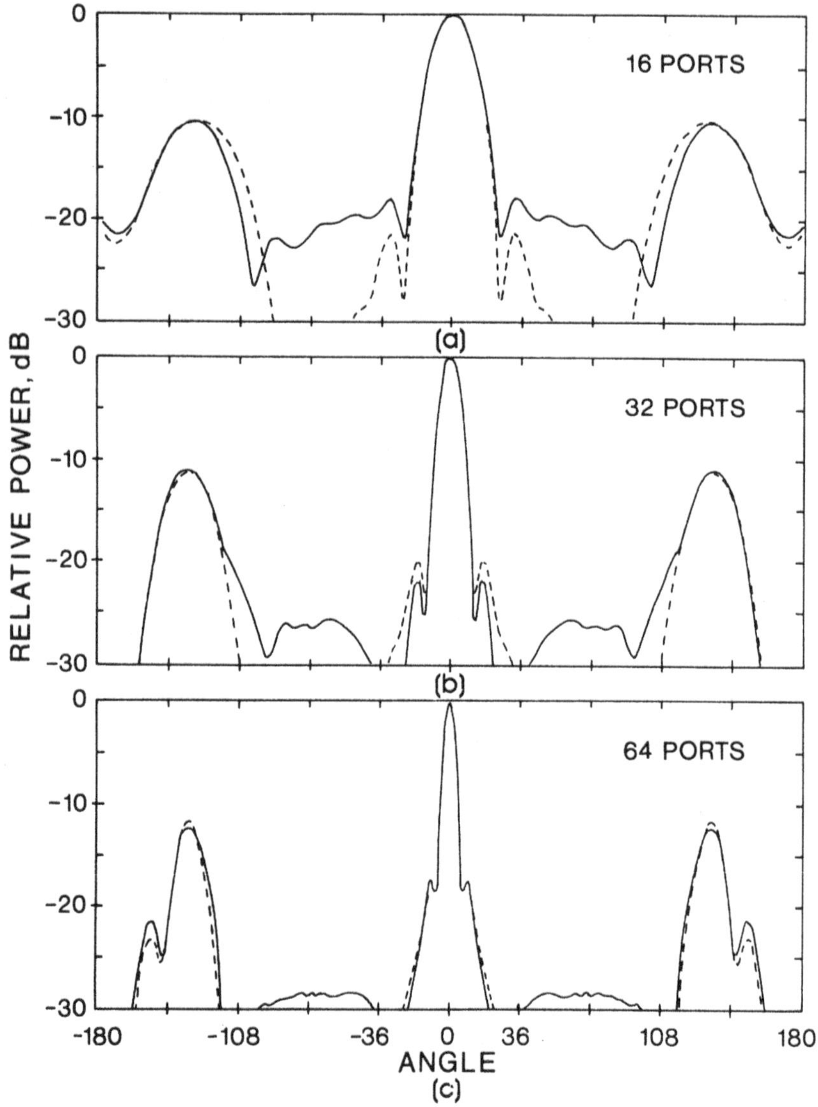

Figure 6.41 Simulated Radiation Patterns of RKR Lenses of Various Sizes: (a) 16 ports; (b) 32 ports; and (c) 64 ports

The three devices that were discussed in this section, the R2R, RKR, and Luneburg lenses, are unique among constrained lenses because their focusing is a result of their geometry rather than the application of degrees of freedom to enforce path length equality. They are a natural choice for feeding cylindrical arrays, as the Rotman lens is the typical choice for linear arrays. The R2R and RKR lenses can be implemented in printed circuit form, similar to the printed Rotman lens [9].

Program RKRLENS.FOR can be used to simulate cylindrical array antennas fed by RKR and R2R lenses. The launchers' reflection coefficients are assumed to be $1 - \cos^x(\theta)$, with x being a user-defined parameter. This program was used to calculate the patterns shown in Figure 6.41.

6.8 THREE-DIMENSIONAL BOOTLACE LENSES

This final theory section discusses the most challenging application of constrained lenses as two-axis scanning antennas. The goal is to use an array of feeds (beam ports) distributed over a concave, bowl-shaped surface to form beams in all directions over a large, conical region. Switching between feeds will accomplish scanning in both the azimuth and elevation planes. The general structure of such an antenna is illustrated by Figure 6.42. It is a three-dimensional structure, in contrast to Rotman, R2R, and RKR lenses, which are two-dimensional and restricted to scanning in one plane only.

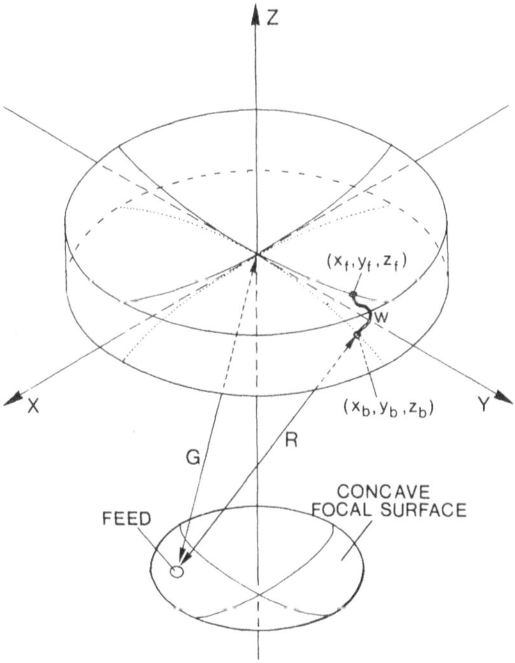

Figure 6.42 Three-Dimensional Lens Structure

The general design procedure for 3D bootlace lenses is no different than for 2D lenses. Path length equality is enforced for the selected foci to determine the element locations and line lengths, and an optimum focal surface is found (numerically or analytically) to establish the beam port positions. One important difference, which was briefly discussed earlier, is that the 3D design may be symmetric in ϕ, or it may have up to four perfect foci. The difference in these two approaches will be explored further in this section. Some practical designs will also be described.

6.8.1 Multifocal *versus* Symmetric

A simple example illustrating the difference between multifocal and symmetric lenses was presented in Section 6.4.2. It was shown that two degree of freedom waveguide lenses could be designed by revolving a 2D design (resulting in no perfect focal points), or by resolving the path length equality conditions in three dimensions (two perfect foci). Those two designs could cover roughly the same number of beamwidths for a given rms error level. It is not clear whether the same would be true for three or four degree of freedom lenses. For example, a quadrufocal lens with one focus in each quadrant might have a larger scan region (in terms of total beamwidths) than a four degree of freedom ϕ-symmetric lens. Unfortunately, design equations for the former are not available. The only quadrufocal designs derived thus far are by Rao [16], who chose to place them on a straight line, and Rappaport and Zaghloul [17], who oriented them on a circle. In both cases, the four foci are all in the same ϕ plane, and scanning in the orthogonal plane suffers as a result. Cornbleet [18] gives equations for the alternative, ϕ-symmetric case (see Figure 6.43):

$$\frac{R_4 - R_1}{\sin \alpha} = \frac{R_3 - R_2}{\sin \beta} \tag{6.43a}$$

$$z_f = \frac{w_0 + (R_2 + R_3 - 2G) - (R_1 + R_4 - 2F)}{2(\cos \beta - \cos \alpha)} \tag{6.43b}$$

$$\rho_f = \frac{R_4 - R_1}{2 \sin \alpha} \tag{6.43c}$$

$$w = \frac{w_0 + \cos \alpha(R_2 + R_3 - 2G) - \cos \beta(R_1 + R_4 - 2F)}{2(\cos \beta - \cos \alpha)} \tag{6.43d}$$

The first of the above contains no unknowns other than the back face element coordinates ρ_b and z_b, so it defines the inner contour. The second

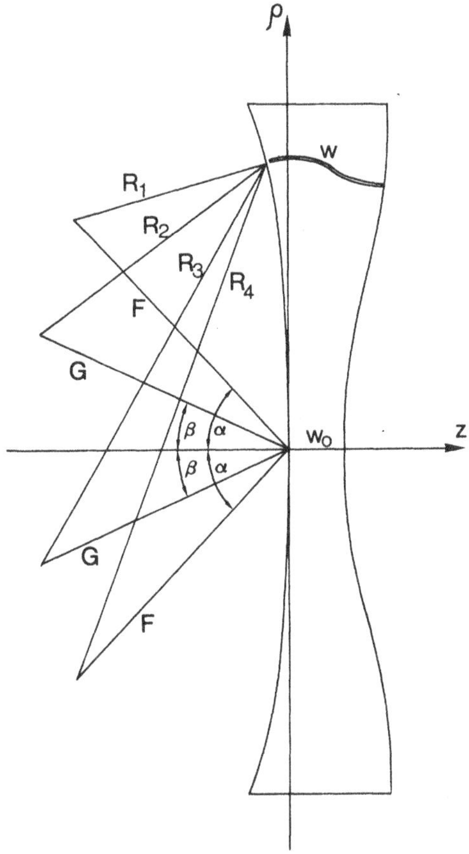

Figure 6.43 Quadrufocal Lens Geometry in Two Dimensions

and third equations give the outer contour and element positions, and the fourth gives the transmission line lengths. A simplified design results from choosing $G = F$, i.e., a circular focal arc. The outer contour in that case will be nearly planar.

Rappaport and Zaghloul also derived a case with a circular focal arc and a planar front face, except it has four perfect foci. The design equations are

$$z_b = -x_f^2 \frac{\cos \beta + \cos \alpha}{2F}; \quad z_f = 0 \tag{6.44a}$$

$$x_b^2 = x_f^2 \left\{ 1 + \frac{1}{F^2 - x_f^2} \left[y_f^2 - x_f^2 \cos \alpha \cos \beta + x_f^4 \left(\frac{\cos \beta + \cos \alpha}{2F} \right)^2 \right] \right\} \tag{6.44b}$$

$$w = F(1 - x_b/x_f) \tag{6.44c}$$

The added condition that $y_b = y_f$ is imposed since, according to the authors, "scan angles α and β are in the x-z plane," and so "there is no need to distort the mapping in the y direction."

In either case above, the path length equality condition for an arbitrary feed point at a radial distance H from the origin and angles θ, ϕ is $H + s = R + w$ and the path length error is $\Delta l = H + s - R - w$, where

$$s = x_f \sin \theta \cos \phi + y_f \sin \theta \sin \phi + z_f \cos \theta \tag{6.45}$$

and R is the distance from the feed to a back face element.

Figure 6.44 shows two curves of rms path length error versus scan angle (*dashed*) for symmetric lenses and a set of curves (*solid*) for a quadrufocal. For both of these, $\beta = 10°$ and $\alpha = 20°$. As expected, the quadrufocal performs better in the $\phi = 0$, or x-z plane than in the orthogonal plane.

The data in Figure 6.44 were generated by program QFOCAL.FOR, which simulates both types of four degree of freedom lens discussed above. In addition to calculating radiation patterns, it can either find the path length error for an arbitrary feed point location, or find the optimum focal length for a given scan angle.

In spite of their very wide angle scanning characteristics, four degree of freedom lenses are extraordinarily difficult to build. At least one face must be curved, with radiators placed in precise locations with transmission lines connecting the two surfaces. This difficulty is the probable reason that development of 3D bootlace lenses (and their application as multibeam antennas) has lagged behind that of phased arrays and reflectors. The remainder of this section discusses two 3D bootlace lens designs that solve some of the fabrication problems. One uses a stack of two-dimensional trifocal lenses to focus a cylindrical aperture onto three line sources, while the other uses two interconnected planar printed circuit arrays to focus in the direction normal to the circuit.

6.8.2 Cylindrical Lens Array

Newman, Hannan, and Ruskowski [19, 20] constructed a three-dimensional lens from a stack of two-dimensional lenses, as illustrated in

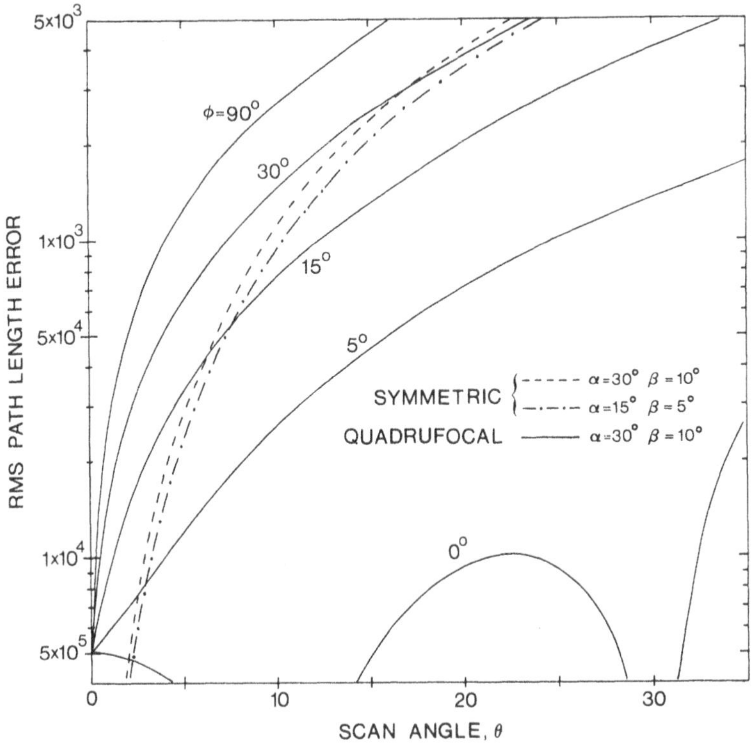

Figure 6.44 RMS Path Length Error for Four Degree of Freedom Lenses

Figure 6.45. Each shelf-like layer is a printed circuit board similar to the one shown in Figure 6.46. The dipole elements are at the very edge of the board. Metal strips oriented vertically and parallel to the board edges act as a ground plane behind the dipoles and as spacers between the circuit boards. The meandering sections of the microstrip transmission line join the front and back face dipoles.

The theoretical design was based on a four degree of freedom concept reported by Franchi, Kernweis, and Rotman [21]. The front and back face curvatures and element displacement in the x direction (horizontal, E-plane) are used to obtain perfect focus at three vertical line-source feeds. The fourth degree of freedom, line lengths in the lens, could be used for elevation focusing. In this experimental model, the line lengths were all the same so that the focusing would be independent of elevation.

The quadrufocal derived by Rappaport and Zaghloul could be implemented in this type of printed circuit construction, since it has no element displacement in the y direction. A significant problem with this form of lens is that the transmission lines tend to be quite long. They must be long enough to join the edge elements, which leaves a large length near the center that must be "meandered" to take up the slack. The electrical lengths of the several mitered bends required is a potential source of path length error.

Figure 6.45 Cylindrical Lens Array [19, 20]

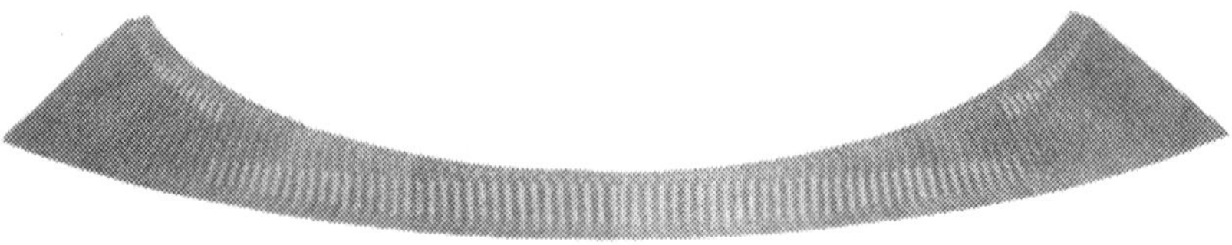

Figure 6.46 One Printed Circuit Layer of the Cylindrical Lens Array

6.8.3 Microstrip Constrained Lens

The microstrip constrained lens [22] is the most lightweight and structurally the simplest 3D constrained lens. Its entire structure is comprised of two printed circuit boards. Figure 6.47 shows one version in a test fixture that includes a pyramidal horn feed. The front and back faces are arrays of microstrip patches that have a common ground plane separating their dielectric substrates. In the two versions reported, the pairs of elements were connected by a pin coupler [Figure 6.48(a)] or a slot coupler [Figure 6.48(b)].

This lens is based on a two degree of freedom design [23]. The experimental versions were ϕ-symmetric, although two perfect foci are possible ([22] shows that their scan regions are comparable in size). Since both faces must be flat, the only features available are line lengths and element displacement, but moderately wide angle scanning is possible nonetheless. Scanning to ± 6 beamwidths off-axis has been demonstrated, with ± 18 beamwidths theoretically possible.

Figure 6.47 Microstrip Constrained Lens [22]

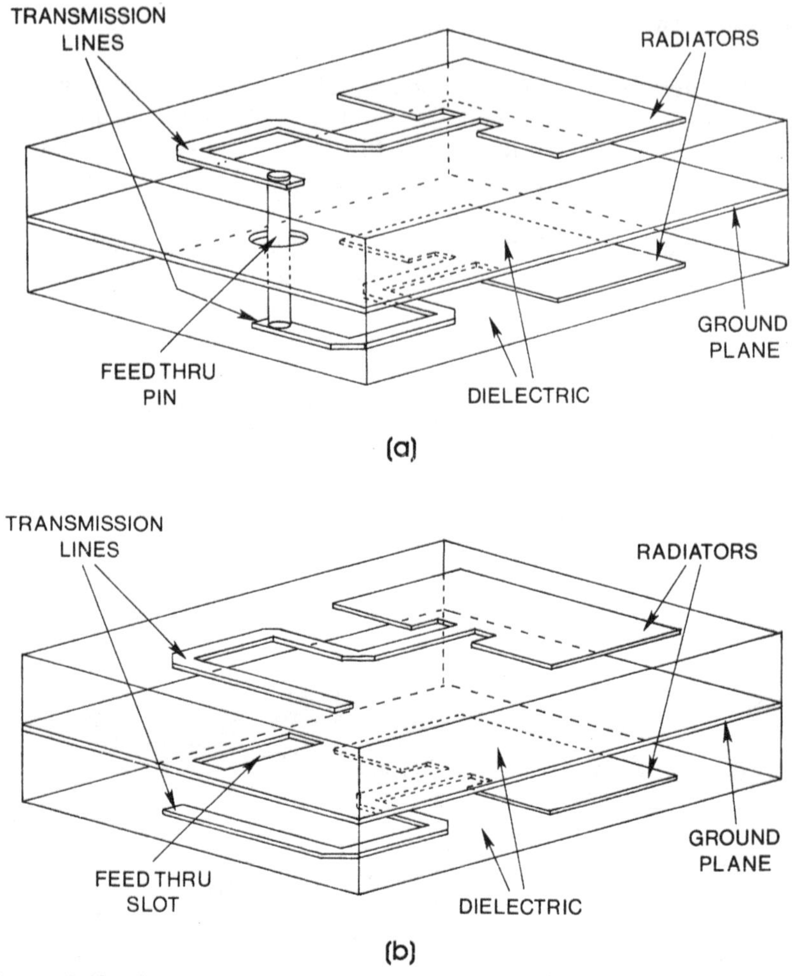

Figure 6.48 Element Connections in the Microstrip Constrained Lens: (a) Pin Coupler; (b) Slot Coupler

6.9 SUMMARY

This chapter has described a large variety of constrained lenses, from the earliest metal plate structures to more recent three-dimensional antennas. They all have in common the general structure of two interconnected antenna arrays. Their theoretical scanning properties are governed by how effectively they use a number of possible degrees of freedom.

Their actual performance depends on controlling electromagnetic effects such as array impedance, element amplitude pattern and phase center, and internal reflections, because those factors determine how closely the lens approaches its idealization as a ray optic device.

The technique of designing a constrained lens geometry follows a predictable, step-by-step approach: first, path length equality conditions are solved simultaneously for equations relating the lens features; second, an optimum feed surface is found (usually numerically); third, and finally, the lens' scanning ability is characterized by evaluating its path length error for a range of scan angles. The geometry thus derived is generic and suited to a wide range of applications, but the particular application determines what kinds of feed radiators, lens radiators, and transmission lines are appropriate.

References

[1] Brown, John, *Microwave Lenses,* Methuen's Monographs on Physical Subjects, London: Methuen, 1953.

[2] Kock, W.E., "Metal Lens Antennas," Proc. IRE, Vol. 34, Nov. 1946, pp. 828–836.

[3] Cohn, Seymour B., "Lens Type Radiators," Chapter 14, *The Antenna Engineering Handbook,* 1st Ed., New York: McGraw-Hill, 1961.

[4] Ruze, J., "Wide Angle Metal Plate Optics," Proc. IRE, Vol. 38, Jan. 1950, pp. 53–59.

[5] Peeler, George D.M., "Lens Antennas," Chapter 16, *The Antenna Engineering Handbook,* 2nd Ed., New York: McGraw-Hill, 1984.

[6] Gent, H., "The Bootlace Aerial," Royal Radar Establishment Journal, Oct. 1957, pp. 47–57.

[7] Jones, S.S.D., H. Gent, and A.A.L. Browne, "Improvements in or Relating to Electromagnetic-Wave Lens and Mirror Systems," British Provisional Patent Specification No. 25926/56, Aug. 1956.

[8] Rotman, W., and R.F. Turner, "Wide Angle Microwave Lens for Line Source Applications," IEEE Trans. Antennas and Propagation, Vol. AP-11, Nov. 1963, pp. 623–632. *See also* AFCRL-TR-62-18, same title and authors, Jan. 1962.

[9] Archer, D.H., "Lens-Fed Multiple Beam Arrays," Microwave Journal, Sep. 1984, pp. 171–195.

[10] Amitay, N., V. Galindo, and C.P. Wu, *Theory and Analysis of Phased Array Antennas,* New York: John Wiley and Sons, 1972, p. 22.

[11] Tomasic, Boris, and Alexander, Hessel, "Arrays of Coaxially-Fed Monopole Elements in a Parallel Plate Waveguide," Proc. Phased Arrays Symp., RADC-TR-85-171, Aug. 1985, pp. 223–250.

[12] Southall, Hugh L., and Daniel T. McGrath, "An Experimental Completely Overlapped Subarray Antenna," IEEE Trans. Antennas and Propagation, Vol. AP-34, Apr. 1987, pp. 465–474.

[13] Clapp, Roger, E., "Extending the R2R Lens to 360°," IEEE Trans. Antennas and Propagation, Vol. AP-32, July 1984, pp. 661–671.

[14] Luneburg, R.K., *Mathematical Theory of Optics,* Berkeley: Univ. of California Press, 1964.

[15] Peeler, G.D.M., and D.H. Archer, "A Two-Dimensional Microwave Luneburg Lens," IRE Trans. Antennas and Propagation, Vol. AP-1, July 1953, pp. 12–23.

[16] Rao, J.B.L., "Multifocal Three-Dimensional Bootlace Lenses," IEEE Trans. Antennas and Propagation, Vol. AP-30, Nov. 1982, pp. 1050–1056.
[17] Rappaport, Carey M., and Amir I. Zaghloul, "Optimized Three-Dimensional Lenses for Wide-Angle Scanning," IEEE Trans. Antennas and Propagation, Vol. AP-33, Nov. 1985, pp. 1227–1236.
[18] Cornbleet S., *Microwave Optics,* New York: Academic Press, 1976, pp. 73–77.
[19] Hannan, Peter W., and Edward M. Newman, "Study and Design of a Cylindrical Lens Array Antenna for Wideband Electronic Scanning," RADC-TR-83-128, Dec. 1983.
[20] Newman, E.M., and C.M. Ruskowski, "Cylindrical Lens Array for Wideband Electronic Scanning," Proc. Phased Arrays Symp., Aug. 1985, pp. 207–222.
[21] Franchi, Peter, Nicholas Kernweis, and Walter Rotman, "Wideband Microwave Lens Antenna for Tactical Radar Applications," Proc. Antenna Applications Symp., Sep. 1980.
[22] McGrath, Daniel T., "Slot-Coupled Microstrip Constrained Lens," Proc. Antenna Applications Symp., Sep. 1987.
[23] McGrath, Daniel T., "Planar Three-Dimensional Constrained Lenses," IEEE Trans. Antennas and Propagation, Vol. AP-34, Jan. 1986, pp. 46–50.

Chapter 7
MULTIBEAM AND SCANNING REFLECTOR ANTENNAS
C. J. Sletten

7.1 INTRODUCTION

There are three reflectors commonly used for wide-angle beam-forming antennas: the paraboloid, the sphere, and the parabolic torus, which is a shape generated by rotating a parabolic curve. The paraboloid has limited angular coverage with feed displacement from the focus, and therefore is often overlooked for wide-angle scanning. The offset paraboloidal section is capable, however, of producing good patterns many beamwidths from its axis. The offset subreflector-fed paraboloid has a broad focal surface for beam scanning.

Both the concave spherical mirror and the parabolic torus antenna can be scanned with little pattern degradation over very wide angular sectors. All patterns on these reflectors, however, show some aberration effects, and aperture phase errors must be corrected to obtain good performance at the higher frequencies of operation. Different portions of the sphere or torus are generally used for producing beams in different directions, unlike the paraboloid where the whole surface is generally illuminated for all beam angles. The choice of scanning reflector depends on the effectiveness of the beam-forming techniques (including those presented in this chapter), the needed angular coverage, the wavelengths, and the required sidelobe and pattern characteristics for each application.

The scanning performance of each reflector type can be improved by using aberration correcting subreflectors or diffraction correction by multiple-feed clusters. Fan-shaped search patterns can be generated with extended sources using information about focal surfaces. The quality of the wide-angle multibeams and their economy of generation with reflector antennas rivals, and often surpasses, the capabilities of lenses or phased arrays.

The presentation of each reflector-antenna design technique in this chapter is done with the aid of computer programs. The simpler and more tutorial programs are listed in the text for study and possible use by those readers with access to computers. All software mentioned in the text is available in an optional package with a user's manual explaining how to operate the code for the example antennas. A list of all the computer programs developed for multibeam and scanning reflector antennas may be found at the end of the chapter. The programs indicated by an asterisk are printed in Appendix 7A. To aid the reader in following the codes, unsubscripted variable designations are found occasionally in the text.

7.2 THE FOCAL REGION OF THE PARABOLOID

With the paraboloidal reflector we are assured of an axial pattern of good quality with point source feeds, even at very short wavelengths, provided the mirror is constructed to tolerances of $\lambda/16$ or better. What happens to pattern performance when the feed is moved from the focus for beam scanning? To find the best paraboloidal section and the best focal surfaces to locate the source feeds for the best radiation patterns, we need to examine the caustics of the mirror.

Optical studies of the paraboloid caustics, or focal surfaces where rays reflected by the surface are tangent, were done for light images [1], and these results were further developed by N.G. Parke, III, for antenna applications [2]. The general caustic surface equations for the paraboloid are derived and illustrated in Reference [3]. The caustic surface for an incident plane wave inclined with respect to the reflector axis has in general a focal surface of two sheets, one corresponding to rays focusing in the plane of incidence and the other corresponding to rays focusing in the orthogonal plane. Ridge lines or traces of the caustic on the principal coordinate planes are helpful in finding the best feed locations.

A simple derivation of the "ridge-line" equations below ([3], pp. 58–59) using an incident plane wave results in a parametric set of equations that locate the coordinates Y_0 and Z_0 of the caustic trace on the y-z plane.

$$Y_0 = \frac{-\sin_\theta (x^2 + y^2 + 4f^2)}{4f \cos \theta - 2y \sin \theta} \tag{7.1}$$

$$Z_0 = \frac{\cos \theta (x^2 + y^2 + 4f_2)}{4f \cos \theta - 2y \sin \theta} - 2f + z \tag{7.2}$$

where f is the paraboloid focal length, x, y, z are points on the reflector surface, and θ is the inclination of the plane wave with the z-axis. These ridge lines are plotted in Figure 7.1 for focal length $f = F = 1$ and θ identified with elevation angles. These equations are programmed in CAUSTIC.BAS which prints out the points Y_0, Z_0 for paraboloid parameters for θ angles of interest. The curves are generally plotted for the reflector strip along the y-axis where x approaches zero. The loops produced when x is not zero, however, are not large. These curves are useful for constructing phased-line sources that improve pattern quality for beams far from the antenna z-axis direction. These curves are not especially useful for determining best focal point coordinates, X_0, Y_0, Z_0, although the point of best azimuth focus corresponds to the point for $Y = Y_c$ and $X = 0$.

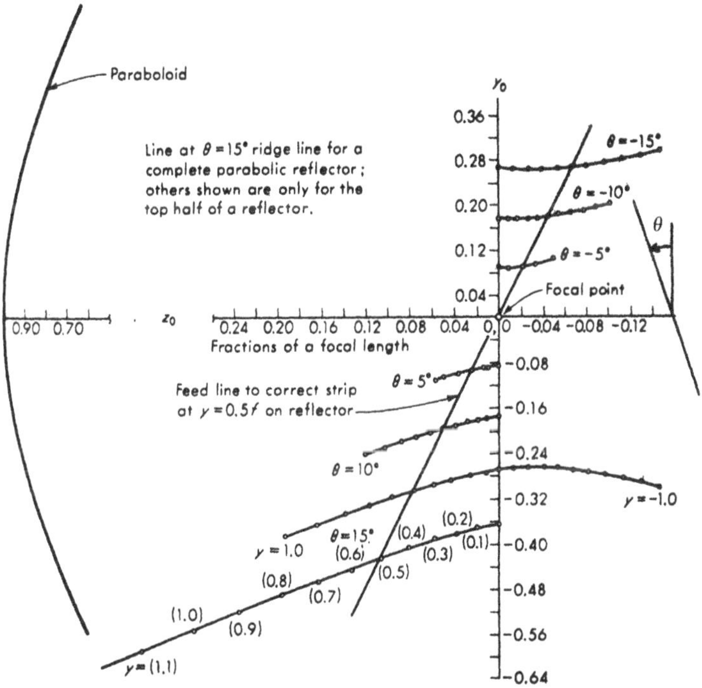

Figure 7.1 Ridge Lines Produced by Plane Waves Incident on Paraboloid Section Inclined at Angle θ
(From Collin and Zucker, Part II [3]. Reprinted with permission.)

By using the general paraboloid caustics, trace curves can be derived that locate curves for the best θ-plane (elevation plane) patterns and for the orthogonal plane patterns referred to as the azimuth plane patterns. Although these focal curves are simple straight lines and circles, their derivation is involved and found in [3], Chapters 16 and 17. These curves are shown in Figure 7.2 and their equations are,

$$Y_0 = -2Z_0/y \quad \text{(for best azimuth angle focus)} \tag{7.3}$$

$$Y_0^2 + Z_0^2 - (1 - 3/4y^2)Z_0 + y/8(y^2 - 12)Y_0 = 0 \tag{7.4}$$

(for best elevation angle focus; normalized to $f = 1$)

The intersection of these two curves at F_0 and at another point F_2 locate the two "focal points" of an offset paraboloid section. Note that these two curves are quite close to each other in the region above the focus, F_0, and they diverge below that point. The elevation loci lie on a circle and the best azimuth focal curve is a straight line.

The significant information is that, for the offset paraboloid, by locating feed sources above the true focal point, F_0, of Figure 7.2, better patterns over a wider elevation angle sector can be realized than when feeds are positioned below the focal point. The second approximate focal point can be located by simultaneous solutions of Equations (7.3) and (7.4), and the elevation angle at which this occurs is given by

$$\tan \theta_{A=0} = -y/z \tag{7.5}$$

where y, z are reflector coordinates of midpoint. $A = 0$ means zero astigmatism.

The second focal point may be computed using LOCI.BAS, based on Equations 7.3 and 7.4, and it is seen to be at $Y_0 = Y_c$ on the straight line curve (7.3). LOCI.BAS prints coordinates Y_0, Z_0 of the best azimuth focal line and then the best elevation focal circle for an offset paraboloid section with focal length F and center Y_c. Because for most applications the focal positions in the y-z plane are most useful, the analysis and the software thus far is limited to these focal traces. The program LOCI.BAS gives values of Y_0, Z_0, the focal points, on the $X = 0$ plane for selected values of Y_c (for different offset heights of paraboloidal sections). The data corresponds to the general curves shown in Figure 7.2 which are normalized to $f = 1$. (Note the elevation angle, θ, as shown in Figure 7.2, is opposite in sign to that in Figure 7.1). As the circle solution is quadratic, care must be taken to keep Y_c inputs inside this focal circle.

The programs CAUSTIC.BAS and LOCI.BAS are approximations for viewing the best focal surfaces of the paraboloid. However, to find the best focal spot for both elevation and azimuth beam directions for the

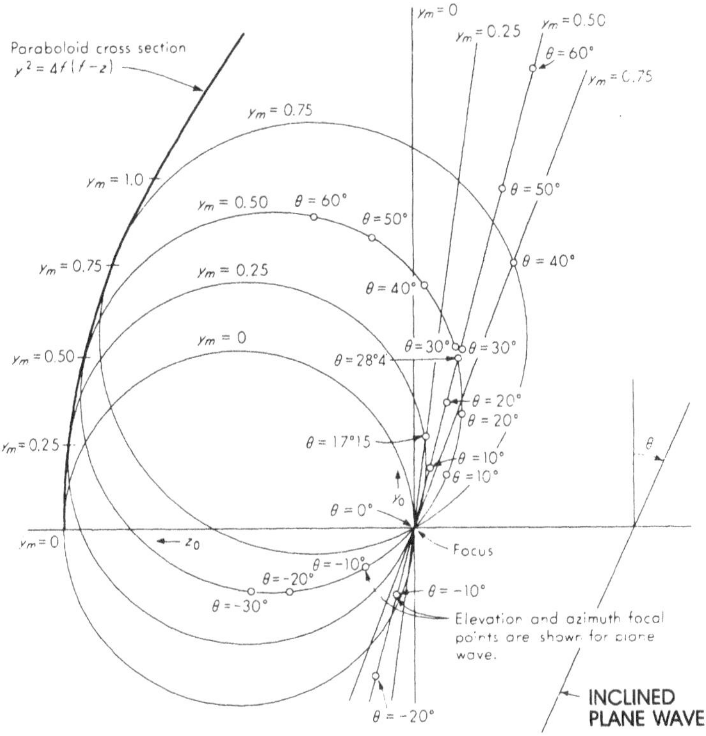

Figure 7.2 Best Feed Loci for Azimuth and Elevation Focusing for Paraboloid Section with Midpoints Y_m
(From Collin and Zucker, Part II [3]. Reprinted with permission.)

prime focus fed offset (or a symmetric antenna with $Y_c = 0$) paraboloid, the program PRIME.BAS and the graphics version PRIMEG.BAS are provided. These programs treat all rays reflected from the paraboloid surface, not just a central strip. A circular-edged aperture for the paraboloid section is assumed, although by simple modification of the programs the reflector edge points of elliptical or square-shaped apertures can be traced.

The program PRIME.BAS gives focusing information to select the best focal points for pointing a beam in a direction EL $= A_2$ (elevation angle, θ, as in Figure 7.1) and AZ $= A_5$ (azimuth angle measured clockwise about the $+y$-axis). It is programmed to find the best azimuth focal point. A parameter DELZ, however, is provided to move a focal "screen," which is oriented perpendicular to the central ray, in and out from the best azimuth location. Usually DELZ is set to zero, for best azimuth plane focusing, and the focal spot observed by studying the spot coordinates on the screen generated by the rim rays from the paraboloid. This curve is generally a nearly vertical line when graphics are available (PRIMEG.BAS), indicating that rays are focused well in azimuth but not so well in elevation. The parameter DELZ can be used to move along the central ray using the direction of the best elevation focus curve as a guide (Figure 7.2) until the smallest focal spot is obtained for the rim rays. The rays reflected from the area near Y_c, however, may not lie inside the focal spot for the dish rim rays. The focal trace for any point on the paraboloid can be found, however, by making the R (radius of aperture of interest) smaller than the radius of the dish aperture. The focal spot for small aperture circles is always near the center of the focal screen.

The derivation of the equations used in PRIME are obtained by writing:

$$\mathbf{s}_0 = \cos A_2 \sin A_5\mathbf{x} - \sin A_2\mathbf{y} + \cos A_2 \cos A_5\mathbf{z} \tag{7.6}$$

for the rays incident on the paraboloid surface (and normal to a tilted plane wave) at angles of AZ = $A5$ and EL = $A2$. The lower-case boldface variables are unit vectors:

$$\mathbf{n} = \frac{x\mathbf{x} + y\mathbf{y} + 2fz\mathbf{z}}{R}$$

$$R = (x^2 + y^2 + 4f^2)^{1/2} \tag{7.7}$$

Equation (7.7) is for the unit normal to the paraboloid of focal length, $f = F$. Snell's law gives us the reflected ray $\mathbf{s}_1$,

$$\mathbf{s}_1 = \mathbf{s}_0 - 2(\mathbf{s}_0 \cdot \mathbf{n})\mathbf{n} = L_1\mathbf{x} + M_1\mathbf{y} + N_1\mathbf{z} \tag{7.8}$$

From the simultaneous solution of the ray as line equation from any point x, y, z on the reflector to the best azimuth focal plane, we can write

$$(X_2 - x)/L_1 = (Y_2 - y)/M_1 = (Z_2 - z)/N_1 \tag{7.9}$$

$$Z_2 = -Y_2Y_C/(2F) \tag{7.10}$$

and obtain the equations for the best focal point X_2, Y_2, Z_2 given in lines 180 to 210 of PRIME, noting that in the BASIC code all variables are in upper case letters (without subscripts) and that $Y_c = Y_3$. The program from lines 210 to 360 sets up a flat "image screen" using, in line 300, the focal adjustment DELZ = Z_5. The derivation of the screen orientation follows from the equation of the plane normal to the central ray through Y_c which is,

$$B_1X + B_2Y + B_3Z + D = 0 \tag{7.11}$$

where $B_1 = L_1$, $B_2 = M_1$, $B_3 = N_1$ are direction cosines of the central ray and D is a constant.

When DELZ or AZ are not zero, the focal point printed out as X_F, Y_F, Z_F is different from X_2, Y_2, Z_2 for best azimuth focus. The X_0, Y_0 traces plotted by the graphic version programs or printed out on the line printer give the focal spot coordinates or the *circles of least confusion* for a circular dish aperture of radius = R. DELZ should be varied also to "focus" for the best feed position. The relative diameters (or areas as the traces are not circles) of the best focal spots give a qualitative estimate of the radiation patterns that can be expected for a given beam angle, EL, AZ. The smaller focal spots mean better focused radiation patterns with lower sidelobe levels. The diffraction patterns depend too, of course, on the diameter = $2R$-to-wavelength (D/λ ratio) of the antenna. These diffraction effects can be studied using the best focal coordinates X_F, Y_F, Z_F in the program PARAP.FOR, discussed in the next section, which computes the antenna aperture scalar fields for any feed position X_0, Y_0, Z_0. This aperture amplitude and phase distribution is then used by PATT.FOR (see Chapter 1) to compute the diffraction patterns with the main beam peaks in directions very nearly the input angles, EL and AZ.

To run PRIME.BAS start with these simple example data. At the prompts enter $F = 10$ and $Y_c = 5$. Try elevation angle, EL = 0, AZ = 0, R, radius of the aperture, $R = 5$, and DELZ, the focusing parameter, DELZ = 0. Then try EL = 2, AZ = 2, $R = 5$, DELZ = 0.2 for more interesting numbers.

There are many papers discussing best focal curves using Seidel aberration calculations [4, 5] to predict the beam peak position and sidelobe levels for paraboloids. Most of these approaches are limited to small scan angle and large f/D antennas. The techniques presented here are not so limited and illustrate that the offset paraboloid can be scanned over wide angles when feeds are located on the best focal surfaces.

7.3 TECHNIQUES FOR COMPUTING ANTENNA PATTERNS FOR FEEDS DISPLACED FROM THE FOCUS OF A PARABOLOID ANTENNA

As most numerical integration methods use a regular grid of aperture points to compute the diffraction pattern from the phase and amplitude aperture distribution, a method is developed here that accurately finds the pathlengths to a given aperture point X_1, Y_1, Z_1 located on a simple (r,ϕ) grid. The pattern gain, beamwidth, and near-in sidelobes, for a feed displaced to a focal point X_0, Y_0, Z_0 (in the focal region of an offset or symmetrical paraboloidal reflector) can be computed using PARAP.FOR and PATT.FOR. It is found that the focal points of the circles of least confusion obtained from the program PRIME.BAS give the best radiation patterns.

Let the coordinates X_1, Y_1, Z_1 be given to a point on a tilted aperture plane with equation,

$$X_1 \cos(\text{EL}) \sin(\text{AZ}) - Y_1 \sin(\text{EL}) + Z_1 \cos(\text{EL}) \cos(\text{AZ}) + D = 0 \tag{7.12}$$

This plane is normal to the incident parallel ray bundle of equation (7.6), where $D = 0$ or a constant. The unit vectors for the tilted aperture plane axes are,

$$\begin{aligned} \mathbf{x}_P &= \cos(\text{AZ})\mathbf{x} - \sin(\text{AZ})\mathbf{z} \\ \mathbf{y}_P &= \sin(\text{AZ}) \cos(\text{EL})\mathbf{x} + \cos(\text{EL})\mathbf{y} + \cos(\text{AZ}) \sin(\text{EL})\mathbf{z} \\ \mathbf{z}_P &= \cos(\text{EL}) \sin(\text{AZ})\mathbf{x} - \sin(\text{EL})\mathbf{y} + \cos(\text{EL}) \cos(\text{AZ})\mathbf{z} \end{aligned} \tag{7.13}$$

To find the coordinates X_1, Y_1, Z_1 (in the same coordinates X, Y, Z as the paraboloid) of the aperture points X_P, Y_P, Z_P Equation (7.13) can be used to make the three-coordinate transform. The aperture coordinates are,

$$\begin{aligned} X_1 &= X_P \cos(\text{AZ}) + Y_P \sin(\text{AZ}) \sin(\text{EL}) + Z_P \cos(\text{EL}) \sin(\text{AZ}) \\ Y_1 &= Y_P \cos(\text{EL}) - Z_P \sin(\text{EL}) + Y_C \\ Z_1 &= -X_P \sin(\text{AZ}) + Y_P \cos(\text{AZ}) \sin(\text{EL}) + Z_P \cos(\text{EL}) + Z_{11} \end{aligned} \tag{7.14}$$

The regular grid of aperture points used by PATT.FOR are

$$X_P = R \cos \phi$$

$$Y_P = R \sin \phi$$

$$Z_P = 0 \tag{7.15}$$

$$0 \leq R \leq A_P$$

Both R and ϕ are incremented; $-\pi \leq \phi \leq \pi$.

Equations (7.14) and (7.15) are seen in the code (listed in Appendix 7A) for PARAP.FOR, where, because $Z_P = 0$, Equation (7.14) is simplified. The constant Z_{11} allows the aperture plane to be moved back away from the paraboloid for small f/D antennas or for large feed displacements. The parameter $A_P > A$ where A is the aperture radius is needed for tilted apertures. Only rays reflected from the circular paraboloid section of radius A are used.

The ray pathlengths from the aperture points X_1, Y_1, Z_1 to points X, Y, Z on the paraboloid are given by

$$R_1 = [(X - X_1)^2 + (Y - Y_1)^2 + (Z - Z_1)^2]^{1/2} \tag{7.16}$$

and the ray pathlengths from the paraboloid to the displaced feed point by

$$R_0 = [(X_0 - X)^2 + (Y_0 - Y)^2 + (Z_0 - Z)^2]^{1/2} \tag{7.17}$$

where X_0, Y_0, Z_0 is the focal point where the center of phase of the feed is located. These points may be found with PRIME using the X_F, Y_F, Z_F printout. Using Sncll's law from Equation (7.8) and R from Equation (7.7) we can write the equations for the direction cosines of the incident and reflected rays as

$$L_0 = (X_0 - X)/R_0;\ M_0 = (Y_0 - Y)/R_0;\ N_0 = (Z_0 - Z)/R_0$$

$$L_1 = (X - X_1)/R_1;\ M_1 = (Y - Y_1)/R_1;\ N_1 = (Z - Z_1)/R_1 \tag{7.18}$$

$$L = X/R;\ M = Y/R;\ N = 2F/R$$

Defining

$$B = LL_1 + MM_1 + NN_1$$

we can write three nonlinear equations which when solved give the X, Y, Z coordinates on the reflector for any source point X_0, Y_0, Z_0 and aperture point X_1, Y_1, Z_1. The three equations used in PARAP are

$$Z = F - (X^2 + Y^2)/(4F)$$

$$L_0 = L_1 - 2BL \tag{7.19}$$

$$M_0 = M_1 - 2BM$$

These three nonlinear equations are solved using Numerical Recipes software for the solution of sets of nonlinear equations with programs MNEWT, LUDCMP, and LUBKSB linked to PARAP with subroutine USRFUN developed for Equations (7.16) to (7.19). As partial derivatives are needed for (7.19), 12 equations, ALPHA(3, 3) and BETA(3), must be evaluated in USRFUN. This subroutine returns X, Y, Z coordinates for the paraboloid reflection point that connects the point X_1, Y_1, Z_1 on the aperture plane with the focal point X_0, Y_0, Z_0. Approximate trial values of X, Y, Z must be entered first and the numerical procedure finds X, Y, Z coordinates that satisfy Snell's law to good accuracy. The pathlengths R_0 and R_1 can then be calculated from (7.16) to (7.18) to give the aperture phase distribution.

Rather detailed descriptions of PARAP.FOR are given, because the computational techniques are repeated in several programs that follow. The source code for PARAP.FOR is printed in Appendix 7A. Program PARAP accepts input data from file PARD.DAT, containing the best focal point X_0, Y_0, Z_0 from PRIME and the elevation tilt angles EL and AZ.

Phase PHAS is calculated from the path lengths from

$$\text{PHAS} = (R_0 + R_1)2\pi/(\lambda) \tag{7.20}$$

and amplitude feed distribution is simply tapered, in the example case programmed in PARAP assuming the "point source feeds" are pointed at Y_c and produce a radial aperture distribution governed by

$$\text{AMP} = [1 - (1 - \text{DEL})r^2]R_{00}/R_0$$

where r is the fraction of dish aperture at the aperture point, DEL is a parameter used to control the edge taper, and R_{00} is the normalizing central ray pathlength, R_0. When the parabolic radial distribution contains DEL = 0.5 there is about a −6 dB edge taper. An enlarged aperture radius

AP is needed to accommodate the tilted geometry and the aperture edges are filled with zeros to give only the illuminated aperture radiation pattern, using a conditional that requires all the reflected rays to be from a dish of radius A.

To run PARAP.FOR (actually PARAP.EXE) the data file PARD.DAT must be prepared using EDLIN, for instance. An example file is given in the software and uses numbers computed by PRIME.BAS. The data is read back by the program; if not correct, use a Control C to abort and then edit the file PARD.DAT. As seen in the FORTRAN code of PARAP, the parameters are A = radius of paraboloid section, LAM = wavelength in units of radius, Y_c = height of aperture center point, F = paraboloid focal length, Z_{11} = Z coordinate of aperture center where phase, amplitude is measured, EL = elevation angle in degrees, AZ = azimuth angle in degrees. On the second line the coordinates of the center of phase of the feed, X_0, Y_0, Z_0 are first entered and then AP = (a larger aperture than A) to accommodate a tilted wave front. DEL is a constant that controls the aperture distribution and is related to the directivity of the feed. DEL = 0.5 is used in the example. The path length attenuation is also taken into account in the program. Finally, on line 3 of the data file PARD.DAT, we enter the number of radial points for aperture integration, NRAD, and NANG, the number of ϕ angle cuts over the dish aperture. The point spacing on the aperture edge should not be much greater than two wavelengths for computing beams a few degrees from the beam axis to avoid grating lobe effects.

The program pauses to verify the data and also to show the center point on the aperture which for displaced feed points is not at X_c, Y_c, Z_c but usually near to that point. For gaining familiarity, aperture amplitude and phase values are shown on the screen as the program computes them over the tilted aperture of the dish. Generally this display can be made into a "C" or comment statement to save computer run time. About ten minutes are needed to run the example program which has the X_0 and Y_0 feed point located more than a wavelength from the focus. The skewing makes the aperture distribution asymmetric. To compute the radiation pattern from the data file PHAMP.DAT (unformatted) the program PATT.EXE is used. It asks for THETAI (INITIAL), THETAF (FINISH), and DTHETA (INCREMENT). Because the aperture is tilted to bring the peak of the beam near to $\theta = 0$ enter 0.0,4.0,0.2 to get a cut over 4° at 0.2° intervals. The program requests PHI, the ϕ-cut angle. PHI = 0.0 is an azimuth cut, and PHI = 90° is an elevation angle cut. The example data give a gain of 39 dBi and azimuth sidelobes about −20 dB from the peak. The pattern data can be read by TYPE CJS.DAT listing the contents of the data file CJS.DAT. The pattern may be plotted using the graphics program PLPATT.BAS. The G_0 value for the 15-unit radius is about 39.5 dBi, but spillover is not included in PARAP's calculations. See Chapter 3 for spillover and cross-polarization numbers.

For those running the program PARAP.FOR, it must of course be compiled and linked with the Numerical Recipes object modules MNEWT, LUDCMP, and LUBKSB. The data file PARD.DAT for the example antenna has the numbers entered as below:

```
15.0,1.0,15.0,30.0,0.0,2.0,2.0

1.12289,-1.12337,.281275,15.5,0.5

21,105
```

The capability to compute radiation patterns for large optimized feed displacements from the paraboloid focus is a powerful and unique tool for dish designers. An offset, prime-focus-fed paraboloid can be scanned many beamwidths by feed displacements.

7.4 FOCAL SURFACES AND WIDE-ANGLE RADIATION PATTERNS FOR OFFSET CASSEGRAIN AND GREGORIAN ANTENNAS

Best focal surfaces for feeding a Cassegrain or Gregorian antenna are not easy to find analytically, especially for beams scanned many beamwidths from the paraboloid axis. Again we will consider mainly optimally tilted (optimum for radiated beams in the Z-axis direction only) offset-fed paraboloid sections with the subreflectors either hyperboloids or ellipsoids. Antenna dimensions and parameter are readily available using the programs CASS.BAS or GREG.BAS described in Chapter 1. We will always use the entire paraboloid surface by enlarging the subreflector surfaces as necessary to generate a beam at some angle EL or AZ. The computer technique for mapping the best focal surfaces and computing the diffraction patterns for offset dual reflector antennas proceeds parallel to the development for the offset prime-focus paraboloid.

Although both offset Cassegrain and Gregorian antennas will be treated, the hyperboloid subreflector antenna of Figure 7.3 (a and b) will be used as the primary example.

To find the *circles of least confusion* or best focal spot for the offset Cassegrain antenna the constants for the main reflector and subreflector as printed out by CASS.BAS are edited into the BASIC program CLCC.BAS in line 60 (or entered interactively in line 70). Next the scan

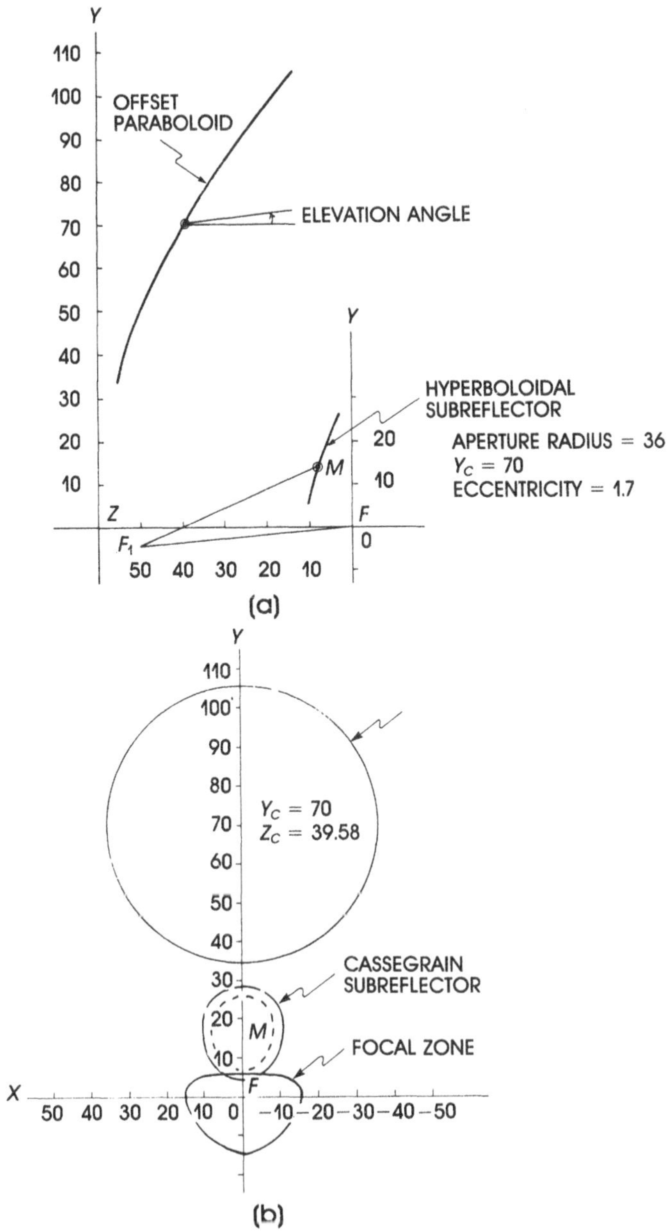

Figure 7.3 (a) Profile of Offset Cassegrain Antenna; (b) Front View of Offset Cassegrain Antenna

angles EL and AZ are entered with the radius, R, of the paraboloid's circular aperture. Usually R is the dish rim radius, although it is well to examine focal traces for $R < R_{rim}$. The focusing variable $Z_{focus} = Z_4$ moves the computer "image screen" (a plane perpendicular to the central ray passing through Y_C) on which the rays intersect to form a trace or spot. It is well to start with EL = 0, AZ = 0 and $Z_{focus} = Z_4 = C \cos(A_1)$ where C is the distance between the foci of the hyperboloid subreflector and $A_1 = \beta$ the tilt angle between the subreflector axis and the paraboloid axis. With all constants properly entered the program CLCC prints the trace coordinates, X_0, Y_0, relative to the subreflector focus at $X_0 = 0$ and $Y_0 = C \sin(\beta)$, a point trace. (The program permits the subreflector to be misaligned in the y-z plane by variables H and K, which are generally set to zero.)

To identify variables in CLCC.BAS (and the graphics program CLCCG.BAS), the direction cosines L_1, M_1, N_1 in lines 180–200 are for the reflected rays $\mathbf{s}_1$ of Equation (7.8) after reflection from the paraboloid. The intersection points of these rays, after reflection, on the subreflector surface have coordinates X_1, Y_1, Z_1, in lines 280–310. The rays reflected from the subreflector have the direction cosines L_2, M_2, N_2 in lines 360–380.The focal screen is set up perpendicular to the initial central ray through Y_c and this work is done in lines 560–610. The graphics version of CLCCG is much preferred for quickly finding the best focal point for a given beam direction.

To study the focal surfaces of offset Gregorian antennas only the input constants from GREG.BAS need to be entered on line 50 and the sign before SQR in line 280 changed from + to −. This program change is done in CLCG.BAS. The equations for the program CLCG.BAS are are found also in [6]. Again the program CLCGG.BAS should be used for PCs with BASIC graphics capabilities.

The programs CLCC and CLCG also print out to screen (not the image screen referred to earlier) the edge points ("footprint") for a circle points on the offset dish edge at 30° angular intervals in ϕ over the antenna aperture. The focal trace or spot is hardprinted on the line printer so that comparison can be made as the in-and-out focal variable Z_4 is varied. These data are plotted (automatically using graphics with CLCCG or CLCGG BASIC programs) and the smallest spot of a vertical or horizontal line trace chosen for running patterns using the diffraction programs MULTE.FOR or MULTH.FOR which are discussed next.

Once the best focal surfaces for a dual-reflector antenna are known, the diffraction patterns for feeds in the preferred focal spots can be computed using the programs MULTE.FOR and MULTH.FOR which are written for ellipsoidal and hyperboloidal subreflectors respectively. The best focal surfaces need not be at the smallest focal spots or *circles of least confusion*, but when best azimuth sidelobes are desired the Z_4 value corresponding to the vertical focal trace can be chosen. Likewise, the horizontal trace or astigmatic line can be selected using focal coordinates X_4, Y_4, Z_4 for best elevation plane patterns.

The development of the code for MULTE.FOR or MULTH.FOR proceeds similarly to PARAP.FOR, resulting in the aperture distribution for scanned patterns. Here no set of nonlinear equations is employed, however, and the algebra of the code for computing ray-path lengths is more complex than in PARAP. These distributions are again summed by PATT.FOR to compute radiation patterns.

An incident ray bundle $\mathbf{s}_0$ is used again with a regular grid of aperture points X, Y. The points where these rays strike the paraboloid is used for integration on a tilted aperture plane as in PARAP. The receive-ray tracing process is similar to that used in CLCC.BAS to find the subreflector focal point labelled in the code as FX, FY, FZ. Then a ray is defined coming from the best focal point F_X, F_Y, F_Z back to the subreflector points XPR, YPR, ZPR (in the subroutine PHSAMP of MULTE and MULTH). Using Snell's law twice, the tilted aperture intercept points X_1, Y_1, Z_1 are found and the pathlength from the feed point F_X, F_Y, F_Z to the point X_1, Y_1, Z_1. This pathlength PLT is computed and multiplied by $2\pi/(\lambda)$ to give the phase at the aperture points X_A, Y_A. There is a small error in this procedure, as X_1, Y_1, Z_1 are not exactly the points where the $\mathbf{s}_0$ rays intercept the tilted aperture plane. The aperture integration by PATT.FOR can be carried out with good approximation from the X, Y, Z points on the offset paraboloid surface using the programs MULTE and MULTH. The focal surfaces of typical Gregorian and Cassegrain antennas are given in the figures that follow.

7.4.1 Some Example Focal Surfaces and Radiation Patterns for Dual-Reflector Antennas

To illustrate mapping of dual-reflector focal surfaces using CLCC.BAS and CLCG.BAS and computing the verifying radiation patterns using MULTE.FOR and MULTH.FOR, some examples are provided.

The first example is an offset Cassegrain of circular aperture with the following antenna constants: For the main reflector $F = 60$ (where length units can be centimeters, inches, etcetera), $Y_c = 70$, $R =$ diameter/2 $= 36$; for the subreflector eccentricity $= 1.7$, $C = 50$, $D = -16.34948$, $B = 451.193$, $\beta = 4.947447°$ and $\alpha = -18.9210$. When the antenna is fed by a horn with semiflare angle of about 9.12° then the antenna has a gain of about 45.8 dB and half-power beamwidth of about 1°. For these calculations, frequency is 12 gHz and length units are in inches (a six-foot diameter dish).

Figure 7.3(a) shows the antenna profile and Figure 7.3(b) gives a frontal view. The hyperboloidal subreflector is made large enough so that the traces of the dish rim rays, the "footprints," all strike the subreflector for the elevation and azimuth scanning sectors of interest. These footprints for EL = AZ = 0° and for a 4.0° azimuth scan (from the direction along the paraboloid axis) are shown in Figure 7.4. These traces of footprints can be obtained from CLCC.BAS by plotting the X_1, Y_1, Z_1 points on the subreflector surface as the program sequentially enters the paraboloid rim points. The full paraboloid aperture is always illuminated and only the footprint sections of the subreflector are used to compute a scanned beam in a given angle direction, EL, AZ (and its near-in sidelobe pattern), using MULTH.FOR.

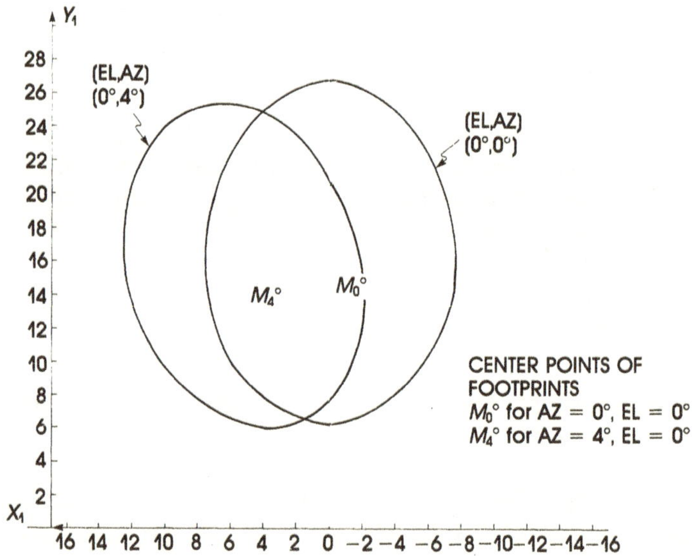

Figure 7.4 "Footprints"—Illuminated Portions of Cassegrain Subreflector for Scan Angles Shown

The best focal surfaces for this example that are shown in Figures 7.5(a) and (b) are typical for offset Cassegrains. Figure 7.5(a) shows the

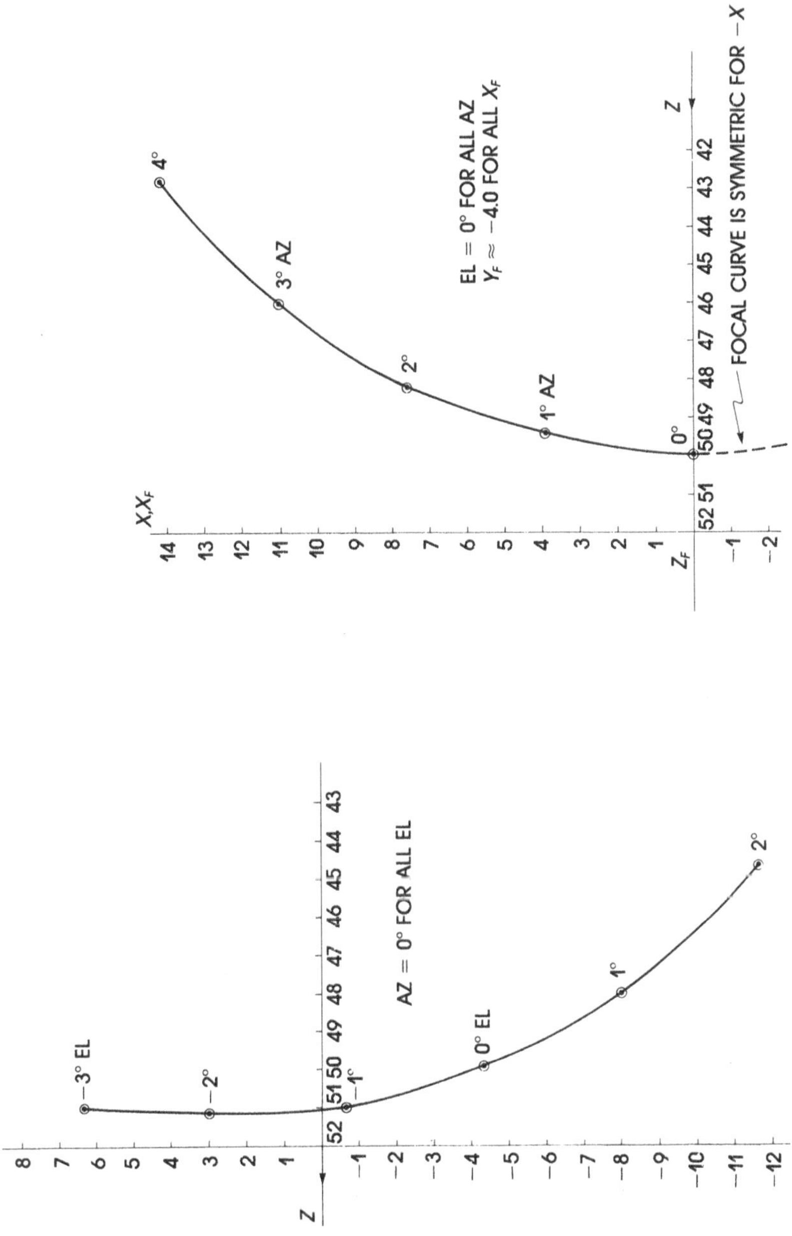

Figure 7.5 (a) Elevation Plane Focal Curve; (b) Azimuth Plane Focal Curve

focal curve for the smallest circles of least confusion for elevation angle scanning when AZ = 0°. Figure 7.5(b) shows the best azimuth scanning curve when EL = 0°. Figure 7.6 presents a view of the focal surface area as seen looking along the Z-axis for both elevation and azimuth scanning with the $Z_F = Z_4$ depth of focus data given on the figure. It is this variable, Z_4, entered into CLCC (or preferably in CLCCG.BAS for graphics) that is varied at each EL and AZ scan position to find the preferred focal spot.

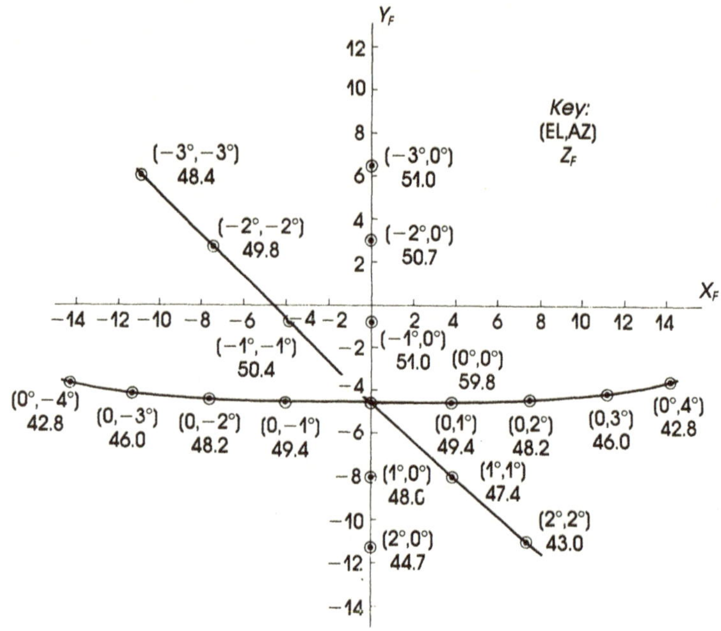

Figure 7.6 Focal Surface Area for Offset Cassegrain Antenna

Radiation patterns were computed using MULTH.FOR and PATT.FOR for EL = AZ = 0° and the scanned patterns at AZ = 4.0°. This pattern is shown in Figure 7.7 for a $P = \cos^N \theta$ feed pattern with $N =$ 324.5. Note that there is little gain or sidelobe deterioration for the 4.0° scanned pattern when λ is approximately 1 (unit).

Although offset-fed Gregorian antennas have generally smaller angular scanning sectors, they can also be scanned using feed displacements. Figure 7.8 shows the elevation plane focal arc for an offset ellipsoidal subreflector antenna with these parameters [7]: the paraboloidal reflector of dimensions in inches; $F = 70$, $Y_C = 62$, $R = 48$; for the subreflector eccentricity = 0.538, $D = 42.96074$, $\beta = 5.14239°$, $\alpha =$ 17.0046. When fed by a corrugated horn with semiflare angle = 12° the

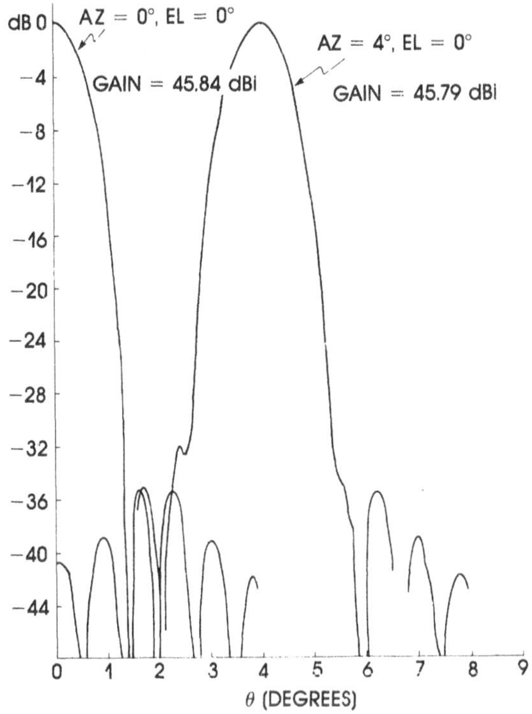

Figure 7.7 4.0 Degree Azimuth Scan Pattern

antenna had a gain of about $G = 0.8(2\pi R/\lambda)^2$ and beamwidth$_{1/2}$ = $1.2(\lambda/2R)$ radians. Figure 7.9 illustrates both the azimuth and elevation focal arcs with a frontal view when the circle of least confusion focal points are found varying Z_4 in CLCG.BAS. In Figures 7.10(a and b) patterns were taken using MULTE and PATT to show the patterns on both vertical and horizontal astigmatic focal traces produced by CLCGG. The patterns verify that the patterns have good azimuth plane sidelobes when the vertical trace is selected as the focus and good elevation plane patterns when the horizontal focal trace is chosen.

To run MULTH.FOR and MULTE.FOR input data files are needed called DATINH.DAT and DATINE.DAT, respectively. They ask for A = radius, LAM = wavelength, B = the transform constant printed out by CASS.BAS or GREG.BAS, Y_C = dish center Y value, and F = paraboloid focal length. On line 2, the input data files enter K (subreflector alignment factor) = 0.0, H = 0.0, BE = beta for CASS or GREG, E = eccentricity, D = subreflector constant from CASS or GREG. The programs interactively ask for N the power coefficient of the $\cos^N \theta$ pattern of the horn illuminating the subreflector. You should be able to run the

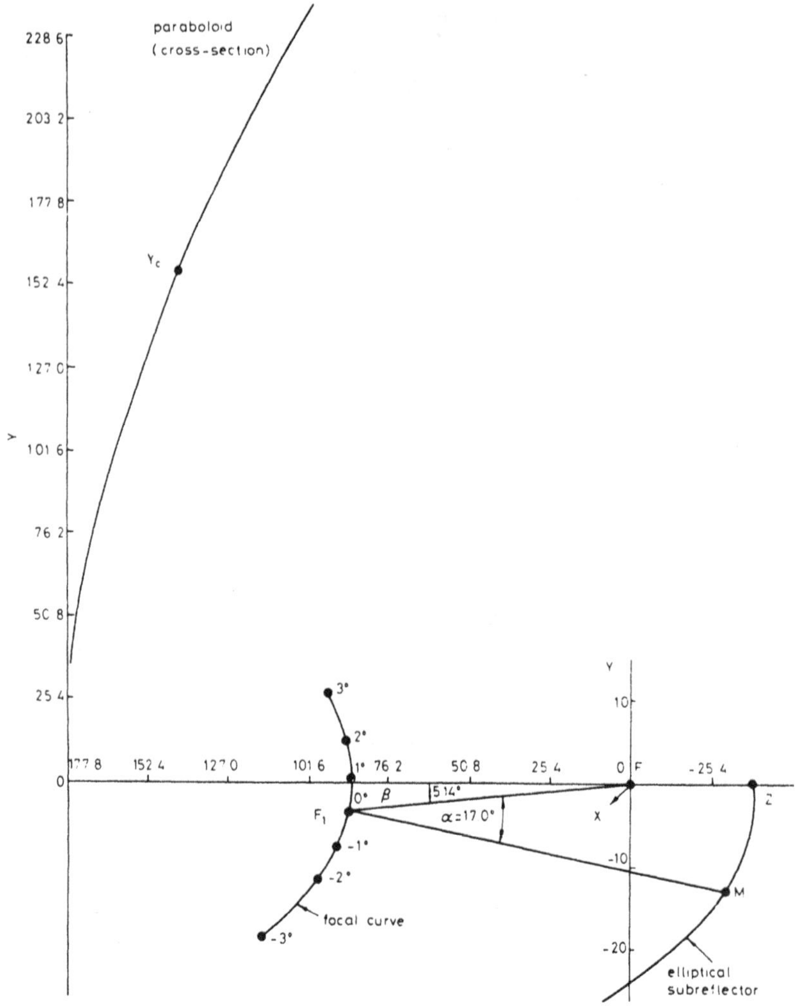

Figure 7.8 Best Elevation Plane Focal Curve for Gregorian Antenna (From Sletten and Shore, *IEE Proc.* [6]. Reprinted with permission.)

example data for MULTH.FOR and then employ the programs for your antennas, using the optional software package with user's manual and example booklet.

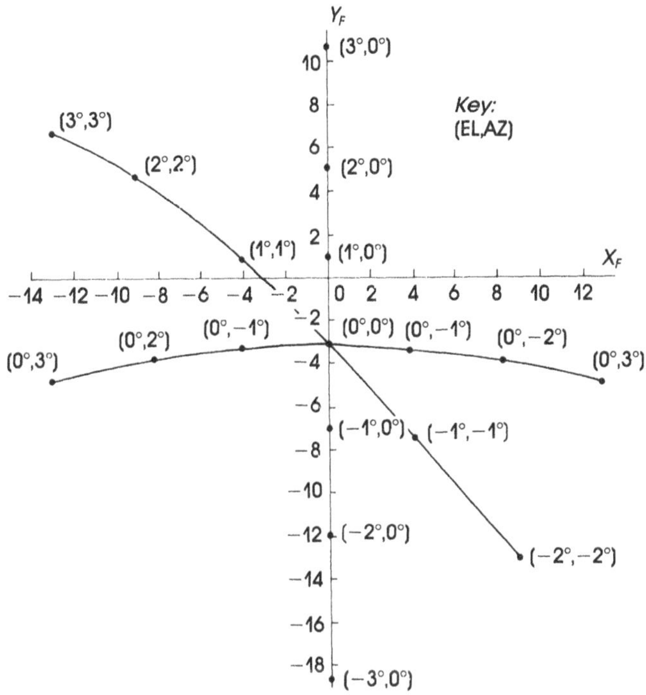

Figure 7.9 Focal Surface $x - y$ Plot for Gregorian Antenna

7.5 SHAPING SUBREFLECTORS TO INCREASE THE WIDE-ANGLE COVERAGE OF OFFSET DUAL-REFLECTOR ANTENNAS

Although the offset Cassegrain antenna especially can be scanned many beamwidths (by feed displacements) when large f/D geometry is used with an enlarged hyperboloidal subreflector, excellent patterns can be achieved at even larger scan angles by shaping subreflector footprint areas. Because the focusing of the antenna is good, the shaped subreflector sections are nearly hyperboloids. One of the simpler applications of this shaping technique is to shape the footprint of the subreflector area that generates the beam scanned furthest from the antenna axis. The edge pattern can be made free from phase errors and with characteristics approximating the on-axis pattern.

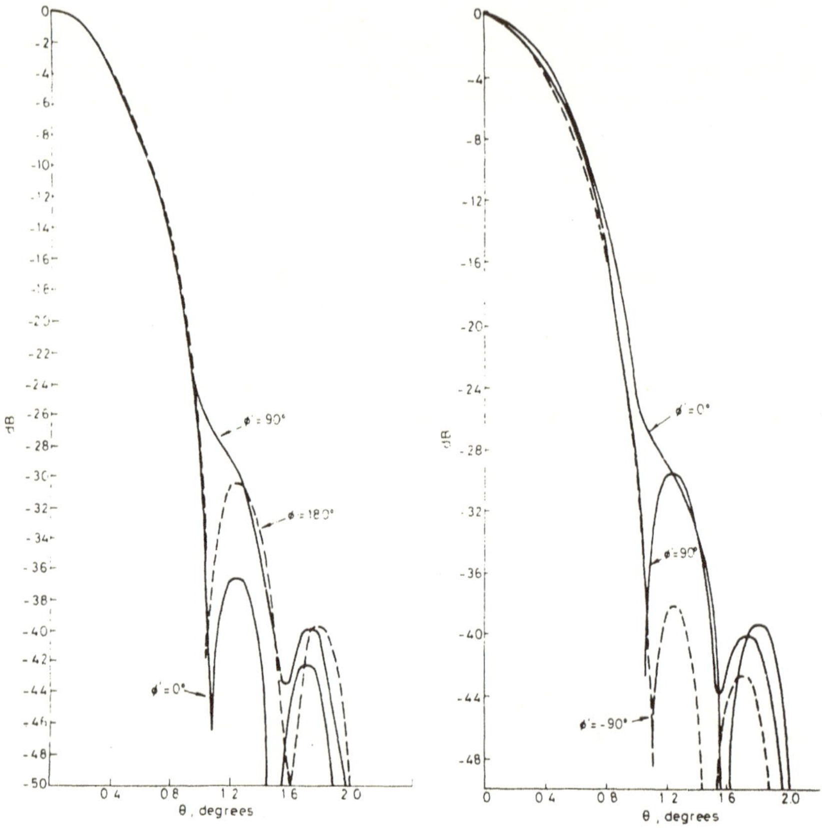

Figure 7.10 (a) Diffraction Pattern Scanned 3° in Azimuth Focused at Best Azimuth Focus; (b) Diffraction Pattern Scanned 3° in Elevation Focused at Best Elevation Focus
(From Sletten and Shore, *IEE Proc.* [6]. Reprinted with permission.)

Following the surface shaping methods developed in Chapter 3 for 4XSTUP.BAS, a program SHAPE.FOR for shaping offset Cassegrain subreflectors (see source code for SHAPE.FOR in Appendix 7A) begins by computing start-up values (three back derivative on X-cuts) near the edge of the footprint. The subreflector may be unshaped in the central portions when the remote scanned angle footprint is at the edge of the enlarged subreflector. For the shaping of the footprint for EL = 0°, AZ = 4° the footprint as seen in Figure 7.4 covers the central portion of the hyperboloid subreflector. This Cassegrain antenna scanned 4.0° in azimuth is the example case presented here, but wider angles can be easily computed using SHAPE. The best focal points, for the end pattern to be improved by subreflector tip shaping, must first be found using CLCC.BAS. SHAPE then synthesizes a subreflector surface that correctly focuses all rays from the look direction, EL, AZ, to the focal point found by entering the antenna configuration parameters and EL, AZ in CLCC.BAS.

Program SHAPE.FOR uses a function CASS (X_1, Y_1) to calculate points on the reference Cassegrain for computing the start-up data needed for the Adam's predictor-corrector [8] and to compare the shaped surface with the hyperboloidal surface. For axis-normal plane wave incidence, EL = AZ = 0°, the synthesis produces the original hyperboloid. The true direction cosines of the rays reflected from the paraboloid intercepted at any point on the subreflector are computed by subroutine USRFUN which needs Numerical Recipes programs MNEWT, LUDCMP, and LUBKSB to solve the three nonlinear equations for X, Y, Z on the paraboloid based on Snell's law. The shaped subreflector surface coordinates X_1, Y_1, Z_1 are printed by SHAPE along with DEL, the difference of the Z_1 values from the true hyperboloid. The three-element array, beta, in USRFUN defines the negative of the three nonlinear functions used to compute X, Y, Z (main reflector points) and the nine-element array contains the partial derivatives of the beta functions. In the program the direction cosines L_{11}, M_{11}, N_{11} are for the rays incident on points projected by the numerical Adam's predictor-corrector. The variables K_8 and K_9 are the slopes of the shaped subreflector (refer to Chapter 3 for details of the shaping method) and are calculated by inverting Snell's law. Constants of different antenna configurations can be entered into SHAPE using data file SHAP1.DAT.

For the reader having the Numerical Recipes subroutines and the program SHAPE.FOR, as listed in Appendix 7A, the data file SHAP1.DAT with numbers as printed below is for shaping the footprint area for a beam at 4.0° in azimuth, 0.0° in elevation. The numerical data is arranged in this form in SHAP1.DAT:

```
1.70,60.0,50.0,36.0,70.0,0.0,4.0

4.947447,-16.349488,-36.0,70.0

14.20405,-3.676753,42.8

-2.0746,1.0746,13.63144,15.63144
```

The program SHAPE.FOR identifies these variables when the program is initiated. They are

E,F,C,R0,YC,EL,AZ

A,D,XI,YI

X0,Y0,Z0

U0,U1,V0,V1

where E = reference Cassegrain eccentricity, F = paraboloid focal length, C = distance between subreflector foci, R0 = radius of dish, YC = midpoint of main dish, EL = elevation angle of beam, AZ = azimuth angle of beam. In the second line of SHAP1.DAT A = β angle of reference subreflector, D = subreflector constant, XI = dish X-edge point where shaping starts, YI = dish Y-edge point where shaping starts. On line three, the best focal point coordinates, X_0, Y_0, Z_0, for the reference subreflector obtained by CLCC.BAS are entered. The rectangular zone of the subreflector X_1 and Y_1 are entered in line four. U0 and V0 are the starting values of X_1 and Y_1 on the subreflector surface respectively and they must correspond to XI and YI on the paraboloid. The values of U0 and V0 are the upper bounds of X_1 and Y_1 defining a rectangularly shaped area on the subreflector.

The example used to illustrate this shaping for wider scanning is the Cassegrain antenna shown in Figure 7.3(a) and (b) with footprint shown in Figure 7.4 at $AZ = 4.0°$. Using CLCC.BAS the best focal point, X_0, Y_0, Y_0, was chosen to be at $F_X = 14.20405$, $F_Y = -3.67675$, and $F_Z = 42.8$. We wish to shape the subreflector in the area of the footprint so that all the rays incident from AZ = 4.0°, EL = 0° focus perfectly to F_X, F_Y, F_Z. This is done in program SHAPE.FOR.

As mentioned earlier, this shaping method can be used to shape the ends of an elongated subreflector to make beams far from the antenna's Z-axis well focused. When this is done the central portion of the subreflector can be kept hyperboloidal to insure a well focused axis beam. To make sure that only the tip footprint area is shaped, a conditional is programmed into SHAPE. An oversized rectangular tip zone of the end of the subreflector footprint can be entered, and the program will only shape the oval footprint for the EL,AZ end beam, using only incident rays (from the tilted plane wave arriving and departing the main paraboloid aperture) that are intercepted by the main reflector surface. This means that only the portion of the enlarged subreflector used by the edge beam will be shaped.

The program SHAPE.FOR is tutorial and simplified to introduce the shaping method. In Chapter 8 alternate offset subreflector shaping codes are developed for Cassegrain-fed paraboloids. We have described the computer code for SHAPE in detail, as it is a prototype of the more complex shaping programs to follow.

7.6 SPHERICAL REFLECTOR ANTENNAS

The concave spherical mirror offers wide-angle scanning possibilities when multiple or movable antenna feeds are positioned in the focal region of the fixed reflector. The hemispherical reflector surface appears the same when viewed from any radius (in the central sectors) due to its rotational symmetry. The optical focusing properties of the sphere are useful approximations at microwave and millimeter wavelengths and can be studied using caustic surfaces as was done with the paraboloid mirror.

Figure 7.11(a) and (b) shows a useful construction [3] of incident rays (assuming an incident bundle of parallel rays representing a plane wave normal to the radius of the sphere, OV) of which AB is typical. These rays do not focus to a point, but those reflected near the vertex region V are concentrated near the paraxial focus F which is located at $f = 0.5R$ from the point O at the center of the spherical reflector of radius R. The caustic curve labeled in Figure 7.11(b) is a surface of revolution and has the cross section shown. When we take coordinates with origin at O, Z toward V, Y toward the vertical, and θ defined as shown in Figure 7.11(a), then the parametric equations for the caustic profiles are,

$$\begin{aligned} Z_c &= R(3 \cos \theta - \cos 3\theta)/4 \\ Y_c &= R(3 \sin \theta - \sin 3\theta)/4 \end{aligned} \tag{7.21}$$

There is another degenerate branch of the caustic which lies along the line (radius) FV of Figures 7.11(a) and (b). It is along this line that phased-line sources can be placed to intercept all the rays in a 60° cone measured at the sphere center about the radius normal to the incident plane wave. By proper control of amplitude and phase along this caustic line the electromagnetic energy incident on this spherical cap can be approximately brought into phase coherence (as is done optically with a paraboloidal mirror with a point source at its focal point). Feeds to accomplish this line source correction of spherical aberrations are discussed in Section 7.8 and in [3,9].

Subreflectors offer a wideband technique for focusing the rays reflected from a spherical cap to a point of phase where a feed horn can be located. Implicit (parametric) functions for the subreflector or the Gregorian type can be derived using raypaths and Snell's law at the spherical

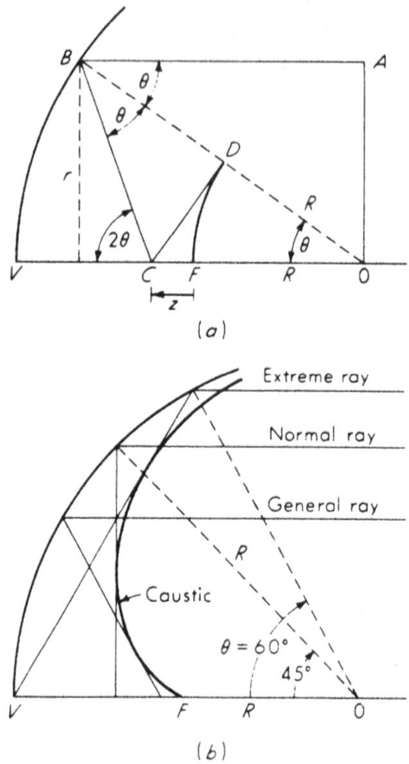

Figure 7.11 Geometry of (a) Spherical Cap Antenna Used to Compute Pathlengths and (b) Rays that Form Caustic
(From Collin and Zucker, Part II [3]. Reprinted with permission.)

surface [10]. Because of the axial symmetry the analysis and construction can be done in two dimensions and the subreflector curve rotated about the Z-axis to produce the Gregorian subreflector in three dimensions. These equations of the Gregorian generating curves are from [3,10],

$$u = Z/R = c - \frac{[c^2 - 4(1 - n)c - (1 + p^2) + (2 + p - 2n)^2](2c^2 - 1)}{4(pc^2 - c - n + 1)]}$$

$$v = Y/R = \pm \frac{(2cu - 1)s}{2c^2 - 1} \tag{7.22}$$

where $c = \cos\theta$ and $s = \sin\theta$

These equations are programmed in SGREG.BAS which produces the curves shown in Figure 7.12. The shape of representative Gregorian

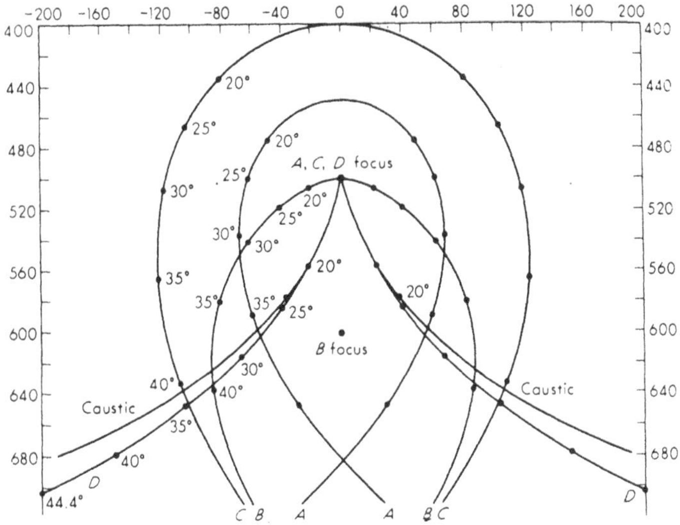

Curve	Corrector	
	Vertex *n*	Focus *p*
A	0.45	0.5
B	0.50	0.6
C	0.40	0.5
D	0.50	0.5

Figure 7.12 Gregorian Corrector Contours for Spherical Reflector Antennas (From Collin and Zucker, Part II [3]. Reprinted with permission.)

subreflectors are seen with vertex of subreflector generally near the paraxial focus and focal positions for point source feeds at p locations generally between F and V (see Figure 7.11) on the radius. The caustic cusp (7.21) is shown also in the figure. The simple program SGREG requires input of the sphere radius, R, and the subreflector focus and vertex positions normalized to R. For first try use $R = 10$, $F = 0.6$, and $V = 0.4$ to explore this simple program.

Designers can use an offset sector of the subreflector to illuminate an offset section of the sphere to avoid feed and subreflector blocking. The subreflector tends to concentrate the aperture energy near the edges of the illuminated spherical cap, causing what is known as an inverse taper when horns are used to feed the Gregorian subreflector. This problem also affects offset feeding when the shaped subreflector is used as a line source phase corrector by constructing the curves of (7.16) as two-dimensional pillboxes rather than as surfaces of revolution.

Improved subreflector correction methods for scanning spherical reflectors have been reported [11,12,13]. The tri-reflector technique of

Reference [11] is especially successful for feeding offset sections of spherical reflectors. (Offset angle performance is generally poor for the spherical antenna unless correcting subreflectors are used.) By shaping two subreflectors using numerical computer techniques, tapered aperture distributions with low cross polarizations are realized. Methods to compute the effect of subreflector edge diffraction using GTD methods on multiple subreflector systems for the sphere are treated in [12,13].

Spherical reflectors can be more simply fed (prime-focus fed with point-source feeds) with directive horns pointed along radii illuminating small symmetrical θ-caps to produce relatively good patterns. These feeds can be scanned by moving them on the surface of a sphere of radius R_f slightly larger than F with origin at O of Figure 7.11. Good patterns with beamwidths of one degree or greater over wide-angles can be obtained by feeding on any radii at selected points inside the paraxial focus. It is difficult to estimate the pattern gain, polarization, and sidelobe characteristics for feed horns of different directivity operating at different frequencies and at different positions along a feed radius. Program SPHAP.FOR is provided to give designers this information and to study trade-offs by computing radiation patterns using different parameters. The simple spherical antenna is not free of cross polarization and therefore the program allows both principal and cross-pol patterns to be computed. A preliminary look at the cross-pol aperture fields on the sphere can be made first using the program SPHPOL.BAS.

To run SPHPOL.BAS (listed in Appendix 7A) to examine aperture cross polarization, we select a sphere radius and for our first trial we use $R = 10$ units. This value has been edited into line 50. The program asks for, Z_0, the position of the feed along the feeding radius, and MM a power coefficient to make the feed directive. Try $Z_0 = 5.25$ and $MM = 5$. The program prints P_0, the co-polarized power in dB, and P_x, the cross-polarized power. The P_z component gives an indication of the nonparallel reflected rays. To see aperture fields without a directive feed set $E_0 = 1$ in line 260.

The analytical method for computing aperture fields and radiation patterns for the spherical reflector is similar to technique used in PARAP.FOR. A directive point source feed is located at Z_0 and pointed along a radius of the sphere. The origin is taken at F, the paraxial focus and the equations for sphere and its normals are

$$X^2 + Y^2 + \left(Z + \frac{R}{2}\right)^2 = R^2$$

$$-\mathbf{n} = \frac{X\mathbf{x} + Y\mathbf{y} + (Z + R/2)\mathbf{z}}{(X^2 + Y^2 + (Z + R/2)^2)^{1/2}} \tag{7.23}$$

We can compute ray paths and hence aperture phases assuming a wavelength and a feed pattern. A corrugated horn is assumed to have a $\cos^N \theta$ pattern with Y-polarized Huygen source phase directed along the feeding radius. The conical scanning interval of a hemispherical reflector for a corrugated horn with semiflare angle θ_h is approximately, $\pi - \theta_h$.

Running SPHAP.FOR requires a data file, SPH.DAT, with A = illuminated radius of spherical cap (not the whole cap), LAM = lambda, AP = radius larger than A, R = sphere radius, and Z_0 = the feeding point on the radius. No provisions for feeding off-radius are included because one can always rotate the aperture to a new central radius and position the feed to a Z_0 location. The second line of the data file SPH.DAT inputs the X_1, Y_1, Z_1 points of the center of the aperture and MM a field (not power) coefficient of the $\cos^{MM} \theta$ illumination function. Aperture phase and amplitude data are stored in the PHAMP.DAT file that can be summed by PATT.EXE to give the gain and sidelobes of the pattern for the illuminated spherical cap. By reducing MM, larger surface areas of the sphere can be illuminated but with larger (pathlength) phase errors. Also the position of Z_0 along the sphere radius is critical in this optimization with Z_0 positioned near the paraxial focus $Z = R/2$ for more directive feeds.

7.7 PARABOLIC TORUS ANTENNAS

The hybrid antenna, the parabolic torus, can be scanned many beamwidths along a circular arc in one angular plane. It is generally less aberrated than the sphere and can produce quite narrow beams in the microwave frequency bands. First the prime focus performance is explored with TORPOL.BAS to show the cross-polar aperture components for symmetric and offset feeding. Then antenna pattern computations to display gain and near-in sidelobe characteristics are developed in TORAP.FOR. The diffraction patterns of a prime-focus-fed torus antenna can be computed using the same numerical solution of three nonlinear equations as done in PARAP.FOR and SPHAP.FOR. A technique for shaping an offset subreflector to correct the aperture phase errors of the parabolic torus antenna is developed in TORCA.FOR, again using nonlinear solutions for an accurate representation of Snell's law, as done in SHAPE.FOR.

The parabolic torus antenna as generated by rotating a parabola of focus F in a circular arc of radius $R = R_0$ as seen in Figure 7.13 and discussed in [14] is now considered. Often the parabola focal length F is chosen to be about $0.49 \times R_0$. The length $\rho = R_H$ in (7.24) is then about $0.51 \times R_0$. The equations of the parabolic torus surface are

$$R_0 = F + R_H$$

$$X^2 + (Z + R_H)^2 = (R_0 - Y^2/(4F))^2 \tag{7.24}$$

The unit normal (inward or outward) vector to the surface is,

$$\pm\mathbf{n} = \frac{X\mathbf{x} + (R_0 - Y^2/(4f))Y/(2F)\mathbf{y} + (Z + R_H)\mathbf{z}}{R1}$$

$$R_1 = [X^2 + (R_0 - Y^2/(4F))Y^2/(4F^2) + (Z + R_H)^2]^{1/2} \tag{7.25}$$

To become familiar with the characteristics of the torus two BASIC programs are available, TORPOL.BAS and TORRAY.BAS. TORPOL explores offset feeding by computing the principal and cross-polarization components in the antenna aperture assuming a linear polarized Huygens source illumination with a directive pattern variable with NN of the $\cos^{NN}\theta$ pattern. TORRAY illustrates that the reflected rays, even when the plane wave incidence is normal to the Z-axis, do not pass through the focal point F except for points on the generating curve lying along the Y-axis.

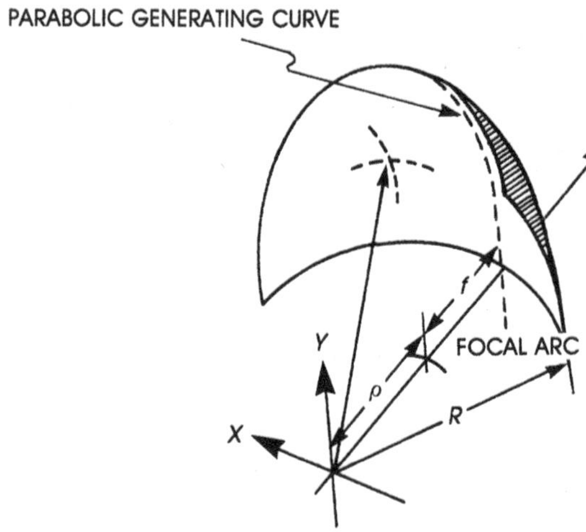

Figure 7.13 Parabolic Torus Antenna Geometry
(From Collin and Zucker, Part II [3]. Reprinted with permission.)

The parabolic torus antenna can be offset-fed like a paraboloid with some sacrifice in cross-pol performance. The program TORPOL.BAS calculates the aperture fields for an offset torus with radius of 10 (as for the sphere examined with SPHPOL). A value YC = 4 is edited in the program and the aperture co- and cross-pol fields and power levels displayed for a horn tilted toward YC. By responding to requests for XAP, and YAP (which are points on the torus surface relative to the reflector center point at YC), aperture power levels are printed as with SPHPOL.

Generally large f/D configurations are used to reduce the torus aperture phase errors and cross polarization. The program TORRAY.BAS has, edited in the code for the example, configuration parameters $F = 75$ and $R = 153.0612$ units with YC = 30. This program prints out the points X_1, Y_1, where incident rays parallel to the Z-axis intersect the plane, Z_1 = (a constant). By entering $Z_1 = 0$ (a plane normal to the torus axis at the focus) it is seen that in-coming rays striking points X, Y on the reflector surface do not focus to F, but generally cluster about F. The printout K_X, K_Y, E_X, E_Y are the line coefficients for these rays that can be used to check if the shaped subreflector truly focuses such rays to a point by using the program TRACE.BAS (See Chapter 8 and the software code manual for more details). The computation of torus aperture fields and radiation patterns is done nearly the same in TORAP.FOR as for the sphere in SPHAP.FOR, except that provisions for off-axis (X_0, Y_0, Z_0) focal locations are included. Again the solutions of three nonlinear equations are needed (or another mathematical procedure based on minimum pathlengths) to compute accurate patterns at all wavelengths.

7.8 A SHAPED CASSEGRAIN SUBREFLECTOR FOR CORRECTING APERTURE PHASE ERRORS ON THE PARABOLIC TORUS ANTENNA

A synthesis technique to shape Cassegrain subreflectors that corrects the aberrations of the offset parabolic torus antenna is now described as provided by TORCA.FOR. (Cluster feeds [15] can also be used to reduce near-in sidelobes on the torus. These are discussed later under distributed feeds.)

The shaping method to correct the torus aperture phase errors makes use of the numerical phase-correcting (surface-shaping) techniques explained in Chapter 3 and again in Chapter 8. Figure 7.3(a) shows in profile the offset Cassegrain geometry used although much longer focal lengths are used. (The configuration program CASS.BAS is used to compute the reference Cassegrain parameters. The example antenna in the optional software package with user's manual and example booklet uses the configuration constants edited in the program TORRAY.BAS.) In the $Y - Z$ plane the profiles of the torus subreflector and the paraboloid

(Cassegrain) subreflector are the same. For general rays reflected from the torus, (7.24) and (7.25) are used to shape the correcting subreflector. Only incident rays for beam angles $EL = 0°$, $AZ = 0°$ are considered, as scanning is done by feed positioning of the horn and shaped subreflector on the many identical torus radii and surface sectors of Figure 7.13. Beam scanning is done on one circular arc only.

Nonlinear equations are again used assuming a plane wave normal to the Z-axis to find the ray direction cosines for reflected rays intersecting any point on the shaped subreflector. The optimum tilting relations for an offset paraboloid are used for obtaining start-up numerical data for the shaping synthesis. First four X-cuts are made starting from the hyperboloidal subreflector that would feed the reference offset paraboloid. These cuts use the hyperboloid points only for a very small area at the midpoint of the subreflector and computed shaped X-cuts under the control of the rays reflected by the torus, using solutions to the nonlinear equation based on Snell's law of reflection on the torus surface. Shaping is achieved from the ray condition that all rays reflected from the subreflector focus to the reference subreflector focus.

TORCA.FOR is a rather large and complex program. The optional software package with user's manual and example booklet explains how to synthesize a Cassegrain subreflector and provide input data for the file SHAP.DAT. The contours of the shaped subreflector are computed along Y-cuts using the four X-cut data along a central strip of the shaped subreflector as inputs for the Y-cut shaping. The coordinates of the shaped Cassegrain are stored in file YU.DAT for the upward Y-cuts and in YD.DAT for the downward Y-cut data.

The input data file SHAP.DAT has the parameters needed to define the subreflector synthesis procedure. Similar to SHAPE.FOR, on line one of the file is entered E = eccentricity of reference subreflector, F = focal length, R_0 = radius of rotation of the torus, $RH = \rho = R_0 - F$. On line two is entered A = radius of torus reflector illuminated, D = hyperboloid constant, XI and YI are center points of the illuminated aperture, D_1 = increment of numerical shaping, and D_2 = interval of filing data points for constructing the shaped subreflector. SH is for shortening the X-cut length for trials. Line three enters the reference subreflector focus for EL = AZ = 0°. On line four the boundaries on the shaped area are defined U0 and U1 for X and V_0, V_1, V_B for Y. The program displays numbers as the 4-X strips are first made and then the YCUTU (up) and YCUTD (down) subroutines generate data files that can be used to machine the shaped surface or to check using TRACE if reflected rays computed by TORRAY are really focused to X_0, Y_0, Z_0. It is noted from the DEL values, displayed when TORCA runs, comparing the point (Z_1 coordinates) deviations of the shaped subreflector points from those on the reference (hyperboloidal) subreflector that the Y cuts near the Y axis have the smallest DEL values.

By using a large f/D configuration for the reference paraboloidal Cassegrain system, DEL values are smaller. Also the optimized tilt, based on results of Chapter 1 in CASS.BAS, of the offset Cassegrain subreflector with respect to the offset reference paraboloid reduces the cross polarization on the torus aperture. The torus antenna with correcting subreflector can be used at very short wavelengths depending on construction accuracy of the reflecting surfaces. Scanning over wide angles can be done with very narrow, high gain beams by moving a directive feed horn and the subreflector along the circular focal arc of the torus. Also, the configuration parameters and horn feed angle can be chosen to make the subreflector small, allowing multiple feeds to be placed closely along the circular feed arc.

The capability to correct aperture phase errors on the offset parabolic torus antenna provides the designer with a very high gain, low sidelobe antenna that can be scanned over a large angular arc by feed (and subreflector) positioning.

7.9 DISTRIBUTED FEEDS FOR REFLECTOR ANTENNAS

Phased line sources composed of strings of slot or dipole radiators are used to generate fan-shaped beams and to correct for reflector aberrations. In particular, line sources can lie along the best azimuth focal line of the offset prime-focus-fed paraboloid to radiate a $\csc^2 \theta$ coverage pattern for radar applications. The most familiar use of line sources, perhaps, is to correct for reflector aberrations on the spherical reflector where distributed sources are excited and phased to produce a high-gain, pencil-beam pattern. Usually at microwave frequencies slots on waveguides are the feed elements. The designer needs to know how to couple the slots (or dipoles) to the feed line to achieve the proper amplitude and power radiation along the line source. The phasing along the radiating line source is critical, and it can be realized usually by varying the radiator element spacings along the waveguide and using phase reversals (180° phase shift by reversing, for example, the polarization of dipoles or alternating the transverse location of longitudinal shunt slots from side to side on the wide surface of a rectangular waveguide).

7.9.1 Line Sources in the Paraboloid Focal Region

The analytical and computer techniques needed to design phased line sources can be illustrated for the important application of producing a shaped radiation pattern with a midpoint line source feed [3] lying along the best azimuth focus straight line of the prime-focus-fed paraboloid that was described in Section 7.2. The excellent wide angle characteristics of the offset paraboloid section computed by PRIME.BAS and PARAP.FOR can be exploited, for example, to produce a $\csc^2 \theta$ pattern by properly coupling and spacing radiators along a rectangular waveguide. This line source feed is called a midpoint corrector because the multiple radiating elements are all phased to focus to the center ($y_m = Y_c$ in Figure 7.14) of the offset paraboloid which is for the coordinates used here at the middle point Y_c, Z_c ($X_c = 0$).

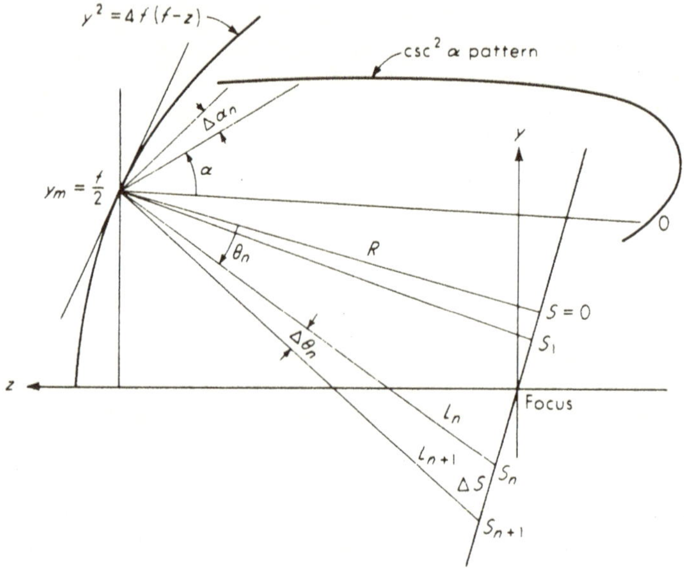

Figure 7.14 The Midpoint Line Source for Synthesizing Fan-Shaped Patterns
(From Collin and Zucker, Part II [3]. Reprinted with permission.)

The geometry in cross section is shown in Figure 7.14, and the designer has a choice of placing his phased line source below the focus or above the focus along the best azimuth focus line,

$$Y = -2fZ/Y_c$$

The best patterns are obtained above the focus, although some aperture blocking will result. If the above focal locus is chosen, the antenna must be turned over to make use of the $\csc^2 \theta$ pattern above the surface of the earth, because θ is now negative.

To focus the line source radiation on Y_c the longitudinal spacings along the rectangular waveguide of uniform cross section with $\mu = (\lambda_0/\lambda g)$ radiator spacings, S_n, must be such as to bring the radiation in phase at the midpoint, or

$$\mu(S_n - S_{n+1}) + \lambda_0/2 = L_{n+1} - L_n$$

where λ_0 and λ_g are wavelengths in free space and in the waveguide respectively, S is the radiator spacing, and L_n is the distance to the midpoint as seen in Figure 7.14 for radiators numbered n and $n + 1$. Some algebra is done in [3] which can be seen in the programs LINSOU.BAS (feeding above the focus F) and LINSOD.BAS for feeding below the surface. The peak of the $\csc^2 \theta$ pattern should be generated by a radiator at the paraboloid focus for maximum pattern gain in both cases. The origin of the spacing variable, S, is taken at a point perpendicular to the line source from the midpoint. The programs first print out the element spacings assuming $\lambda = 1$ and $\mu = 0.7$. We have entered the parameters $F = 30$ and YC = 15 in line 70 for the example that prints out without prompts. First a table of slot spacings, DELS at points S_0 appear for the θ pattern angles. The programs LINSOU and LINSOD are available in the optional software package with user's manual and example booklet.

The second parts of the programs LINSOU and LINSOD are used to find the power couplings to the waveguide to achieve a $\csc^2 \theta$ pattern. When the PAUSE is broken with the CR key a table of decibel couplings appears, assuming a traveling-wave line source array that dumps 10 dB of the power in a matched load. Information about distance to the paraboloid focus measured along the line source and the power radiated by the slot to achieve the csc pattern is printed. These couplings can be computed for longitudinal shunt slots displaced from the center line of the broad side of the rectangular waveguide (see [16], pp. 587–620). The coupling in dB for dipoles can be measured by experimentally increasing the probe depth and noting the dB drop as a function of probe depth. In the example here, a traveling-wave array is assumed and the power remaining at the end terminal is 10 dB. The program shows the couplings needed for the $\csc^2 \theta$ power values also displayed. The equations used in these programs for calculating transmission line couplings can be found in [3, p. 69].

These line-source amplitude and phase parameters for the elemental radiators at their focal locations can be entered into the program PARAP.FOR. The compound aperture distributions can be obtained by successively summing the PHAS and AMP data in the file PHAMP.DAT. This summation of the contributions of all the line source elements with their amplitude, phase, and locations taken into account can then be integrated by PATT.FOR to get the shaped antenna radiation pattern. The losses in gain for points away from the paraboloid focus are then taken into account and sidelobe levels computed. This technique permits a computer method for shaping fan beams.

When the phases of received radar echos are compared at the two ends of the midpoint corrector consisting of a traveling-wave array, a phase versus angle function is measured that can be used to extract elevation angle data from the shaped fan beam.

The phasing condition is simple in the case of the midpoint corrector for offset paraboloids, but the same procedure can be used to phase curved line sources to correct the phase along a ridge line of a paraboloid (using CAUSTIC.BAS).

The phase (ray path error) function for a correcting line source lying along a radius of a spherical reflector [3,9] is

$$\delta = Z[1 - 2f/(f + Z)]$$

where Z is measured from the paraxial focus of the sphere toward the spherical reflector as shown in Figure 7.11(a), and $f = R/2$, where R is the radius of the sphere.

When the required phase change along the line source is for an "end-fire" pattern, as is the case near the end of the correcting line source for the spherical reflector to illuminate at very small θ angles, then the above method using uniform rectangular waveguide fails and waveguide loading may be needed to vary μ to achieve end-fire radiation.

7.9.2. Transverse Correcting Techniques

When pencil beam patterns have high sidelobes due to aberrations it is possible to reduce the near-in lobes by using a cluster of small radiators as feeds. This technique reported in reference [15] for the torus antenna is also called the seven-feed cluster, as a central feed surrounded by six similar radiators can be used to reduce all the near-in sidelobes. The computed patterns with the phase of the sidelobes are used to position, phase, and intercouple the six side feeds to the main central illuminator feed. The coupling is at about the sidelobe level and the phase is the conjugate of the radiation pattern phase (see Chapter 8). The wide angle patterns computed by MULTE.FOR and MULTH.FOR, for example, can be plotted with phase and angular position of sidelobes data. The angular positions of sidelobes can be easily converted to physical distance as the metrics of the focal surfaces are known from CLCC.BAS and CLCG.BAS. Thus correcting cluster feeds, or, better, the coupling of adjacent radiators on a focal plane phased array, can be designed to reduce sidelobes on far-out patterns. This may be a better technique than shaping the ends of the subreflector, as each pattern can be adjusted separately (subreflector shaping influences all patterns and not always favorably).

7.9.3. Beam Pointing Using a Tilting Plane Reflector

Sometimes it is desirable to keep the transmitter, receiver, and feeding waveguides fixed, as well as the main reflector, but to scan over a limited sector by moving a plane mirror. The idea was suggested by Henning Olesen, of GTE, and reported in [17]. This technique is useful for both prime-focus-fed and dual-reflector antennas, although because prime-focus-fed paraboloids or the parabolic torus have smaller focal areas, the method is perhaps most useful for them in following the small elevation angle variations of satellites in geostationary orbits. The equations of the mirror location to radiate a fixed horn's pattern along a best focal arc are provided here. The plane mirror must be translated to place its center location X, Y, Z on a line between the center point of the main or subreflector (which point moves, too, with scan angle) and the focal point where normally the horn or feed center of focus would be located. The mirror must also be rotated so that the mirror image of the horn or feed radiation pattern appears to originate on the antenna focal arc. It would be desirable to keep the feed horn fixed, but generally some rotation about the feed center of phase is required to point the horn axis at the mirror center. The plane mirror size depends on where the horn is placed and must be large enough to reflect the horn pattern down to about the −30-dB levels. A plane mirror does not depolarize, although the mirror image changes any asymmetries the horn pattern may have with respect to the antenna aperture distribution.

A general program TILT.BAS, supplied in the optional software package with user's manual and example booklet, performs the optical imaging calculations. The first requirement is that the pathlength from the horn-phase center to the mirror center be the same as the pathlength from the best focal point on the focal arc to the mirror center. This condition focuses the pattern. The mirror center must lie on the line between the best focal point and the center of the main or subreflector being illuminated.

We need as input data the coordinates of the dish center, the arc focal point, and the location of the horn-phase center. For the dual-reflector systems the programs CLCC.BAS or CLCG.BAS give the first points assuming the designer has found the smallest circle of confusion or astigmatic focal points if narrow elevation or azimuth patterns are preferred.

The location of the horn requires some trials to find a location for a small mirror size and the smallest angular motion of the feed horn axis. The orientation of the mirror and horn are given in direction cosines (vector components) which for engineering need to be changed to angles in some convenient frame. Here all angles and positions are referred to the coordinate frame shown in the Figure 7.15.

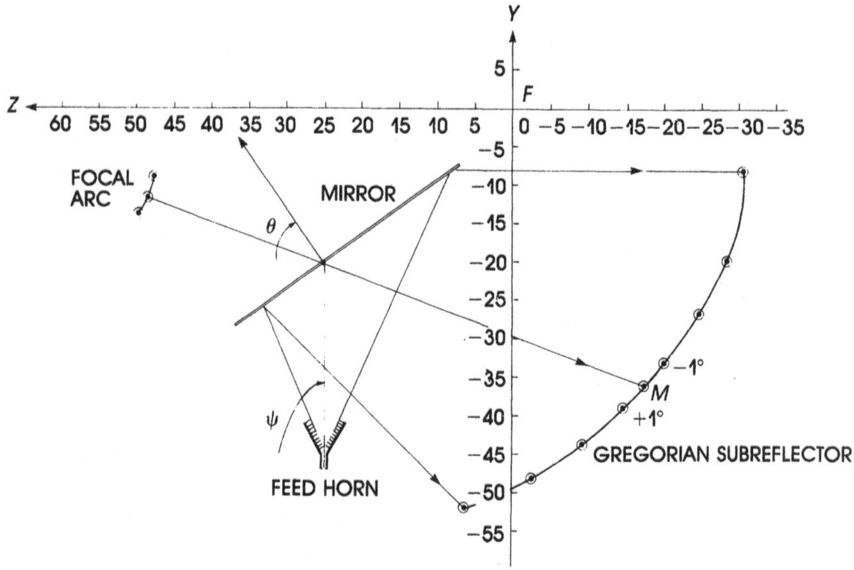

Figure 7.15 The Tilting Plane Mirror Scanning Technique

The derivation of the correct location of the tilting mirror can be followed in TILT.BAS. The constants (using computer program notation) C, D_1, D_2, and D_3 are defined to simplify the equation of the mirror center obtained by setting the distance squared from the mirror center to the focal point X_F, Y_F, Z_F and the feed phase center X_0, Y_0, Z_0 equal. Then the condition that the center point X,Y,Z lie on the line between the reflector center X_C, Y_C, Z_C and X_F, Y_F, Z_F are obtained from the equation of that line in terms of its direction cosines L_1, M_1, N_1. By solving for Y, we get the expressions written as NUM(numerator) and DEN(denominator). The angular tilt on the flat plate is obtained from Snell's law of reflection to get the mirror normal vector, and the horn-tilt direction cosines follow directly from the geometry.

For operational use, the burden of entering data points into the program can be alleviated by making TILT a subroutine of the programs that compute the focal arcs of the antenna and the location of the sub-reflector or main reflector middle points. In the high power or millimeter-wave (with lossy feed lines) feeding of the prime-focus-fed offset paraboloid or parabolic torus antennas this tilting mirror technique may be useful.

References

[1] Eisenhart, L.P., *A Treatise on the Differential Geometry of Curves and Surfaces,* New York: Dover Publications, 1960.
[2] Parke, N.G., III, "Calculations of the Caustic Surface of a Paraboloid of Revolution for an Incoming Plane Wave of 20 Degrees Incidence." MA: Parke Math. Lab., Concord, May 1952.
[3] Collin, R.E., and F.J. Zucker, *Antenna Theory,* Part II, Chapters 16 and 17. New York: McGraw-Hill, 1969.
[4] Ruze, John, "Lateral-feed Displacement in Paraboloid," IEEE Trans. AP, Sept. 1965.
[5] Afifi, M.S., "Aberration and Dispersion of the Focus of a Paraboloid," IEEE Antennas and Propagation Symp. 1971, p. 219.
[6] Sletten, C.J., and R.A. Shore, "Focal Surface of Offset Dual-Reflector Antennas," IEE Proc., Vol. 129, Pt. H, No. 3, pp. 109–115, June 1982.
[7] Faigen, I.M., C.F. Reichert, C.J. Sletten, and R.A. Shore, "Reflector Antennas with Low Sidelobes, Low Cross Pol and High Aperture Efficiency," Chu Assoc., Littleton, MA for NASA Lewis Space Center, NASA Rep. Cr-174170, March 1984.
[8] Press, W.P., B.P. Flannery, S.A. Teukolsky, and W.T. Vetterling, *Numerical Recipes,* "The Art of Scientific Computing," pp. 569–572. Cambridge Univ. Press, 1986.
[9] Spencer, R.C., C.J. Sletten, and J.E. Walsh, "Correction of Spherical Aberration by a Phased Line Source," Air Force Camb. Res. Labs., Bedford, MA, Rep. E-5069, May 1961.
[10] Holt, F.S., and E. Bouche, "A Gregorian Corrector for Spherical Reflectors," IEEE Trans. Vol. AP-12, No. 1, Jan. 1964.
[11] Watanabe, F., and Y. Mizugutch, "An Offset Spherical Tri-Reflector Antenna," Trans. of IECE of Japan, Vol. 66, No. 2, Feb. 1983.
[12] Kildal, Per-Simon, "An Asymptotic Transition Region Theory for Calculation of Diffraction Losses in Multiple Reflector Antennas,"

IEEE AP-S Int. Symp. Digest, Vol. I, Philadelphia, June 1986.

[13] Baker, Lynn, "A Dual Reflector Shaping Method and Its Application to Feeds for Spherical Main Reflector," IEEE AP-S Int. Symp. Digest, Vol. I, Philadelphia, June 1986.

[14] Kreutel, R. "A Multiple-beam Torus Earth Station Antenna, IEEE-ICC'80, June 1980.

[15] Mavroides, W.G., and J.H. Provencher, "Off-axis Sidelobe Reduction in Elliptic Torus Reflectors," AFCRC-TR-58-199, Bedford, MA, Dec. 1958.

[16] Collin, R.E., and F.J. Zucker, "Antenna Theory Part I," pp. 587–620. New York: McGraw-Hill, 1969.

[17] Sletten, C.J., and S.E. Carrillo, "Scanning Multibeam Communication Antennas," IEEE AP-S Int. Symp. Digest, Vol. 1, pp. 474–477, June 1984.

Appendix 7A
List of Chapter 7 Computer Programs and Source Code Listings for Selected Programs

In this appendix the names of computer programs developed especially for Chapter 7 are listed. The source code for some of the simpler, basic programs are included here. All programs are discussed in the text and brief operating instructions are given for the tutorial programs listed in this appendix. By studying and running this code detailed information on the design software is clearly illustrated. The programs are written in Microsoft BASIC and FORTRAN-77 for IBM PC and compatible computers.

The program names are written in the first few lines of the source code. The suffix .FOR indicates a FORTRAN program and .BAS a basic program; .DAT indicates a data file.

SOFTWARE LIST, CHAPTER 7

PROGRAM NAME: ASSOCIATED PROGRAMS, FILES AND SUBROUTINES:

CAUSTIC.BAS
LOCI.BAS
*PRIME.BAS
PRIMEG.BAS; CIRC.BAS, LINEPL.BAS
*PARAP.FOR; PRIME.BAS, PARD.DAT, MNEWT, LUDCMP, LUBKSB, PHAMP.DAT, ALAM.DAT, PATT.FOR, CJS.DAT, PLPATT.BAS
*CLCC.BAS; CASS.BAS
CLCCG.BAS; CIRC.BAS, LINEPL.BAS
CLCG.BAS; GREG.BAS
CLCGG.BAS; CIRC.BAS, LINEPL.BAS
MULTH.FOR; CLCC.BAS, DATINH.DAT, PHAMP.DAT,

ALAM.DAT, PATT.FOR, CJS.DAT, PLPATT.BAS
MULTE.FOR; CLCG.BAS, DATINE.DAT, PATT.FOR, PHAMP.DAT, ALAM.DAT, CJS.DAT, PLPATT.BAS
*SHAPE.FOR; CLCC.BAS, SHAP1.DAT, MNEWT, LUDCMP, LUBKSB, TIP.DAT
*SGREG.BAS
*SPHPOL.BAS
*SPHAP.FOR; SPH.DAT, MNEWT, LUDCMP, LUBKSB, PHAMP.DAT, ALAM.DAT, PATT.FOR, CJS.DAT, PLPATT.BAS
*TORPOL.BAS
TORRAY.BAS
TORAP.FOR; TORD.DAT, MNEWT, LUDCMP, LUBKSB, PHAMP.DAT, ALAM.DAT, PATT.FOR, CJS.DAT, PLPATT.BAS
TORCA.FOR; SHAP.DAT, MNEWT, LUDCMP, LUBKSB, YD.DAT, YU.DAT
LINSOD.BAS
LINSOU.BAS
TILT.BAS

* These programs are printed in this appendix.

All programs can be listed and operated from the source codes in ASCII in the optional computer software package with user's manual and example booklet. Programs MNEWT, LUDCMP, and LUBKSB are Numerical Recipes subroutines in FORTRAN that must be linked to the driver program named. The data files must be in the directory to run the programs. Unformatted files PHAMP.DAT and ALAM.DAT are generated by the FORTAN pattern programs PARAP.FOR, SPHAP.FOR, and TORAP.FOR. PHAMP.DAT may be large. These are used by PATT.EXE and must be available. *Do not abort* (*CtrlC*) *these pattern programs in progress* (run them to termination). CtrlC deletes the unformatted files needed for PATT.

```
10 REM Computes Focal Spots for Prime Focus Feeding of Offset Parab.
20 REM Sets up Screen perpendicular to Central Ray.
30 PRINT "PRIME.BAS"
40 PRINT
50 INPUT "F, Yc, ", F,Y3
60 LPRINT "F=";F;"Yc=";Y3
70 INPUT "EL, AZ, RADIUS, DELZ ", A2,A5,R,Z5
80 LPRINT "EL=";A2;"AZ=";A5;"R=";R;"DELZ=";Z5
90 I=1 : PI=3.141592654# : D=0
100 A2=A2*PI/180 : A5=A5*PI/180
110 X=0 : Y=Y3
120 Z=F-(X*X+Y*Y)/4/F
130 R3=2*(X*COS(A2)*SIN(A5)-Y*SIN(A2)+2*F*COS(A2)*COS(A5))/(X*X+Y*Y+4*F*F)
140 L1=COS(A2)*SIN(A5)-R3*X
150 M1=-SIN(A2)-R3*Y
160 N1=COS(A2)*COS(A5)-R3*2*F
170 IF I=0 THEN GOTO 310
180 Y2=((-Z)*M1/N1+Y)/(1+Y3*M1/2/F/N1)
190 X2=L1/M1*(Y2-Y)+X
200 Z2=-Y2*Y3/2/F
210 LPRINT "X2=";X2;"Y2=";Y2;"Z2=";Z2
220 T1=L1/N1 : A7=ATN(T1)
230 T2=M1/N1 : A6=ATN(T2)
240 B1=L1 : B2=M1 : B3=N1
250 Z4=(B1*T1*Z+B2*T2*Z-(D+B1*X+B2*Y))/(B3+B1*T1+B2*T2)
260 Y4=T2*(Z4-Z)+Y
270 REM PRINT "Y4=";Y4;"Z4=";Z4
280 Z3=Z2-Z4
290 Z6=Z3/COS(A6)^2
300 D=-(Z6+Z5)*B3
310 D1=B3+B1*L1/N1+B2*M1/N1
320 Z1=(B1*L1/N1*Z+B2*M1/N1*Z-(D+B1*X+B2*Y))/D1
330 X1=L1/N1*(Z1-Z)+X
340 Y1=M1/N1*(Z1-Z)+Y
350 REM PRINT "X1=";X1;"Y1=";Y1;"Z1=";Z1
360 IF I=0 THEN GOTO 420
370 X5=X1 : Y5=Y1
380 LPRINT "XF=";X5;"YF=";Y5;"ZF=";Z1
390 PRINT "XF=";X5;"YF=";Y5;"ZF=";Z1
400 LPRINT
410 IF I=1 THEN GOTO 510
420 X0=(X1-X5)/COS(A7)
430 Y0-(Y1-Y5)/COS(A6)
440 REM LPRINT "PH1-";A8*180/PI;"X0=";X0;"Y0=";Y0
450 PRINT "PH1=";A8*180/PI;"X0=";X0;"Y0=";Y0
460 IF A8>359*PI/180 THEN GOTO 70
470 A8=A8+30*PI/180
480 Y=Y3+R*SIN(A8)
490 X=R*COS(A8)
500 GOTO 120
510 I=0 : Y=Y3 : X=R : A8=0
520 GOTO 120
530 END
```

```
      PROGRAM PARAP
C     Computes phase and amplitude distribution on offset circular
C     paraboloid section for point source feed in location X0,Y0,Z0
      PARAMETER(TOLX=1.0E-6,N=3,TOLF=1.0E-6,NP=15)
      DIMENSION X(NP),ALPHA(NP,NP),BETA(NP)
      COMMON X1,Y1,Z1,X0,Y0,Z0,F,R0,R1
      REAL LAM
      DATA EP,PI/.00001,3.141592654/
      ZP=0.0

      OPEN(3,FILE='PARD.DAT',STATUS='OLD')
      REWIND 3
      READ(3,'(7F8.3)') A,LAM,YC,F,Z11,EL,AZ
      READ(3,'(5F11.6)') X0,Y0,Z0,AP,DEL
      READ(3,'(2I4)') NRAD,NANG
      TWOPI=2.0*PI
C     TH=ATAN(-Y0/(F-Z0))
C     THETA=TH*180/PI
C     ZD=TAN(TH)*YC

      WRITE(*,8) A,LAM,YC,F,Z11,EL,AZ
8     FORMAT(' A=',F8.3,' LAM=',F8.3,' YC=',F8.3,' F=',F8.3,
     *  ' Z11=',F8.3,' EL=',F8.3,' AZ=',F8.3)
      WRITE(*,10) X0,Y0,Z0,AP,DEL
10    FORMAT(' X0=',F11.6,' Y0=',F11.6,' Z0=',F11.6,' AP=',F11.6
     *  ' DEL=',F11.6)
      WRITE(*,12) NRAD,NANG
12    FORMAT(' NRAD=',I4,' NANG=',I4)
      OPEN(4,FILE='PHAMP.DAT',STATUS='NEW',FORM='UNFORMATTED')
      OPEN(7,FILE='ALAM.DAT',STATUS='NEW',FORM='UNFORMATTED')
      WRITE(7) AP,LAM
      WRITE(7) NRAD,NANG
      PAUSE'FILES OK?'

      EL=EL*PI/180.0
      AZ=AZ*PI/180.0
      ZC=F-YC**2/4.0/F
      Y1=YC
      Z1=Z11
      X(1)=0.0
      X(2)=YC
      X(3)=ZC
      PLT0=0.0

C     Finding initial ray pathlength PLT0
      DO 13 J=1,5
              CALL MNEWT(1,X,N,TOLX,TOLF)
              CALL USRFUN(X,ALPHA,BETA)
13    CONTINUE

      R00=R0
      PLT0=R1+R0
      WRITE(*,15) X(1),X(2),X(3),PLT0
15    FORMAT(' X=',E15.6,' Y=',E15.6,' Z=',E15.6,' PLT0=',E15.6)
      XM=X(1)
      YM=X(2)
      ZM=X(3)
      PAUSE'CENTER POINT'

      DELPHI=TWOPI/FLOAT(NANG-1)
      DELRAD=1.0/FLOAT(NRAD-1)
```

```
      PHI0=-PI
      RA=0.0

      DO 101 I=1,NANG

      X(1)=XM
      X(2)=YM
      X(3)=ZM
      PHI=PHI0+(I-1)*DELPHI

              DO 100 K=1,NRAD
              R=RA+(K-1)*DELRAD
              XP=AP*R*COS(PHI)
              YP=AP*R*SIN(PHI)
      X1=XP*COS(AZ)+YP*SIN(AZ)*SIN(EL)
      Y1=YP*COS(AZ) + YC
      Z1=-XP*SIN(AZ)+YP*COS(AZ)*SIN(EL) + Z11

      DO 98 J=1,4
              CALL MNEWT(1,X,N,TOLX,TOLF)
              CALL USRFUN(X,ALPHA,BETA)
98    CONTINUE

              PHAS=(R0+R1-PLTO)*TWOPI/LAM
              AMP=(1.0-(1.0-DEL)*(R*AP/A)**2)*R00/R0
      RT=SQRT(X(1)**2+(X(2)-YC)**2)
      IF(RT.GT.A) THEN
      AMP=0.0
      PHAS=0.0
      ENDIF

      WRITE(4) AMP,PHAS
      WRITE(*,96) I,K,AMP,PHAS
96    FORMAT(' I=',I4,' K=',I4,' AMP=',F12.6,' PHAS=',E13.4)

100   CONTINUE
101   CONTINUE
      STOP
      END



      SUBROUTINE USRFUN(X,ALPHA,BETA)
      PARAMETER(NP=15)
      DIMENSION ALPHA(NP,NP),BETA(NP),X(NP)
      COMMON X1,Y1,Z1,X0,Y0,Z0,F,R0,R1


      A0=X0-X(1)
      B0=Y0-X(2)
      C0=Z0-X(3)
      A1=X(1)-X1
      B1=X(2)-Y1
      C1=X(3)-Z1
      R=SQRT(X(1)**2+X(2)**2+4.0*F*F)
      R1=SQRT(A1**2+B1**2+C1**2)
      R0=SQRT(A0**2+B0**2+C0**2)
      BR=A1*X(1)+B1*X(2)+C1*2.0*F
      F1=2.0*X(1)*BR
      F2=R**2*R1
      F3=2.0*BR+4.0*X(1)**2-2.0*X1*X(1)
      F4=2.0*X(1)*R1+R**2/R1*A1
      F5=2.0*X(2)*R1+R**2/R1*B1
      F6=2.0*X(2)*BR
      F7=2.0*BR+4.0*X(2)**2-2.0*Y1*X(2)
```

```
      ALPHA(1,1)=X(1)/2.0/F
      ALPHA(1,2)=X(2)/2.0/F
      ALPHA(1,3)=1.0
      ALPHA(2,1)=-(B0**2+C0**2)/R0**3-(B1**2+C1**2)/R1**3+
     *  (F2*F3-F1*F4)/F2**2
      ALPHA(2,2)=A0*B0/R0**3+A1*B1/R1**3+
     *  (F2*(4.0*X(1)*X(2)-2.0*X(1)*Y1)-F1*F5)/F2**2
      ALPHA(2,3)=A0*C0/R0**3+A1*C1/R1**3+
     *  (F2*4.0*F*X(1)-F1*R**2/R1*C1)/F2**2
      ALPHA(3,1)=B0*A0/R0**3+A1*B1/R1**3+
     *  (F2*(4.0*X(2)*X(1)-2.0*X(2)*X1)-F6*F4)/F2**2
      ALPHA(3,2)=-(A0**2+C0**2)/R0**3-(A1**2+C1**2)/R1**3+
     *  (F2*F7-F6*F5)/F2**2
      ALPHA(3,3)=B0*C0/R0**3+B1*C1/R1**3+
     *  (F2*4.0*X(2)*F-F6*R**2/R1*C1)/F2**2

      BETA(1)=F-X(3)-(X(1)**2+X(2)**2)/4.0/F
      BETA(2)=-A0/R0+A1/R1-F1/F2
      BETA(3)=-B0/R0+B1/R1-F6/F2

      END
```

```
10 REM Computes focal spots for offset Cassegrain antenna
20 REM using constants from CASS.BAS and El, Az angles.
30 PRINT "CLCC.BAS"
40 PI=3.141592654#
50 REM Parameters of Main Reflector and Hyperboloid Subrefl.
60 A11=4.947447  : F=60  : E=1.7  : Y3=70  : C=50  :D=-16.34948
70 REM INPUT "Beta,E,F,YC,C,D ",A11,E,F,Y3,C,D
80 Z4=C*COS(A11*PI/180)
90 PRINT "For A2 & A5=0 then Zfocus= ";Z4
100 INPUT "EL, AZ, RADIUS, Zfocus ",A2,A5,R,Z4
110 LPRINT "EL=";A2;"AZ=";A5;"R=";R;"Z4=";Z4
120 H=0 : K=0: I=0 : A2=A2*PI/180 : A5=A5*PI/180
130 A1=A11*PI/180 : SN1=SIN(A1) : CS1=COS(A1)
140 X=0 : Y=Y3 : J=1 : SN2=SIN(A2) :CS2=COS(A2) : SN5=SIN(A5)
150 CS5=COS(A5)
160 Z=F-(X*X+Y*Y)/4/F
170 R3=2*(X*CS2*SN5-Y*SN2+2*F*CS2*CS5)/(X*X+Y*Y+4*F*F)
180 L1=CS2*SN5-R3*X
190 M1=-SN2-R3*Y
200 N1=CS2*CS5-R3*2*F
210 T=L1/M1 : S=X-T*Y : U=N1/M1 : V=Z-U*Y : P=V-K
220 A=T*T + 1 +U*U-E*E*(U*U*CS1^2+SN1^2-2*U*CS1*SN1)
230 B1=2*E*E*(-U*P*CS1^2+H*SN1^2+(P-U*H)*CS1*SN1-D*U*CS1+D*SN1)
240 B2=2*(T*S-H+P*U)
250 B=B1+B2
260 C3=H*P*CS1*SN1+D*P*CS1+D*H*SN1
270 C4=S*S+H*H+P*P+E*E*(-D*D-P*P*CS1^2-H*H*SN1^2-2*C3)
280 Y1=(-B+SQR(B*B-4*A*C4))/(2*A)
290 X1=Y1*T+S
300 Z1=Y1*U+V
305 PRINT "X=";X;"Y=";Y;"Z=";Z
310      PRINT "X1=";X1;"Y1=";Y1;"Z1=";Z1
320 C5=CS1*(Z1-K)-SN1*(Y1-H)+D
330 R5=X1^2+(Y1-H+E*E*C5*SN1)^2 + (Z1-H-E*E*C5*CS1)^2
340 R4=2*(L1*X1+M1*(Y1-H+E*E*C5*SN1)+N1*(Z1-K-E*E*C5*CS1))
350 R6=R4/R5
360 L2=-R6*X1+L1
370 M2=M1-R6*(E*E*SN1*C5+Y1-H)
380 N2=N1-R6*(Z1-K-E*E*C5*CS1)
390 IF J=1 THEN GOTO 560
400 Z0=(L3*L2/N2*Z1+Z1*M3*M2/N2-(D3+L3*X1+M3*Y1))/(N3+L3*L2/N2+M3*M2/N2)
410 X0=L2/N2*(Z0-Z1)+X1
420 Y0=M2/N2*(Z0-Z1)+Y1
430 X5=(X0-X4)/COS(A7)
440 Y5=(Y0-Y4)/COS(A6)
450 REM PRINT "X0=";X0;"Y0=";Y0
460 LPRINT "PHI=";A8*180/PI;"X5=";X5;"Y5=";Y5
470 IF J=0 THEN GOTO 500
480 A8=A8+30*PI/180
490 GOTO 510
500 A8=0
```

```
510 Y=Y3+R*SIN(A8)
520 X=R*COS(A8)
530 J=2
540 IF A8>2*PI THEN A8=0 : : LPRINT : GOTO 100
550 GOTO 160
560 L3=L2 : M3=M2 : N3=N2
570 T3=M3/N3 : A6=ATN(T3)
580 T4=L3/N3 : A7=ATN(T4)
590 X4=L3/N3*(Z4-Z1)+X1
600 Y4=M3/N3*(Z4-Z1)+Y1
610 D3=-(L3*X4+M3*Y4+N3*Z4)
620 J=0
630 IF I=1 THEN GOTO 400
640 LPRINT "XF=";X4;"YF=";Y4;"ZF=";Z4
650 LPRINT
660 I=1
670 GOTO 400
680 END
```

```
10 REM Computes Spherical Gregorian Reflector Contour
20 PRINT "SGREG.BAS"
30 PI=3.14159265359#
40 INPUT "Radius=R,Focus=P/R,Vertex=N/R ",R,P,N
50 INPUT "Theta in Degrees from Center of Sphere ",A
60 A=A*PI/180
70 C=COS(A) : S=SIN(A)
80 NUM=(C*C-4*(1-N)*C-(1+P*P)+(2+P-2*N)^2)*(2*C*C-1)
90 DEN=4*(P*C*C-C-N+1)
100 U=C-NUM/DEN
110 V=-S*(2*C*U-1)/(2*C*C-1)
120 Y=V*R
130 Z=U*R
140 PRINT "Y=";Y;"Z=";Z
150 PRINT
160 GOTO 50
170 END
```

```
      PROGRAM SHAPE
C     Computes shaped Cassegrain surface to correct phase errors
C     for wide angle beam.
      PARAMETER(TOLX=1.0E-6,N=3,TOLF=1.0E-6,NP=15)
      DIMENSION X(NP),ALPHA(NP,NP),BETA(NP)
      DIMENSION SS(400)
      COMMON/CCASS/CAE,CAE2,CAE4,DCA,CASA,ESQ,SASQ,DSA2,DSQ
      COMMON/SLV/X1,Y1,Z1,L0,M0,N0,F,R1,A1,B1,C1
      REAL L0,M0,N0,K3,K4,K5,K6,K7,K8,K9,L11,M11,N11
      DATA EP,PI/.00001,3.141592654/
      DATA J,K,N1,M,K5,K6,K7,K8,K9/1,6,0,6,0.0,0.0,0.0,0.0,0.0/
      D1=0.1
      D2=0.5
      ANGRAD = PI/180.0

      OPEN(3,FILE='SHAP1.DAT',STATUS='OLD')
      REWIND 3
      READ(3,'(7F7.3)') E,F,C,R0,YC,EL,AZ
      READ(3,'(4F11.6)') A,D,XI,YI
      READ(3,'(3F11.6)') X0,Y0,Z0
      READ(3,'(4F11.6)') U0,U1,V0,V1

      WRITE(*,8) E,F,C,R0,YC,EL,AZ
8     FORMAT(' E=',F7.3,' F=',F7.3,' C=',F7.3,' R0=',F7.3,
     *  ' YC=',F7.3,' EL=',F7.3,' AZ=',F7.3)
      WRITE(*,10) A,D,XI,YI
10    FORMAT(' A=',F11.6,' D=',F11.6,' XI=',F11.6,' YI=',F11.6)
      WRITE(*,11) X0,Y0,Z0
11    FORMAT(' X0=',F11.6,' Y0=',F11.6,' Z0=',F11.6)
      WRITE(*,12) U0,U1,V0,V1
12    FORMAT(' U0=',F11.6,' U1=',F11.6,' V0=',F11.6,' V1=',F11.6)
      PAUSE'FILES OK?'

      OPEN(6,FILE='TIP.DAT',STATUS='NEW')

      EL=EL*ANGRAD
      AZ=AZ*ANGRAD
      A=A*ANGRAD
```

```
        CA=COS(A)
        SA= SIN(A)
        CASA=CA*SA
        CASQ=CA*CA
        SASQ=SA*SA
        DCA=D*CA
        DSA=D*SA
        DSA2=2.0*DSA
        DSQ=D*D
        ESQ=E*E
        ESQ2=2.0*ESQ
        CAE=CASQ - 1.0/ESQ
        CAE2=CAE*2.0
        CAE4=CAE*4.0
        L0=COS(EL)*SIN(AZ)
        M0=-SIN(EL)
        N0=COS(EL)*COS(AZ)
        L=INT((U1-U0+EP)/D2)+ 1


        Y1=V0
        X1=U0
        Z1=CASS(X1,Y1)
        ZI=F-(YI*YI+XI*XI)/4.0/F
99      X(1)=XI
        X(2)=YI
        X(3)=ZI

C       Finding initial paraboloid point.
        DO 13 J=1,5
                CALL MNEWT(1,X,N,TOLX,TOLF)
                CALL USRFUN(X,ALPHA,BETA)
13      CONTINUE

        WRITE(*,15) X(1),X(2),X(3)
15      FORMAT(' X=',E15.6,' Y=',E15.6,' Z=',E15.6)
        XI=X(1)
        YI=X(2)
        ZI=X(3)

C---Computes start-up points and slopes.
100     Z1=CASS(X1,Y1)
        P1 = X1/ESQ
        R = CA*Z1 - Y1*SA +D
        F0=Y1/ESQ + SA*R
        F3= Z1/ESQ - CA*R
        K8= -P1/F3
        K9= -F0/F3
        K7=K6
        K6=K5
        K5=K8
        N1 = N1 + 1
        WRITE(*,'(1X,A,2I4,E15.6)')' N1,M,K8=',N1,M,K8
        IF(N1.LE.3) THEN
        X1= X1 + D1
        GOTO 100
        ENDIF
C---Computes slope check CHK
        X1= X1+D1
        Z1=CASS(X1,Y1)
101     P1 = X1/ESQ
        R=CA*Z1 -Y1*SA + D
        F0=Y1/ESQ + SA*R
        F3=Z1/ESQ - CA*R
        K8= -P1/F3
        K9= -F0/F3
```

```
        CHK=K8
102     DO 98 J=1,4
                CALL MNEWT(1,X,N,TOLX,TOLF)
                CALL USRFUN(X,ALPHA,BETA)
98      CONTINUE
        RE=SQRT(X(1)**2+(YC-X(2))**2)
        IF(RE.GT.R0.AND.X1.LT.U1/2.) THEN
        N1=0
        GOTO 100
        ENDIF

        L11=A1/R1
        M11=B1/R1
        N11=C1/R1
        K3=L11/N11
        K4=M11/N11
        T1R= -1.0/SQRT(K3*K3+K4*K4+1.0)
        C6=T1R
        A6=K3*T1R
        B6=K4*T1R
        X0MX1=X0-X1
        Y0MY1=Y0-Y1
        Z0MZ1=Z0-Z1
        T2R=1.0/SQRT(X0MX1*X0MX1+Y0MY1*Y0MY1+Z0MZ1*Z0MZ1)
        A7=X0MX1*T2R
        B7=Y0MY1*T2R
        C7=Z0MZ1*T2R
        T3R=1.0/SQRT(2.0*(1.0 - (A6*A7+B6*B7+C6*C7)))
        A3=(A6-A7)*T3R
        B3=(B6-B7)*T3R
        C3=(C6-C7)*T3R
        K8=-A3/C3
        K9=-B3/C3
C---Uses Adam's projector to find next point on subreflector.
        WRITE(*,'(1X,A,2E15.6)')' K8T K8=',CHK,K8
        X1= X1 + D1
        Z1=Z1+D1/24.0*(55.0*K8-59.0*K5+37.0*K6-9.0*K7)
        K7=K6
        K6=K5
        K5=K8
        ZT=CASS(X1,Y1)
        DEL=Z1-ZT
        WRITE(*,18) X1,Y1,Z1,DEL
18      FORMAT(' X1=',E15.6,' Y1=',E15.6,' Z1=',E15.6,' DEL=',E14.6)

        IF(M.GE.K) THEN
        M=0
        WRITE(6,51) X1,Y1,Z1,DEL
        WRITE(*,51) X1,Y1,Z1,DEL
51      FORMAT(' X1=',F11.6,' Y1=',F11.6,' Z1=',F11.6,' DEL=',E14.6)
        ENDIF
        M=M+1

        IF(X1.GE.U1) THEN
        Y1=Y1+5.0*D1
        M=K
        X1=U0
        N1=0
        GOTO 99
        ENDIF
        IF(Y1.GT.V1) GOTO 105

        GOTO 101

105     CLOSE (6)
        STOP
```

```
      END


      FUNCTION CASS(X,Y)
      COMMON/CCASS/CAE,CAE2,CAE4,DCA,CASA,ESQ,SASQ,DSA2,DSQ

      XSQ=X*X
      YSQ=Y*Y
      B = 2.0*(DCA - Y*CASA)
      C = DSQ - (XSQ+YSQ)/ESQ + YSQ*SASQ - DSA2*Y
      CASS = (-B - SQRT(B*B - CAE4*C))/CAE2

      RETURN
      END
C---Computes true values of main reflector points.
      SUBROUTINE USRFUN(X,ALPHA,BETA)
      PARAMETER(NP=15)
      DIMENSION ALPHA(NP,NP),BETA(NP),X(NP)
      COMMON/SLV/ X1,Y1,Z1,L0,M0,N0,F,R1,A1,B1,C1
      REAL L0,M0,N0


      A1=X1-X(1)
      B1=Y1-X(2)
      C1=Z1-X(3)
      R=SQRT(X(1)**2+X(2)**2+4.0*F*F)
      R1=SQRT(A1**2+B1**2+C1**2)
      BR=L0*X(1)+M0*X(2)+2.0*F*N0

      ALPHA(1,1)=X(1)/2.0/F
      ALPHA(1,2)=X(2)/2.0/F
      ALPHA(1,3)=1.0
      ALPHA(2,1)=B1*A1/R1**3+2.*X(2)*L0/R**2-4.*X(1)*X(2)*BR/R**4
      ALPHA(2,2)=-(A1**2+C1**2)/R1**3+
     *  2.0*(L0*X(1)+2.*M0*X(2)+2.*F*N0)/R**2-4.0*X(2)**2*BR/R**4
      ALPHA(2,3)=B1*C1/R1**3
      ALPHA(3,1)=-(B1**2+C1**2)/R1**3+
     *  (4.*X(1)*L0+2.*M0*X(2)+4.*F*N0)/R**2-4.*X(1)**2*BR/R**4
      ALPHA(3,2)=B1*A1/R1**3+2.*X(1)*M0/R**2-4.*X(1)*X(2)*BR/R**4
      ALPHA(3,3)=A1*C1/R1**3
C---Three nonlinear equations from Snell's law.
      BETA(1)=F-X(3)-(X(1)**2+X(2)**2)/4.0/F
      BETA(2)=M0-B1/R1-2.0*X(2)*BR/R**2
      BETA(3)=L0-A1/R1-2.0*X(1)*BR/R**2

      END
```

```
10 REM *THIS IS A DEMONSTRATION PROGRAM TO SHOW*
20 REM *THE CROSS POLARIZATION ON SPHERE APERTURE*
30 PRINT "SPHPOL.BAS"
40 PI=3.14159265359#
50 R=10 : EP=.000001 : LC=.4342944819#
60 INPUT "Focus on Radius, Z0, and Horn taper constant, MM: ",Z0,MM
70 R0=R-Z0
80 REM The Feed is a Y-polarized Huygen's Source.
90 PRINT : INPUT "Enter x,y: ",X,Y
100 X=X+EP
110 Z=SQR(R*R-(X*X+Y*Y))
120 R1=SQR(X*X+Y*Y+(Z-Z0)^2)
130 L=X/R : M=Y/R : N=Z/R
140 PRINT "L,M,N=";L;M;N
150 T=(Z-Z0)/R1
160 A=ATN(SQR(1-T*T)/T)
170 B=ATN(Y/X)
175 IF X<=0 THEN B=B+PI
180 REM Huygen source polarization
190 L1=-(1-COS(A))*SIN(B)*COS(B)
200 M1=1-SIN(B)^2*(1-COS(A))
210 N1=-SIN(A)*SIN(B)
220 B2=2*(L*L1+M*M1+N*N1)
230 L3=B2*L-L1
240 M3=B2*M-M1
250 N3=B2*N-N1
260 E0=COS(A)^MM*R0/R1
265 REM E0=1
270 PRINT "E0=";E0
280 PRINT "A=";A*180/PI;"B=";B*180/PI
290 PRINT "L1,M1,N1=";L1;M1;N1
300 EAX=E0*L3 : EAY=E0*M3 : EAZ=E0*N3
310 PRINT "EAX=";EAX;"EAY=";EAY;"EAZ=";EAZ
320 P0=10*LC*LOG(EAY*EAY)
330 P1=10*LC*LOG(EAX*EAX+EP)
335 P2=10*LC*LOG(EAZ*EAZ+EP)
340 PRINT "Po=";P0;"Px=";P1;"Pz=";P2
350 GOTO 90
360 END
```

```
10 REM *THIS IS A DEMONSTRATION PROGRAM TO SHOW*
20 REM *THE CROSS POLARIZATION ON OFFSET PARABOLIC TORUS*
30 PRINT "TORPOL.BAS"
40 PI=3.14159265359#
50 F=4.5 : R0=10 : RH=5.5 : EP=.00001 : LC=.4342944819#
60 REM INPUT "Tilt angle of feed, A0, DEGS.=",A0
65 A0=47.925 : YC=4 : R00=5.3889 : MM=5
70 A0=A0*PI/180
80 REM The Feed is a Y-polarized Huygen's Source.
90 PRINT : INPUT "Enter XAP,YAP (From aperture center): ",X,YAP
95 Y=YAP+YC+EP
100 X=X+EP
110 Z=SQR((R0-Y*Y/4/F)^2-X*X)-RH
120 R1=SQR(X*X+(R0-Y*Y/4/F)^2*Y*Y/4/F/F+(Z+RH)^2)
130 L=X/R1 : M=(R0-Y*Y/4/F)*Y/2/F/R1 : N=(Z+RH)/R1
140 REM Rotation of coordinates to tilt, Z1-axis
150 Y1=Y*COS(A0)-Z*SIN(A0)
160 Z1=Y*SIN(A0)+Z*COS(A0)
170 X1=X
180 PRINT "X1=";X1;"Y1=";Y1;"Z1=";Z1
190 PRINT "L,M,N=";L;M;N
200 R=SQR(X*X+Y*Y+Z*Z)
210 T=Z1/R
220 A=ATN(SQR(1-T*T)/T)
230 B=ATN(Y1/X1)
240 IF X<=0 THEN B=B+PI
250 REM Huygen source polarization
260 L1=-(1-COS(A))*SIN(B)*COS(B)
270 M1=1-SIN(B)^2*(1-COS(A))
280 N1=-SIN(A)*SIN(B)
290 L2=L1
300 REM Rotation of polarization vectors back to Z-axis.
310 M2=M1*COS(A0)+N1*SIN(A0)
320 N2=N1*COS(A0)-M1*SIN(A0)
330 B2=2*(L*L2+M*M2+N*N2)
340 L3=B2*L-L2
350 M3=B2*M-M2
360 N3=B2*N-N2
370 PRINT "A=";A*180/PI;"B=";B*180/PI
375 E0=COS(A)^MM*R00/R
380 PRINT "L1,M1,N1=";L1;M1;N1
390 PRINT "L2,M2,N2=";L2;M2;N2
400 PRINT "EAX=";E0*L3;"EAY=";-E0*M3;"EAZ=";E0*N3
410 P0=10*LC*LOG(M3*M3*E0*E0)
420 P1=10*LC*LOG(L3*L3*E0*E0+EP)
425 P2=10*LC*LOG(N3*N3*E0*E0+EP)
430 PRINT "Po=";P0;"Px=";P1;"Pz=";P2
435 GOTO 90
440 END
```

Chapter 8

COMPUTER TECHNIQUES FOR REFLECTOR ANTENNA APPLICATIONS

C. J. Sletten

8.1 INTRODUCTION

In reflector antenna applications and optimization a large variety of design problems are encountered. The authors of the chapters of this book have developed computer tools for calculating and improving the performance of many different radiating structures. In this chapter several examples are given in which the methods of reflector surface shaping are used to achieve wide-angle scanning capability and to reduce both near-in and far-out sidelobes. A special topic, beam waveguides using reflecting mirrors, is also introduced for low-loss feeding of microwave and millimeter wave antennas.

Even with software available and running, the design of reflector antennas with computers still takes much time and effort. Configurations must be weighed and selected, focal surfaces must be mapped, and diffraction patterns must be run to verify the design. For two examples included here some results have been reported [1,2,3], but the computational techniques employed were not described in these papers. The reader now can use the available optional software package with user's manual and example booklet to understand better the potential of the design approaches and perhaps to apply the techniques to new configurations.

In addition, some computer techniques for designing beam waveguides are described. This transmission medium is based on optical concepts similar to those used in reflector and lens antennas. Beam waveguides offer low-loss transmission lines for bringing gigahertz power to the antenna, and for realizing a kind of low-loss rotary joint on large antennas scanned by mechanical rotations.

8.2 WIDE-ANGLE SCANNING WITH OFFSET CASSEGRAIN ANTENNAS

For many applications in communication and radar antennas very narrow beams are required to be generated and scanned by motion of the feed rather than by moving the very large antenna apertures needed for the narrow beams. Often multiple beams are formed by simultaneous illumination of the antenna surfaces with fixed radiators feeding fixed gigantic (in wavelengths) main apertures.

The example Cassegrain antenna chosen here to illustrate the design approach uses an offset paraboloidal antenna with an aperture 240 wavelengths in diameter fed by a subreflector illuminated by conical corrugated horn feeds. This multibeam antenna is capable of generating beams of about 0.3° half-power beamwidths out to approximately 20 beamwidths from the paraboloid axis. The same main aperture is used for all beams and patterns but the Cassegrain subreflector is enlarged to intercept all rays over the angular sector of beam scan or search. The technique developed uses subreflector shaping to improve the gain of the beam furthest from the paraboloid axis, or for generating single beams far from the antenna axis. The multibeam generation can be done in a limited solid-angle sector, although, practically, for low aperture blockage the beam scanning is best done in the azimuth angular sector.

8.2.1 Configuring the Wide-Angle Cassegrain

Because we are interested in very narrow antenna beams with very high gain the main reflector aperture must be many wavelengths in diameter. An offset paraboloid section with circular aperture is chosen to give pencil beam radiation and to avoid feed and subreflector blockage. The focal length (f/D) is made large, which improves the wide-angle performance of the antenna. The antenna configuration developed is shown in Figure 8.1. (See also Fig. 7.3(b) for a frontal view of a similar configuration.) The parameters and dimensions given are computed using CASS.BAS. The main offset paraboloidal antenna selected for this example is 20 feet in diameter and the length units are inches. The wavelength, λ, is approximately one inch.

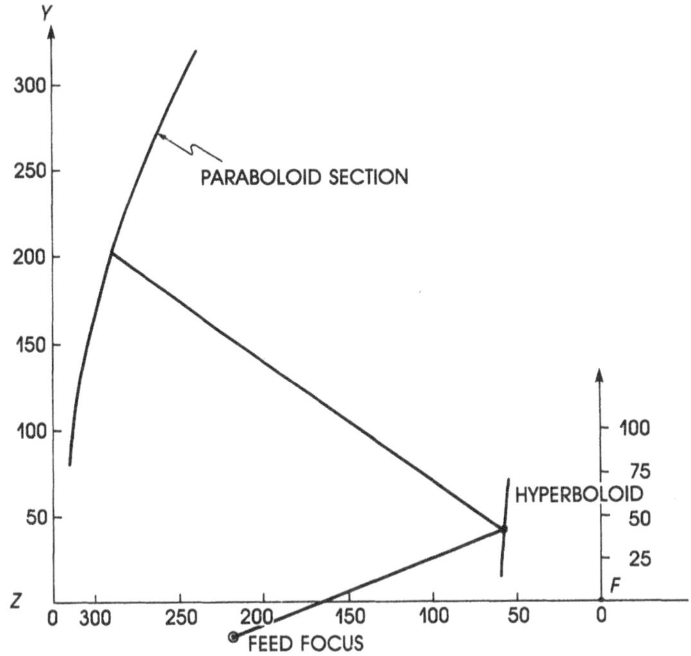

Figure 8.1 **Profile in the Y-Z (X = 0) Plane of the Wide-angle Scanning Cassegrain Antenna**

Main Reflector Dimensions

Offset paraboloid circular aperture radius, R = 120 (units)
Paraboloid focal length, F = 320
Aperture midpoint above z-axis, Y_c = 200

Cassegrain Subreflector Parameters

Eccentricity, E = 2.2
Distance between foci, C = 220
Tilt angle of subreflector axis, β = 5.976°
Horn tilt angle from subreflector axis, α = 15.827
Enlarged subreflector width, about 180 units
Enlarged subreflector height, about 52 units

Conical Corrugated Horn Feed

The horn radiation pattern is approximated by $P = \cos^N \theta$ where $N = 400$. This pattern corresponds approximately to that of a corrugated horn with 8.177° semiflare angle. The rim taper is −17.8 dB with this feed. The central pattern with beam along the z-axis has a gain of 56.5 dBi at 12 GHz computed by MULTH and PATT.

We will call this offset Cassegrain antenna with the hyperboloidal subreflector the *reference antenna*. It produces a conical beam along the paraboloid axis with gain slightly less than computed above due to spill-overs, ohmic and reflection losses, etcetera. Cross polarization is very low. When the antenna's beam and pattern are scanned by moving the feed along the focal arc (of Figure 8.4), however, the beam is no longer axially symmetric and the cross-polarization components in the aperture will increase in magnitude.

An important limitation on the beam-scanning sector is the requirement that the paraboloid's whole caustic sheets (see Section 7.2) must illuminate the subreflector without folding or distortion. We assume the main reflector is receiving a tilted plane wave. We trace the congruence of rays reflected by the paraboloid onto the subreflector. By mapping this converging bundle of rays, using CLCC.BAS, the "footprints" or the subreflector areas for edge rays (parallel incident rays reflected from the paraboloid rim toward the subreflector) are traced on the hyperboloidal surface. Obviously, all the rays must strike the enlarged subreflector without folding or crossing if antenna gains comparable to those of the axis beam are to be realized. The parameter C, distance between subreflector foci, as well as e, subreflector eccentricity, affect how the converging rays are mapped on the footprint.

Figure 8.2 illustrates these effects for the example antenna configuration. The footprint or contours of the dish edge rays on the reference subreflector for EL = AZ = 0 degrees is labeled 0°. Elevation angles are zero for all the footprints shown in the figure. The subreflector footprint used to produce a beam at 8° in azimuth just barely overlaps a small piece of the 0° or axis-beam footprint. The center point of the footprint, corresponding to the ray coming from the center of the main reflector, is shifted in the $+x_1$ direction and is indicated by the circled points labeled for the AZ beams to which they correspond. Note that the footprint for the 12° pattern is quite distorted and the center point is near the edge of the footprint. A feed pattern to illuminate the footprint areas properly should be oval-shaped as shown in Figure 8.3 for the 4° azimuth beam where the 6 dB and 17 dB contours of the (proper) feed pattern are plotted.

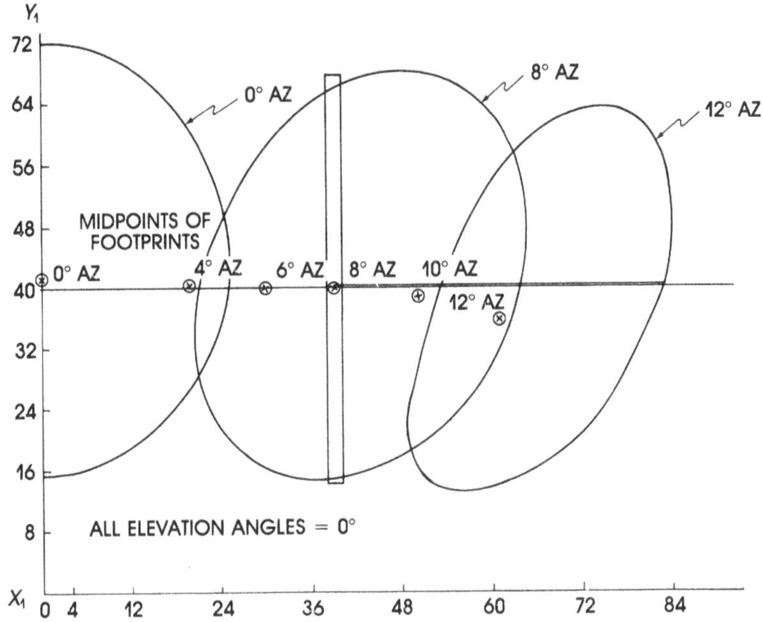

Figure 8.2 "FOOTPRINTS" of Traces of the Main Reflector Edge Rays on the Hyperboloidal (Reference) Subreflector Surface

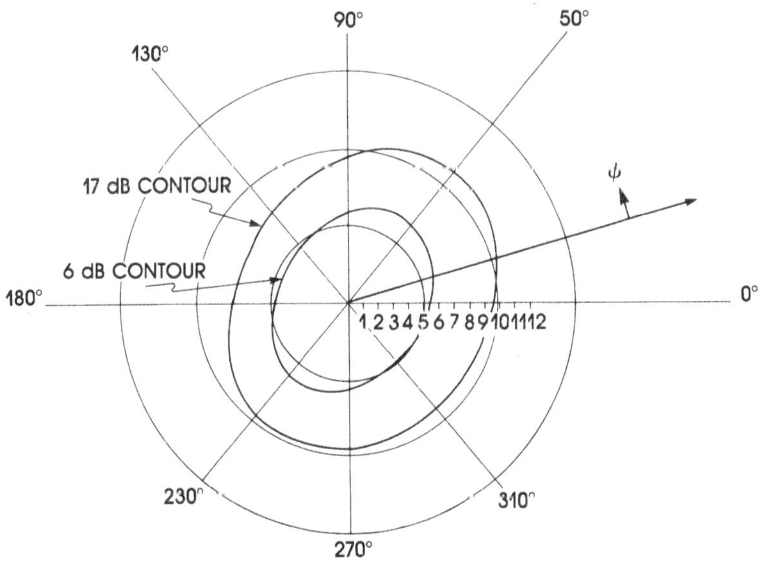

Figure 8.3 Feed Radiation Pattern Amplitude Contours Needed to Illuminate the Footprints Efficiently

8.2.2 Focal Surfaces and Radiation Patterns for the Multibeam Cassegrain Antenna

The next effort is to examine the focusing characteristics of the Cassegrain antenna and to map the best focal curves and surfaces using CLCCG.BAS. Using the smallest circle of least confusion as criteria, the azimuth focal curve is shown in Figure 8.4. The focal arc shown is for azimuth (EL = 0°) scanning only. The midpoints on the Cassegrain subreflector profile are shown for help in directing the feeds (the horn axes) located on the focal arc to illuminate the correct portion (footprint) of the enlarged reference subreflector. The focal curve for negative values of AZ is the mirror image of the +AZ curve in the *y*-*z* plane.

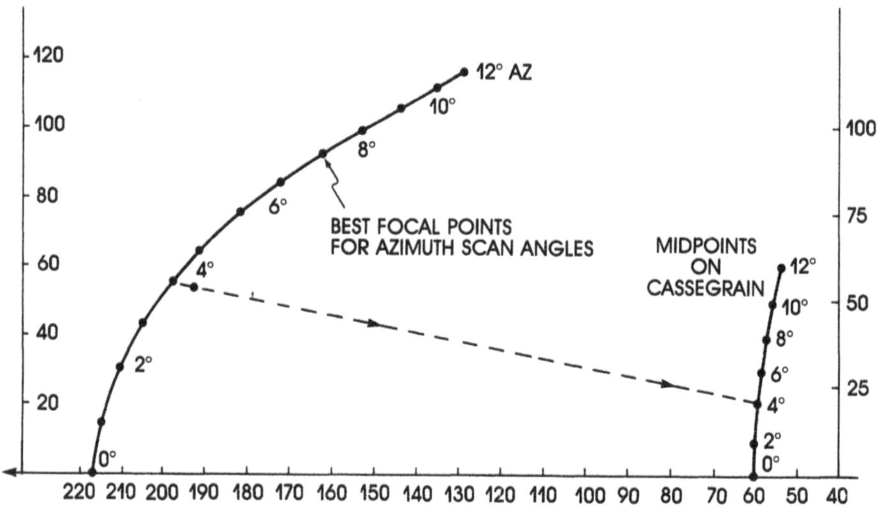

Figure 8.4 Focal Curve (X-Z profile) for Best Azimuth Scan Patterns, and Midpoint Locations of Footprint Centers for Azimuth Angles 0° to 12°

Clearly at the larger scan angles the aberrations are relatively large, and these will cause aperture phase errors and loss of pattern gain when the feeds are conical corrugated horns. As mentioned, the footprints are not circular so the feed pattern should also be oval in shape. We will compute the antenna radiation patterns assuming that the same conical horn generating the axis beam is used for all beams and patterns.

The relation of wavelength to diameter of the circle of least confusion gives some qualitative measure of the highest usable frequency of the antenna for the beam scan. The dish-edge rays can form a focal spot several wavelengths in diameter before the antenna gain loss exceeds 3 dB. The relationship of the size of the circle of least confusion area for rim rays to antenna gain and near-in sidelobe levels can be studied using the CLCCG.BAS traces, and antenna patterns and gains can be computed by MULTH.FOR and PATT.FOR.

Representative radiation patterns are plotted in Figure 8.5 (a, b, c, and d), taken through the beam peaks of the example antenna. There are several influences that make the patterns computed by MULTH.FOR approximate. Cross polarization is neglected, and, also, the computer technique calculates the aperture phase errors correctly but maps the phase and amplitude on the aperture to slightly inaccurate positions. The influence of these effects on gain and sidelobe level is small until scan angles exceed approximately 5°. The radiation patterns shown indicate primarily the influence of phase errors on near-in sidelobes when a conical corrugated horn is moved along the focal arc of Figure 8.4 with its center of phase on these circles of least confusion spots. The axis beam (AZ = EL = 0 degrees) has a beamwidth of about 0.3° at 12 GHz with a computed aperture efficiency of about 78%. The pattern is shown in Figure 8.5(a). Patterns in Figures 8.5(b) and (c) for, 2° and 4°, show little change. At 6° there is a computed loss in gain of about 1 dB (phase error effects only) and a squint of about 0.05° which can be found using MULTH and PATT with different ϕ-cuts. At 10° the pattern (not shown) is badly distorted and the computed gain is −6 dB from the axis directed pattern.

8.2.3 Shaping the Subreflector to Produce a Single Phase-Corrected Wide-Angle Beam

A technique for improving edge beams is developed in Chapter 7 using the subreflector shaping program SHAPE.FOR. The program SHAPE modifies the shape of the subreflector "footprint" for the edge or furthest scanned beam. This is done by shaping the subreflector end footprint area (such as the footprint areas for 8° or 12° in Figure 8.2), blending it into the central hyperboloid section and changing only the end footprint's surface contour.

In this chapter a program, SHAPEC.FOR, is provided that uses a slightly different method to shape a correcting Cassegrain subreflector. SHAPE only allows X-cuts in shaping the surface for perfect focusing of the edge pattern. SHAPEC.FOR permits both X-cuts and Y-cuts and makes it possible to check the data in Y-cut strips for accuracy using TRACXX.BAS (similar to TRACE.BAS of Chapter 3) and CLCCS.BAS. It also makes it possible to use the central $4X$-cut strip (Section 3.12.1)

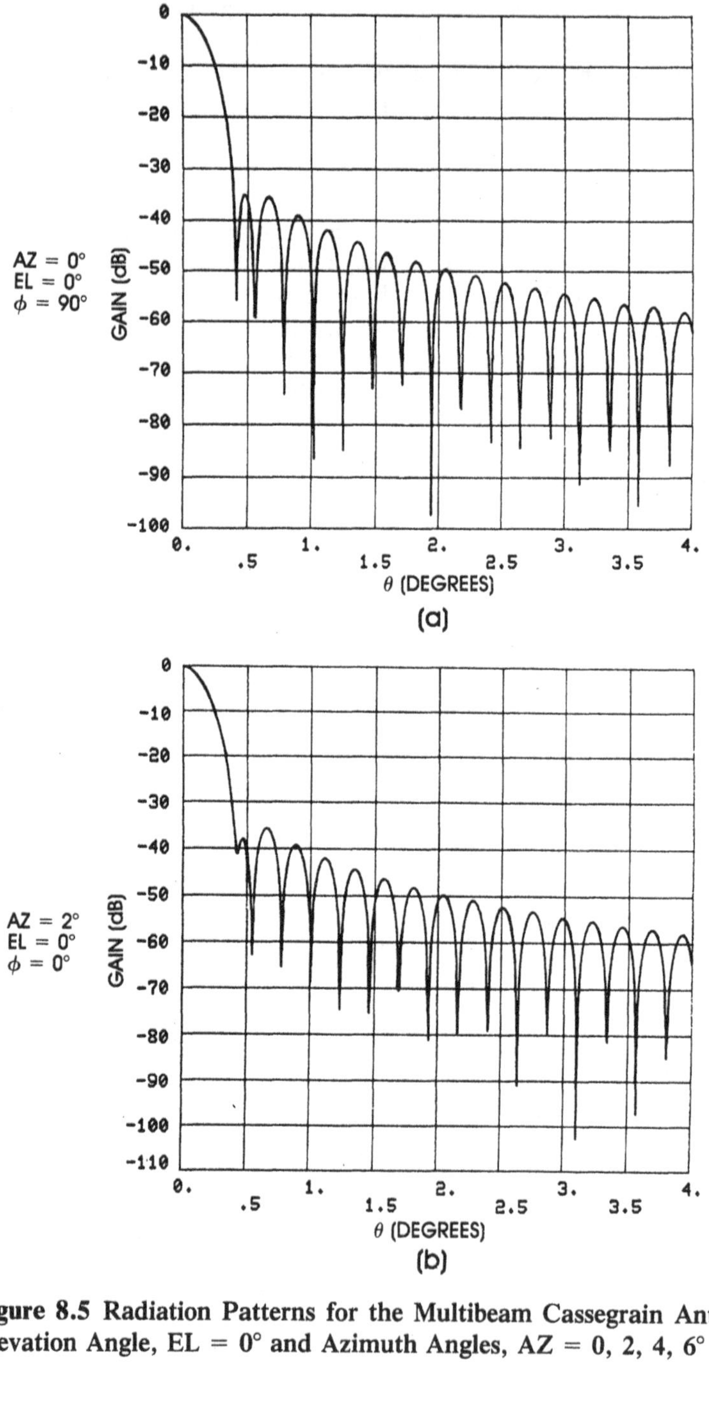

Figure 8.5 Radiation Patterns for the Multibeam Cassegrain Antenna for Elevation Angle, EL = 0° and Azimuth Angles, AZ = 0, 2, 4, 6°

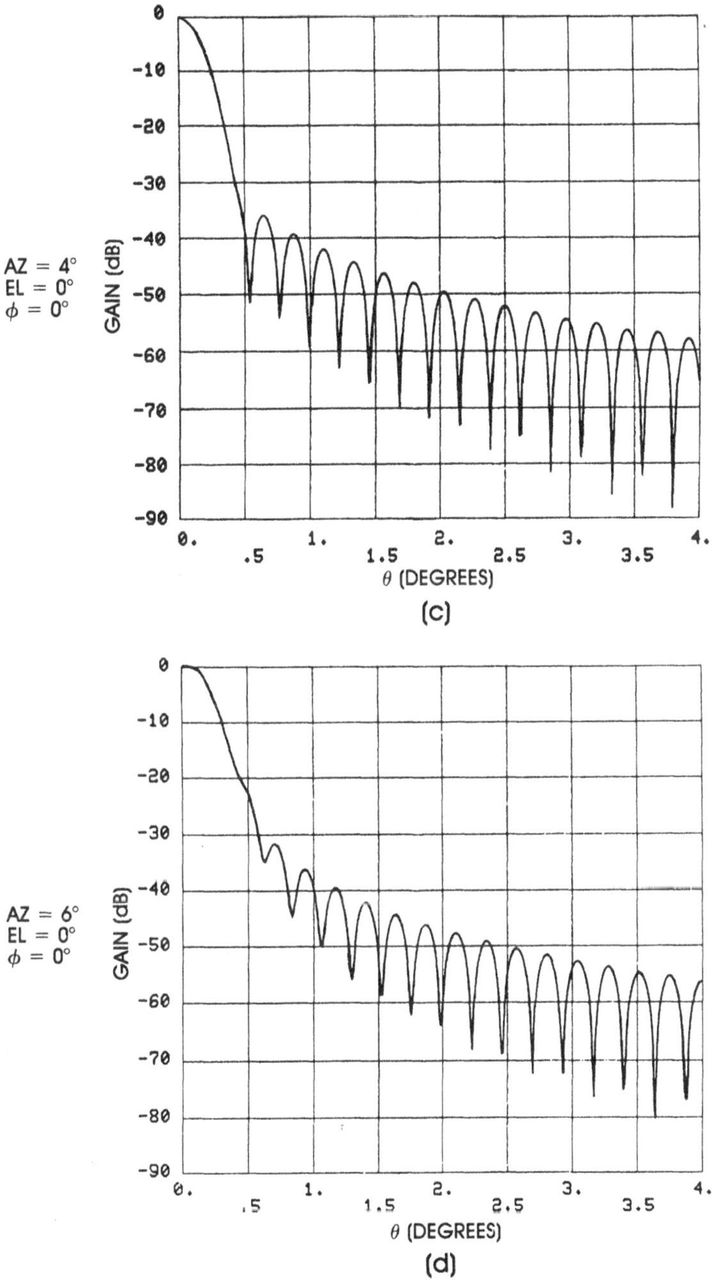

Figure 8.5 (cont'd)

that contains the midpoints of the footprints for continuous focusing to points on the focal curve of Figure 8.4.

The computational procedure used in SHAPEC.FOR is similar to those used in TORCA.FOR described in Chapter 7. Although the program is rather large and complex, the ideas developed are simple. They are reviewed below to help understand the shaping method and for other possible uses of this software.

The procedures employed for shaping a subreflector to refocus rays incident upon it to a point are:

1. Identify the footprint area on the reference hyperboloidal subreflector that intercepts the reflected rays from the main reflector for the EL,AZ beam direction selected. The antenna is assumed to be receiving an inclined plane wave on the paraboloid aperture. The reflected edge rays from the main reflector are mapped to the edge rays on the subreflector footprint using CLCC.BAS.
2. Use the reference (hyperboloid) subreflector surface coordinates and slopes at the edge of the footprint area (to be shaped) as start-up data in the numerical shaping procedure.
3. Find the "true" ray directions and the point, X, Y, Z, on the main reflector for any point, X_1, Y_1, Z_1, on the subreflector surface being shaped, by a numerical solution of three nonlinear equations obtained from Snell's law of reflection on perfect conducting surfaces.
4. Starting from an edge patch of reference points and slopes, use Adam's predictor-corrector [4] under the control of the true ray directions of the rays incident on the subreflector to construct a point-defined subreflector surface that, according to Snell's law, reflects the incident rays to the chosen subreflector focal point.

The analytical steps and computer subroutines for carrying out this shape synthesis are outlined below.

The offset paraboloid reflector with circular aperture is illuminated, in a receive mode sense, by a bundle of rays inclined by angles, EL,AZ, with the paraboloid aperture plane. These incident rays, $\mathbf{S}_0$, in a bundle of parallel rays, all have direction cosines L_0, M_0, N_0 where (bold type indicates unit vectors, and the origin of all coordinates is at the paraboloid focus)

$$S_0 = L_0\mathbf{X} + M_0\mathbf{Y} + N_0\mathbf{Z} \tag{8.1}$$

and

$$L_0 = \cos(\mathrm{EL})\sin(\mathrm{AZ});\ M_0 = -\sin(\mathrm{EL});\ N_0 = \cos(\mathrm{EL})\cos(\mathrm{AZ})$$

The paraboloid of focal length F has coordinates X, Y, Z related by

$$Z = F - (X^2 + Y^2)/(4F) \tag{8.2a}$$

We wish to find the rays, $\mathbf{S}_1$, reflected by the paraboloid toward the subreflector

$$\mathbf{S}_1 = L_1\mathbf{X} + M_1\mathbf{Y} + N_1\mathbf{Z}$$

The reflected ray direction cosines are L_1, M_1, N_1. These are calculated from Snell's law for reflectors:

$$\mathbf{S}_1 = \mathbf{S}_0 - 2(\mathbf{S}_0 \cdot \mathbf{N})\mathbf{N}$$

where $\mathbf{N}$ is the unit normal to the paraboloid. If we know the coordinates, X_1, Y_1, Z_1, of the points on the subreflector where rays from points X, Y, Z on the main reflector intersect it we can always write

$$L_1 = (X_1 - X)/R_1;\ M_1 = (Y_1 - Y)/R_1;\ N_1 = (Z_1 - Z)/R_1$$

where

$$R_1^2 = (X_1 - X)^2 + (Y_1 - Y)^2 + (Z_1 - Z)^2$$

We can find the subreflector point that corresponds to a paraboloid point by using Snell's law to obtain these relations

$$L_1 = L_0 - 2XB \tag{8.2b}$$

$$M_1 = M_0 - 2YB \tag{8.2c}$$

and where

$$B = (L_0X + M_0Y + 2FN_0)/R^2 \text{ and } R^2 = X^2 + Y^2 + 4F^2$$

By solving simultaneously the nonlinear set of three equations, (8.2a), (8.2b), and (8.2c), we can relate the main reflector points to the subreflector points.

These are the principal analytical relations needed to shape the subreflector. The surface shaping process starts by computing three back slopes and points on the reference hyperboloid surface. A regular rectangular grid of points in X_1 and Y_1 subreflector coordinates is used to define the shaped subreflector surface, Z_1 (X_1, Y_1), with a small spacing increment D_1 equal for both X_1 and Y_1 point separations. A larger increment D_2, a multiple of D_1, is used for filing surface coordinates in data files for construction of the shaped surface.

The computed points and slopes are entered into Adam's predictor-corrector which, based on these data, projects to a new point at a distance about D_1 from the last computed point and asks what is the ray direction at this point (ray directions are needed to compute the slope of the shaped subreflector surface). Given the new projected point coordinates, X_1, Y_1, Z_1, and the direction cosines L_0, M_0, N_0 (these are constants) of the ray congruence on the paraboloid, the solution to the three nonlinear equations, (8.2a), (8.2b), and (8.2c), finds the point X, Y, Z on the main reflector and from that the ray direction cosines L_1, M_1, N_1. Snell's law is inverted to give the correct slopes K_8, K_9 to reflect the ray incident on the subreflector to the chosen shaped subreflector focal point, X_0, Y_0, Z_0.

This numerical projection by δD_1 increments, first proceeds along a central strip in X_1 making four X_1-cuts across the footprint area. The data from these four central cuts are stored in the array $S_1()$ and later used as start-up data for Y_1 cuts both upward and downward from the central strip.

The machinery to make the first $4X_1$-cuts spaced D_1 from each other in Y_1 is contained in the main program of SHAPEC.FOR which also defines variables to organize the data and store shaped coordinates in data files. The function CASS is used to compute the start-up data from the Cassegrain reference surface and to continue to compute reference subreflector coordinates to see how much the shaped surface departs from the hyperboloid.

Several subroutines are used in SHAPEC.FOR to carry out the shaping process. KAYS is used to compute the slopes of the shaped subreflector using true ray data. It also calculates the slopes of the reference surface at each new point. USRFUN contains the three nonlinear equations and their partial derivatives. It is used by the *Numerical Recipes* subroutines MNEWT, LUDCMP, LUBKSB in solving the three nonlinear equations to find the X, Y, Z point on the paraboloid where the incident ray is reflected to the point X_1, Y_1, Z_1 on the subreflector. The subroutine YCUTU makes the Y_1 cuts in the upward Y_1 direction and YCUTD does the same for the downward Y_1 cuts. Surface coordinate data are filed in YUP.DAT and YDP.DAT by these last subroutines.

Details for running this program are found in the optional software manual package with user's manual and example booklet. Many of the programming steps are also contained in SHAPE.FOR and can be seen in the source code for that program in Appendix 7A. Input data for shaping the footprint area of the 4° azimuth beam are supplied by the data file SHAP2.DAT with results for this example case given in the user's manual.

8.2.4 Suggestions for Increasing Wide-Angle Coverage of the Offset Cassegrain Antenna

Although the dual-reflector antenna developed above performs well as a limited-scan multibeam antenna, there are several possibilities for improving the angular scanning sectors. In all cases, the first step is to optimize the configuration of components. In general, the optimized tilts calculated by CASS.BAS should be used to guarantee good beam symmetry and low cross polarization on the z-axis or principal pattern (EL = 0°, AZ = 0°), although this configuration does not assure the best performance on all the multibeam patterns. The front-fed Cassegrain [5] configuration offers excellent beam coverage and deserves consideration for dual-reflector multibeam applications.

Instead of using a conical corrugated horn to illuminate the subreflector, cluster feeds made up of closely packed small horns (or a phased array of slots or dipole elements) can be used as the feed for producing beams at wide angles. These elements, about a wavelength in diameter, can be excited using the appropriate phase, amplitude, and polarization to compensate for the antenna's aberrations. The CLCC.BAS plots on the "screen" can guide in constructing the compensating cluster feeds. The program can be modified to give the pathlengths to the focal screen and the polarization components.

The designer has a choice of focal surfaces to use. The focal curves of sharpest focus are in caustic regions, and here the diffraction effects dominate. Here it is necessary to use MULTH and PATT to find the diffraction patterns and the dimensions of the Airy main disk and the spatial location of the sidelobe bands. The pattern phase for each sidelobe relative to the main lobe (and cross-polarization components) are obtained using the diffraction pattern phases computed with the programs similar to PATT or those of Chapter 4. By controlling the phase and amplitude of small radiators [6] (see also Section 7.9.2) arranged in circles about the central or main radiator, the seven-feed cluster correction can be used to reduce sidelobes. Six small feeds arranged in a circle about a central feed at a (transverse focal zone) radial distance that corresponds to the first sidelobe radius are sufficient to reduce near-in sidelobes on all ϕ-cuts through the pattern generated by coupling the auxillary six feeds to the central feed port. The level of coupling of the auxillary feeds in the cluster to the central feed should be at about the dB level of the sidelobes to the main beam. The phase of the coupled energy for sidelobe reduction (cancellation) should be in phase with the main beam pattern as the sidelobe phase (check with the vector diffraction programs of Chapter 4) is nearly 180° out of phase with the main beam.

A somewhat simpler approach to phased-array or cluster-feed correction of aperture antenna aberrations can be developed by moving the correcting plane from the best focal surface to a surface removed from the best focus or caustic zone. For the offset Cassegrain antenna the departures of the ray paths from those that produce a converging spherical wave front can be computed using CLCC.BAS because geometric optics is approximately valid in these regions. The small elements can be phased and excited as complex conjugates [7] of the received phase from the tapered plane wave incident on the main dish aperture. By the optical principle of reversibility, the transmitted fields from the multiple elements will also superimpose in the aperture to a tapered plane wave. The elliptical amplitude pattern can be formed and the phase errors, which are the primary cause of pattern gain loss and sidelobe increases, can be compensated for. The number of feed elements needed is increased, but these can be reused with different phase and amplitudes to form adjacent beams. Beams cannot overlap too closely when they are required simultaneously. Continuous scanning over wide angles, however, limited by the need to get the whole caustic on the subreflector, is theoretically possible. The main reflector aperture can be from 10 to 100 times larger in area than the phased array feeding surface, thus reducing the cost of phased arrays for forming truly high resolution beams with low sidelobes over a limited conical scan sector of perhaps 10°.

It is suggested in the discussion of the program SHAPEC.FOR that subreflector shaping can be used to improve the wide-angle patterns of the offset Cassegrain antenna. Using SHAPE.FOR, the end patterns can be improved by shaping only the ends of the subreflector. High-quality individual patterns can be obtained by using SHAPEC to synthesize a special subreflector for a chosen EL,AZ direction, when close contiguous multibeam patterns are not required. When continuous pattern scanning or an adjacent string of multibeam patterns are needed, however, a scheme for continuously shaping the subreflector surface can be devised.

In Figure 8.4 it is seen that the focal curve in X_0, Y_0, Z_0 coordinates as a function of AZ angle corresponds to another curve in X_1, Y_1, Z_1 points locating the centers of the footprints for these azimuth angles. Figure 8.2 shows that the footprint centers lie nearly on a straight line in X_1, Y_1. It is possible to approximate the focal curve and the line of centers curve by continuous functions. The shaping algorithms of SHAPEC can continuously enter changing X_0, Y_0, Z_0 points as the X_1-cut iterative shaping process moves across the center of the subreflector. When the edge beam angle is reached the X_0, Y_0, Z_0 point is fixed at, for example, 10° to shape the end or edge portion of the subreflector so that this beam is improved. When the Y_1-cuts are made they always start at the center line where the correct focal point can be computer selected. Using this shaping technique the subreflector is always corrected for the central part of the feed patterns near the center of the footprint, but not corrected for the rays for the edges of the footprint. Because the subreflector surface is no longer a true hyperboloid, the z-axis beam (EL = 0°, AZ = 0°) also has some phase errors. A general improvement in pattern quality over a larger scan interval, however, can be realized by this continuous shaping scheme.

Alternately, for larger angular scan sectors the parabolic torus antenna can generate very narrow beams when corrected by the subreflector shaping technique described in Chapter 7. The main parabolic torus reflector, however, must be enlarged, depending on the angular coverage. When the feed and subreflectors are configured to be small, multiple beams can be fairly closely spaced. Again, with the torus, as with the offset Cassegrain antenna in these subreflector shaping examples, the multibeams with very narrow beams are generated in one plane of scan only.

8.3 LOW SIDELOBE LEVEL, OFFSET GREGORIAN ANTENNAS

Reflector shaping techniques can aid in the design of very low sidelobe antennas. We have seen in Chapter 1 that nearly "ideal" aperture amplitude distributions can be realized with conical corrugated horns feeding offset dual-reflector antennas. Blocking sidelobes are avoided by the offset geometry, and the approximately Gaussian-shaped patterns of feed horns produce radial functions on the aperture approaching the curves computed by HANSEN.FOR (or other "ideal" radial distributions) for optimizing gain versus near-in sidelobe performance. There are two problems, however, that tend to limit sidelobe performance. First, the "scalar horn" does not have a true phase center. The edges of the horn pattern phase curve depart significantly from a spherical wave front, causing aperture phase errors. Second, the wide-angle performance of the antenna is largely determined by the main reflector edge illumination and spillover around the rim of the paraboloid. The fields in the vicinity of the main reflector edges are primarily the result of the edge illumination and edge diffraction of the ellipsoidal subreflector.

The edge diffraction from the subreflector can be examined by using the GTD techniques of Chapter 4. Some far-out radiation improvements can be made by increasing or decreasing the diameter of the subreflector to minimize the subreflector pattern in the region of the dish edge. Here, however, another approach is taken. The subreflector surface is both increased in diameter by adding a skirt and shaped (at its edges) to reduce the fields at the edges of the main reflector. The phase errors, due to the feed (horn) pattern phase not being truly spherical, are first corrected by shaping the subreflector edges. Then the ray-ratio method for amplitude control of aperture distributions is used to shape the edges of the subreflector to reduce the edge taper on the main reflector. A special shaped rim or skirt is added to the subreflector to reduce the spillover energy level outside the edges of the main reflector.

8.3.1 Subreflector Shaping for Phase Correction

For very low sidelobes the aperture phase errors caused by horn phase deviations must be corrected. (We have seen in Chapters 2 and 3 that often the horn position can be moved to refocus the feed and partially correct the horn phase errors.) As the phase errors tend to be largest near the edge of the horn pattern where feed amplitude is low, the effect on the antenna pattern is generally small. When very low sidelobe levels are needed, however, and when amplitude pattern changes in the antenna aperture distribution are to be made by shaping the subreflector, the shaping method of horn phase correction described in Chapter 3 can be employed. (Because the feed phase errors require rather large subreflector edge shaping changes from the reference ellipsoidal shape, it is nearly impossible to control the amplitude fields at the edge of the subreflector correctly without first correcting the feed phase errors.)

When low sidelobe levels for both near-in and far-out lobe patterns are desired with a compact, wide-frequency band antenna, then the offset Gregorian configuration has particular merit. In Chapter 3 the principles for Gregorian subreflector shaping to correct for feed phase errors were incorporated in the programs HORNF0.BAS and 4XSTUP.BAS. In these programs the feed phase pattern was represented by a family of discrete rays and their line parameters were used to shape the subreflector to focus all feed rays to the paraboloid focus. For this chapter a FORTRAN double-precision program is developed that performs these functions for feed phase patterns expressed as power series. Explicit computations of horn phase errors, rather than ray interpolation, are used to shape the subreflector. This program is called SHAPH.FOR and can be employed for very wide-angle horn feeds used in constructing compact, wideband antennas.

The Gregorian antenna selected as the example to illustrate the shaping technique is similar to the antenna discussed in [3] and shown in Figure 8.6(a). The conical corrugated horn for this model has a wide semiflare angle of 46° as shown in Figure 8.6(b). The horn phase and amplitude patterns are shown in Figure 8.6(c) and (d). The paraboloid antenna has a focal length of 30 units and a diameter of 30 units with offset center $Y_c = 35$ units (above the z-axis). The subreflector constants $E = 0.2765$ and $C = 4.7$ are chosen to make the volume of the antenna small.

For convenience, SHAPH.FOR computes the feed and subreflector tilt angles, α and β, in the main or driver program. The subroutine FOURX performs the functions of 4XSTUP.BAS by computing four X_1-cuts starting from the center of the reference subreflector. The data calculated are stored in $S_1()$ for use as start-up slopes and points for the Y_1-cuts, downward and upward. The subroutine KAYS computes the subreflector slopes needed to reflect the horn's rays to the paraboloid focal point. The implicit calculations using the horn phase pattern power series are made in KAYS to find the direction cosines L_1, M_1, N_1 of rays striking a subreflector point X, Y, Z (these are the X_1, Y_1, Z_1 points as seen in subroutines FOURX, YCUYD, and YCUTU). The files YCCD .DAT and YCCU.DAT contain the shaped profiles of the Y_1-cuts at D_2 intervals in X_1 and Y_1. These data can be used to construct the shaped subreflector. The subreflector has left-right symmetry, so only the positive X_1 points need to be computed. The differences, DEL, from the ellipsoidal reference subreflector contours are mainly at the edges of the subreflector and are found by comparing the true ellipsoidal points Z_4 (computed by the function GREG) with the shaped points Z_1 over the subreflector footprint found in the output data files. The phase shaping by SHAPH results in a more concave ($Z_1 < Z_4$) Gregorian subreflector.

Subreflector shaping for correcting aperture phase also changes the aperture amplitude distribution. For the Gregorian configuration the large negative phase errors on the edges of the horn phase pattern result in an increased ray density on the edges of the main reflector. This tends to increase the energy density on the subreflector and main reflector edges. (The opposite effect occurs for Cassegrain geometry.) The shaping procedure as performed by SHAPH spreads out the ray density and brings the aperture distribution back to nearly that for a horn with a true phase center located at F_1 in Figure 6.8(a). The shaping removes the aperture phase errors and somewhat overcorrects the aperture amplitude distribution making the edge illuminations somewhat weaker than for a point source horn feed. This drop in edge "taper" is desired for low side lobes and is increased by the deliberate amplitude reduction on the main reflector edges discussed next.

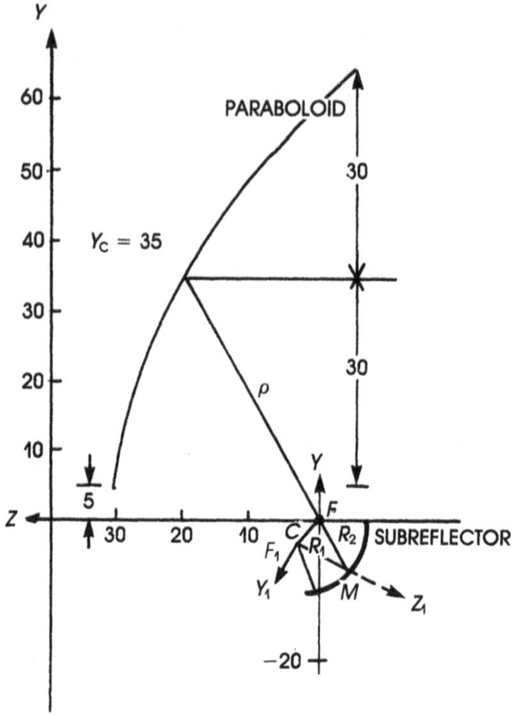

Y
60
50
40
30
20
10
PARABOLOID
30
$Y_C = 35$
30
ρ
Y
5
F
Z
30
20
10
C
R_2
SUBREFLECTOR
F_1
R_1
Y_1
M
Z_1
−20

(a)

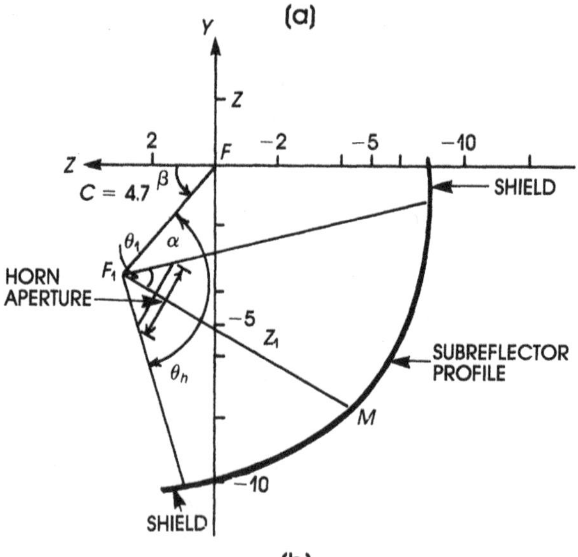

Y
Z
2
F
−2
−5
−10
Z
C = 4.7
β
SHIELD
θ_1
α
HORN
APERTURE
F_1
−5
Z_1
θ_h
SUBREFLECTOR
PROFILE
M
−10
SHIELD

(b)

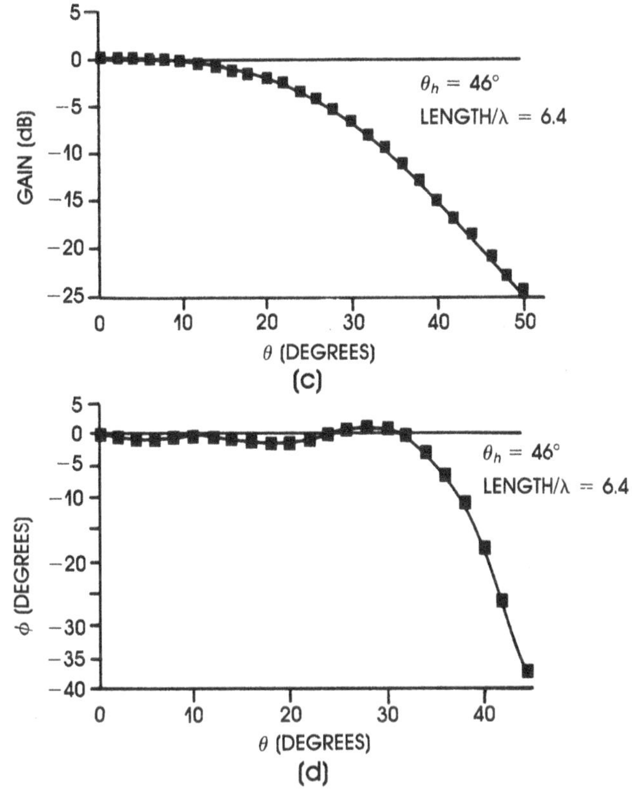

Figure 8.6 (a) Compact Offset Gregorian Antenna; (b) Profiles of Gregorian Feed System; (c) Feed Horn Power Pattern; (d) Feed Horn Phase Pattern (From Sletten, Payne, and Shillue, *Microwave Journal* [3]. Reprinted with permission.)

8.3.2 Shaping for Control of Aperture Amplitude Distributions

Often the far-out lobes caused by (1) rim diffraction by the subreflector and main reflector edges and (2) the spillover caused by the subreflector pattern in the region of the paraboloid edges are the main concerns of the designer. These sidelobes can be reduced by shaping and extending the subreflector edges. The program SHAPM.FOR is written to control the amplitude distribution on the main reflector and on its edges by shaping and enlarging the Gregorian subreflector. The spillover illumination outside the main reflector aperture is reduced using a shaped rim (skirt) or enlarged subreflector surface. The purposes of this skirt are (1) to prevent the subreflector edge diffraction from strongly illuminating the edges of the main reflector and (2) to spread feed rays reflected from the skirt to reduce the level of the spillover radiation beyond the main reflector's rim.

In the ray-ratio subreflector shaping method described in Chapter 3 and developed in RAYRAE.BAS, the operating G-function for the Gregorian containing the ray length, R_1, from the feed point to the subreflector and the ratio of the aperture E-field to the feed E-field is

$$G = R_1 \cos(\theta_f)^2 F_6/K_0 \tag{8.3a}$$

where $\theta_f = A_6$, the feed angle from the feed center of phase to the subreflector point, K_0 is a normalizing constant, and F_6 is

$$F_6 = E_T/E_H \tag{8.3b}$$

where E_T = the desired aperture field at a radial point, R, related to θ_f by the transform relation

$$R = B \tan(\theta_f/2). \tag{8.3c}$$

B is the transform constant obtained from CASS.BAS. E_H is the E-field of the illuminating feed. (The transform (8.3c) relating feed pattern to aperture radial distribution is an approximation that neglects the ϕ dependence in the feed pattern and other factors. Also when F_6 is not equal to (or very close to) one, the relationship of aperture points to subreflector points should be used. They are available in the ray-ratio shaping programs.)

It is the F_6 function that controls the incremental shaping *via* Adam's predictor-corrector [4]. When the subreflector surface is not shaped, i.e., it is the reference ellipsoid, the aperture field is simply the feed pattern transformed by Equation (8.3c). If this aperture distribution is satisfactory we just correct the phase feed errors using SHAPH and accept the diffraction patterns given by MULTE and PATT. We can, however, shape the subreflector surface to obtain, for example, one of the ideal aperture radial distributions described in Chapter 1. When the subreflector surface departs from the ellipsoidal reference shape, there will be phase errors on the aperture, unless the main reflector is shaped. When the transformed feed pattern is nearly that of the ideal aperture distribution, the phase errors are small. This trimming of the feed pattern to realize a more desirable aperture distribution can be done simply by writing

$$E_T = F(r)$$

where the function $F(r)$ is often a power series like those developed by Kouznetsov (see Chapters 1 and 5), for example:

$$F(r) = 1/7 + 6/7(1 - r^2)^2$$

Here $r = R/R_0$ where R_0 is the aperture radius. Using Equation (8.3c) we can obtain E_T as a function of θ_f and also the feed E-field, E_H, as a function of θ_f (a power series filed in COEF.DAT). When only small differences in E_T and E_H exist, F_6 will be nearly one and the subreflector will be close to the reference shape.

In the example programmed here to demonstrate sidelobe control techniques on offset Gregorian antennas the designer is satisfied with the near-in sidelobes produced by the Gaussian-like feed pattern of the 46° semiflare horn radiation. The horn phase errors have been corrected using SHAPH.FOR. Subreflector shaping by SHAPM.FOR is done to reduce the far-out sidelobes in the angular zone of the pattern in the region of the edges of the subreflector pattern.

The approach to far-out sidelobe reduction programmed in SHAPM.FOR is to shape only the edges of the subreflector such that the subreflector radiation pattern drops off abruptly near the edges of the main reflector and beyond. The computer technique drops the edge illumination on the dish enough so that the dish edge diffraction causing far-out lobing and the rear caustic lobe is reduced. The subreflector edges are extended so that the horn pattern is very weak at the subreflector edges. This is done by imposing a terminal shaping function that is smoothly blended with the horn radiation pattern (near the edge of the subreflector). A smooth transition in antenna aperture fields is needed to prevent a discontinuity in pattern or pattern derivative which due to Gibb's phenomena will result in near-in sidelobe increases. A simple algorithm is used to shape the subreflector skirt to spread out the radiation in the dish spillover directions to low levels.

The amplitude control function used for reducing far-out radiation is derived in the steps outlined below and programmed in TERMF.FOR. The purpose of the terminal shaping function is to reduce the dish aperture fields at the edges by from 5 to 10 dB from the edge illumination produced by the horn pattern with the ellipsoidal subreflector.

A smooth terminal function of the form

$$E_t = A + B \cos^M\theta \tag{8.4}$$

can be used where A, B, and M are constants. By substituting $E_t = E_T$ in (8.3b) the control function F_6 achieves the smooth reduction in dish aperture power near its edges. This terminal function is joined to the feed radiation pattern at a feed angle, $\theta = U_1$, where it is given the same value and slope as the feed (horn) pattern. From U_1 to U_2 the terminal function replaces the feed pattern's control of the dish amplitude distribution. E_t

decreases more rapidly than the feed pattern in this terminal angular region of θ. U_2 is generally taken to be the semiflare angle of the horn feed.

The horn pattern, $E_h(\theta)$, is represented as

$$DB_h = A_2U^2 + A_4U^4 + A_6U^6 + A_8U^8 + A_{10}U^{10} \tag{8.5}$$

where U is the feed polar angle, θ, in radians. Imposing the conditions that (1) both the horn and terminal E-fields are equal at U_1, (2) the derivatives with respect to θ are also equal at U_1, and (3) the difference between the two functions is some value (in dB), DB, at U_2, we arrive at the expression

$$C_2 + S_1 \cot(U_1)[\cos^M(U_2)/\cos^M(U_1) - 1]/M = 0 \tag{8.6}$$

The solution to this nonlinear equation gives us M when we substitute the boundary values stated. S_1 is the slope of both the horn pattern and the terminal function at U_1 or

$$S_1 = \mathrm{d}E_h(\theta)/\mathrm{d}\theta|_{U_1}$$

C_2 is obtained from the equation, $C_2 = E_h(U_2) - E_h(U_1)$, where $E_h(U_2)$ is reduced by the dB drop. To obtain (8.6) we differentiate (8.4) and (8.5) with respect to $\theta = U$ to get S_1 and C_2

$$S_1 = -MB \cos \theta^M \tan \theta$$

$$C_2 = B(\cos^M U_2 - \cos^M U_1)$$

where

$$B = -S_1/(M \cos^M U_1 \tan U_1)$$

Equation (8.6) has only one unknown, M, as S_1, C_2, U_1, and U_2 are determined. M can be found from (8.6) with a root-finding program. These equations are written for the DB values in (8.5) into the program TERMF.FOR using Numerical Recipes math subroutines ZBRAK and ZBRENT to find the value of M that makes (8.6) zero.

Caution is needed in introducing the terminal function TERMF. It is best to select the angular interval near the edge of the subreflector with the starting value U_1 about 80% of the semiflare angle, $\theta_h = A_3$, of the horn (or the angle where the feed pattern rays strike the edge of the main reflector). Then set the dB-drop value for the angle $U_2 = A_3$. The feed pattern given by the series may not behave well over θ angles beyond the subreflector edges and should not be used to shape the extended subreflector rim. Also, the terminal function often goes negative beyond U_2. The program SHAPM.FOR uses a value of $F_6 = 0.2$ to shape the subreflector outer border zone for reducing the spillover lobes (by spreading the feed power over a large angular region outside the dish edges).

With a smooth transition function that does not increase the near-in sidelobe levels the program SHAPM.FOR can be used to compute the contours of the shaped subreflector that reduces the fields at the edge of the main reflector (and beyond into the spillover region). First it is assumed that the horn phase error correction contours have been computed using SHAPH.FOR. It is also assumed that the modifications to the subreflector skirts to correct horn phase and to drop the subreflector pattern to reduce far-out lobes can be linearly added. We linearly superimpose the shaped Z_1 values obtained by SHAPH and by SHAPM using the data files these programs produce. The DEL values of the SHAPM computations are added to the Z_1 values calculated by SHAPH for corresponding values of X_1, Y_1. Of course the D_1 and D_2 increments for the data grids must be the same as well as the starting antenna configuration constants obtained from GREG.BAS and recalculated by the main programs in SHAPH and SHAPM.

Program SHAPM.FOR first calculates the reference subreflector parameters. Then in subroutine RAY the four X_1-cuts are made using the ray-ratio method used in RAYRAE.BAS. The Y-cuts up and down are combined into one subroutine, YCUT. The files YCUTB.DAT and YCUTT.DAT are used to store the shaped subreflector profiles (Y_1-cuts at every D_2 units in X_1) for the bottom and top sections of the left (positive X_1) half of the subreflector. The phase shaping data and the amplitude shaping data can be combined for construction convenience by a simple program that reads the data files. The resulting contour of the Gregorian subreflector shaped for correcting horn phase errors and for reducing the far-out sidelobes is shown for a cut in the $Y_1 - Z_1$ plane at $X_1 = 0.1$ in Figure 9 of Reference [3]. Comparing the DEL differences of the shaped surfaces compared with the reference ellipsoid for SHAPH.FOR and SHAPM.FOR shows that for the example case the phase shaping corrections in the YCCD.DAT and YCCU.DAT files are much greater than for the amplitude shaping files YCUTB.DAT and YCUTT.DAT. This means that for the Gregorian subreflector doing only the amplitude shaping without the phase shaping will not be effective in reducing sidelobes. Both the near-in and far-out sidelobes increase with increased main reflector rim illumination.

The calculated reduction in far-out lobing about the edge of the main reflector is seen by comparing Figures 8.7(a) and 8.7(b). These far-out

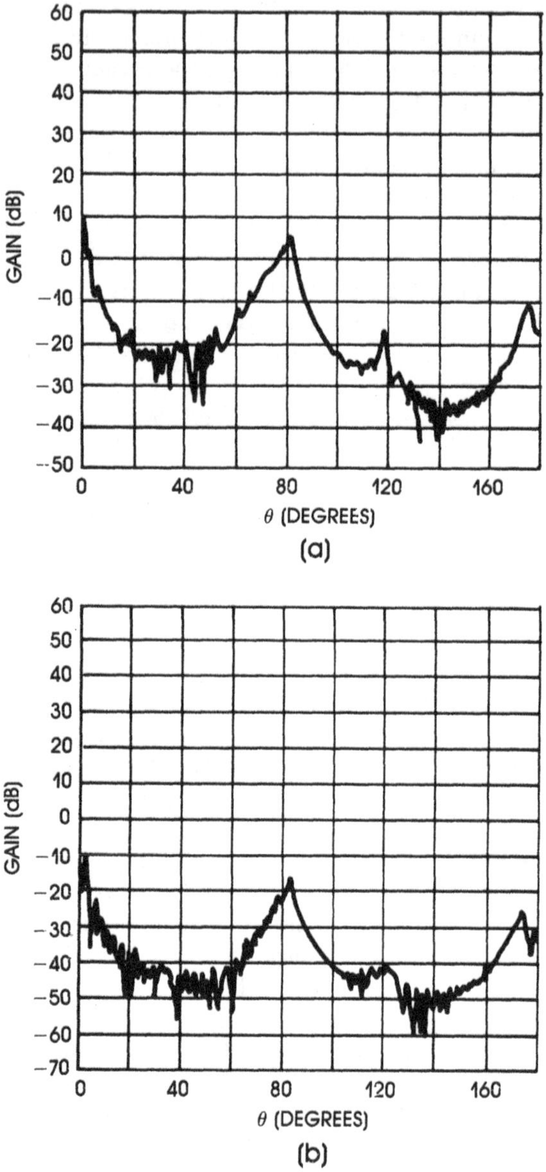

Figure 8.7 (a) Radiation Pattern of the Gregorian Antenna in the Elevation Plane, $\theta = 0°$ to 180°, $\phi = 90°$, Showing Dish Rim Diffraction at $\theta \sim 80°$ and the Back Caustic Lobe at $\theta \sim 120°$; (b) Radiation Pattern of the Gregorian Antenna, $\theta = 0°$ to 180°, Showing Far-Out Lobe Reduction Due to Shaping the Subreflector Edges
(From Sletten, Payne, and Shillue, *Microwave Journal* [3]. Reprinted with permission.)

patterns over a 0° to 180° angular sector in θ are for the $\phi = 90°$ cut where the far-out lobing is most evident. The effect is not seen in the $\phi = 0°$ (azimuth) cut for the offset antenna. These radiation patterns illustrating far-out sidelobe levels were made by using GTD computations for dish edge diffraction and software of the type described in Chapter 4. The technique used to compute the backlobes and far-out sidelobes shown in Figures 8.7(a) and 8.7(b) finds the equivalent feed pattern of the subreflector. The main reflector aperture fields are assumed to be those of the feed pattern out to U_1 and the terminal pattern from U_1 to U_2 transformed by (8.3c). The rim or skirt of the subreflector produces a low-level spillover lobe which is estimated. Then the ray path and angle geometry for the prime-fed offset paraboloid are used to find the equivalent prime-focus feed that would produce the antenna aperture distribution produced by the shaped subreflector illuminated by the feed horn. The program EQFEED.BAS works back from the aperture amplitude distribution to find the asymmetrical prime-focus-feed pattern that would produce the desired aperture distribution. This equivalent feed pattern is entered piecewise into the GTD and aperture-integration program of Chapter 4 to compute 180° patterns in θ. The equivalent "horn feed" illuminates the offset paraboloid exactly like any prime-focus feed making it possible to compute dish-rim diffraction and rear-caustic sidelobes by GTD edge-diffraction programs.

For very low near-in sidelobe it may be desirable to trim the feed pattern to make the antenna aperture fields more closely approximate one of the ideal radial amplitude functions of Chapter 1. This shape synthesis is easy to introduce into SHAPM following suggestions made in this section and in Sections 3.11 and 5.3. The control function $F_6 = E_T/E_H$ can be calculated in the θ-zone of the feed pattern from 0° to U_1, given the ideal E_T aperture function and the feed function E_H. The aperture phase errors caused by subreflector shaping can be displayed to the screen. When the subreflector surface departs appreciably from the reference surface, as evidenced by the aperture phase errors, the main reflector can be shaped to correct for aperture phase errors using SHAPSH.BAS (see Section 3.12.2) to produce the set of discrete rays illuminating the main reflector. These rays can then be entered into PARBUP.BAS to shape the offset main reflector.

The edge shaping of the subreflector for amplitude control to reduce far-out sidelobes introduces some aperture phase errors near the aperture edges. These can also be corrected by main dish shaping. Because the amplitude field on the aperture edges is low, the increase in near-in sidelobes caused by shaping to reduce far-out sidelobes is minimal even with no main reflector shaping.

8.4 AN INTRODUCTION TO BEAM WAVEGUIDES

Beam waveguide (BWG) technology provides low-loss transmission media frequently used by designers of microwave and millimeter-wave antennas. Many of the same optical techniques used in dual-reflector antennas are employed. Introducing frequency dependent Gaussian modes that represent focal fields in caustic regions (like GTD), however, enables the designer to include first order diffraction effects without resorting to boundary value solutions of Maxwell's equations. Goubau and Schwering [8] placed these Gaussian-mode representations on a firm theoretical footing, and the applications to antenna feeding have been published by Bell Laboratory workers [9,10,11]. The fundamental or zero-th order Gaussian mode contains most of the electromagnetic energy and the first-order mode depicts the cross-polarization fields.

Beam waveguide transmission lines for feeding antennas are constructed of a series of lenses or conic section mirrors which periodically refocus the electromagnetic waves propagating from a feed launcher, usually a conical horn, to the focal point (region) of a microwave or millimeter-wave antenna. The BWG feeding system may also serve as low-loss rotating joints by using right angle bends and plane mirrors in the transmission path that rotate with the antenna components for mechanical scanning of antenna patterns. The main design problems are to minimize radiation losses on the open transmission lines, to minimize cross polarization introduced by mirrors or lenses, and to obtain an appropriate amplitude-tapered pattern with a spherical phase front to feed the antenna. In this brief introduction to the subject the governing equations are programmed that allow the antenna designer to configure the refocusing elements and calculate the approximate focal positions and focal fields from the horn launcher to the antenna. Cross-polarization levels and diffraction losses can also be estimated. Reflectors, rather than lenses, are used in the BWG systems described in this text because they have lower propagation losses than lenses and are easier to configure for rotational joint applications.

Figure 8.8 illustrates a typical BWG system used to feed an antenna mechanically rotated in azimuth and elevation planes. Here conic section mirrors, an ellipsoid and a hyperboloid, are used to refocus the electromagnetic fields launched by a fixed conical corrugated horn. The antenna has a Cassegrain subreflector fed by the BWG line at the focus F_c. An azimuth rotating joint is realized by keeping stationary the feed horn that illuminates the first plane mirror and rotating all the remaining BWG and antenna components about the azimuth axis. Likewise, a rotation of the antenna and the second-plane mirror about the elevation axis provides

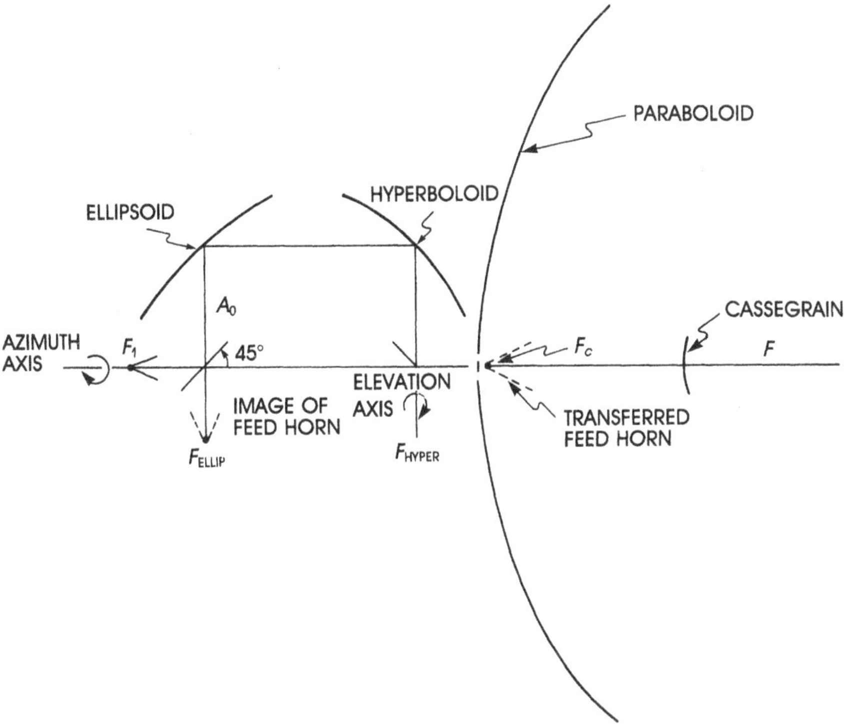

Figure 8.8 Beam Waveguide Feeding System for an Example Cassegrain Antenna

mechanical scanning in the elevation plane. When correctly designed the BWG system effectively transfers the feed-horn pattern with center of phase at F_1, placing its center of phase at the Cassegrain subreflector focus F_c. In [12] Mizusawa and Kitsuregawa show by geometric-optics analysis that when all the BWG curved mirror focal points lie on a straight line and the curved mirrors have the same eccentricity e (or $1/e$), then the beamwidth of the feed pattern and the pattern radiated at F_c will be the same, and an axially symmetric feed horn pattern in θ will remain axially symmetric. This beam symmetry condition also tends to reduce cross polarization caused by the BWG reflectors, although when Gaussian modes are used to compute the cross-polarization levels the vanishing cross-polarization condition is only true at zero wavelengths. Also, the focal points of the mirrors are not located at the geometric-optics focal points in Gaussian-mode analysis. The optimum configurations with focal points on a line, however, are helpful guides in design. Converging mirrors, ellipsoids and paraboloids, generally are used in BWG systems. For example, two paraboloids can be used with their second focal points at infinity to approximate the optimum beam-symmetric configuration given in [12].

In the example BWG feed used to illustrate the design procedure, the components are discussed from a transmitted-wave point of view. A conical corrugated horn illuminates a plane mirror which reflects its pattern onto the first ellipsoidal reflector in Figure 8.8. The plane mirror acts only to make a right-angle bend needed in the rotating joint and its imaging effects are assumed to be perfect in the sense that no cross polarization, diffraction, or ohmic losses are introduced by it. The horn center of phase can as well be located at F_E and the horn axis pointed at the center of the ellipsoidal reflector. The pattern of the conical corrugated horn or other feed can be computed using HORN1-5 or the programs of Chapter 2. Horns do not have true centers of phase, and the ellipsoid could be shaped using SHAPH.FOR to correct for the horn-phase pattern. This kind of correction is usually not done. The actual horn pattern is approximated by a zero-th order Gaussian mode. This is done by finding the $1/e$ or -8.68 dB level and its θ angle on the horn radiation pattern and assuming that the horn pattern can be represented by a Gauss error function. A beam waist radius is defined on the first BWG curved mirror corresponding to the nearly circular curve on the mirror surface intersected by the $1/e$ horn pattern level. Spillover losses around the mirror edges are kept small by making all BWG mirrors large enough to intercept 95% or more of the feed pattern and the subsequent Gaussian propagating modes.

8.4.1 Computing BWG Mirror Surfaces

To construct the first ellipsoidal mirror in Figure 8.8 such that a right-angle bend is accomplished the central ray of the horn pattern is caused to be reflected at right angles from the center point of the mirror in the direction of the second (*GO*) focal point of the ellipsoid. The equations of this ellipsoid can be derived from the geometry shown in Figure 8.9. (Of course, the coordinates of ellipsoids and hyperboloids with any tilt angles can be calculated using the programs GREG and CASS. The special 90° tilt geometry is developed here as a convenience for this important BWG geometry.) It is only necessary to know the lengths, A_0, the distance from the center of phase of the horn to the center of the ellipsoid, and B_0, the distance from the reflector center to the second geometric foci. We can define the ellipsoid by its semimajor and semiminor axes, A

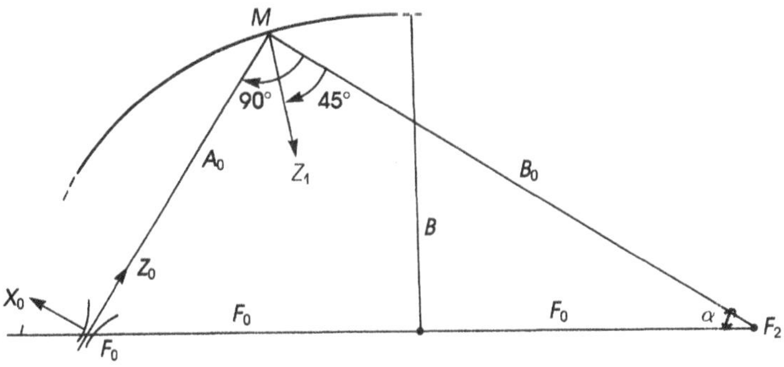

Figure 8.9 Geometry for Computing Ellipsoidal Mirror Surfaces Tilted 90° for BWG Axis

and B, its eccentricity e, and the angle α made by the normal to the horn axis with the ellipsoid axis.

$$A_0 + B_0 = 2A$$

$$F_0 = (A_0^2 + B_0^2)^{1/2}/2$$

$$B = (A^2 - F_0^2)^{1/2}$$

$$e = (A^2 - B^2)^{1/2}/A$$

$$\tan \alpha = A_0/B_0$$

where F_0 is one-half the distance between foci.

To compute the ellipsoidal surface we can use the equation (from Chapter 1):

$$X_0^2 + Y_0^2 + Z_0^2 = e^2(Z_0 \cos \alpha - X_0 \sin \alpha + D)^2$$

where $D = F_0(1 - e^2)/e^2$. For example, we can choose Y_0 and Z_0 on a regular grid with origin at F_0 and compute X_0.

It is advantageous in constructing a tilted ellipsoid as shown in Figure 8.9 to rotate the coordinates 45° making the reflector center the origin and the normal at that point the Z_1-axis. The program BWAVEG.BAS uses the equations listed above to compute the coordinates of points on the ellipsoid reflector. It also rotates the coordinates 45° to make construction easier. The program BWAVEG.BAS allows the rotated X_1, Y_1 coordinates to be entered as uniformly spaced independent variables. BWG-H.BAS is provided for the rather rare cases when a hyperboloidal reflector is used to make a 90° rotating joint. BWG-H.BAS also rotates the coordinates 45° but uses Y_0, X_0 (or Z_0, X_0) variables that result in a construction grid with unequal X_1, Y_1 spacings. (Because of the limits of single precision a compiled double precision BASIC program may be used for greater accuracy in constructing optical BWGs.)

In the example BWG developed we choose an ellipsoid for the first curved mirror with eccentricity near to unity, meaning that it is nearly a paraboloid, and the second focus (at a large distance, B_0, from the mirror center) is remote from the center of the tilted reflector.

8.4.2 Gaussian Mode Representation in BWG Systems

To find accurately the locations of the focal regions along the beam waveguide and compute the radiation pattern illuminating the antenna including the cross-polarization fields we introduce two Gaussian modes. Definitions of two Gaussian modes well suited to antenna feeding are found in Gans and Semplak [9, p. 1332]. Equations used by these authors were derived from [8].

For the fundamental or zero-th Gaussian mode the E-field distributions along the BWG are given by

$$\begin{aligned} E_0 = {} & (H_0 x + V_0 y)\, W_0/w_0 \exp\{-jkz - (r/w_0)^2 \\ & + j[\arctan 2z/(kW_0^2) - kr^2/2R_0]\} \end{aligned} \tag{8.7a}$$

The higher or first-order mode is written

$$\begin{aligned} E_1 = {} & [V_1(x \cos\alpha - y \sin\alpha) \\ & - H_1(x \sin\alpha + y \cos\alpha)]\sqrt{2} r W_1/(w_1)^2 \exp\{-jkz \\ & - (r/w_1)^2 + j[2 \arctan 2z/(kW_1^2) - kr^2/2R_1]\} \end{aligned} \tag{8.7b}$$

The fundamental mode has a maximum on the beam axis and usually represents the major part of the microwave energy propagating in the BWG transmission line. The higher mode has zero intensity on the beam axis and is related to cross-polarization level and beam asymmetry. We will discuss the zero-th order mode first to compute the longitudinal and transverse field distributions in this mode. (The BWG fields are transverse electromagnetic (TEM).) Fortunately the longitudinal positions of beam waists along the direction of beam propagation and the beam waist radii are usually approximately the same for both the zero-th and higher-order modes.

Referring to Equation (8.7a) (the fundamental Gaussian mode), the coordinates used to express the E-field are x, y, z rectangular and r, α, z cylindrical with the $z = 0$ origin at the narrowest beam waist radius, W_0, with w_0 being the transverse radius of the beam at points z not zero. W_0 and w_0 are defined as the beam radii where the principal polarized field drops by $1/e$ from the value on the beam axis. (Some authors use power rather than field or $1/\sqrt{2}e$ as waist radius for the zero-th mode). R_0 is radius of curvature of the beam phase front at a point z on the line of propagation. R_0 is infinite at $z = 0$. k is $2\pi/(\lambda)$ and H_0 means horizontal polarized-beam excitation along x and V_0 means vertically polarized excitation and $r = \sqrt{(x^2 + y^2)}$. The beam waist, w_1, for the higher-order mode with its narrowest beam waist, W_1, has maximum field at radius, $r = W_1/\sqrt{2}$, and decreases to $\sqrt{(2/e)}$ at $r = W_1$. In the derivation of the equations for the two Gaussian modes [8,9] it is shown that both W_0 and W_1 have approximately the same beam waist radii. Also, the radii of curvature of the higher-mode phase front, R_1, is usually equal to R_0. This allows both mode beam waists radii and their longitudinal positions, z, on the BWG system, and the radius of curvature R_0 and R_1 to be computed from these expressions [10, p. 288]:

$$
\begin{aligned}
w &= W\sqrt{(1 + (2z/(kW^2))^2)} \\
R &= z[1 + (kW^2/(2z))^2]
\end{aligned}
\tag{8.8}
$$

where $w = w_0 = w_1$ and $W = W_0 = W_1$ and $R = R_0 = R_1$. These equations can be used to obtain a good approximation for either the zero-th order or higher-order mode.

These equations when solved for W and z allow the beam waist radii and their positions to be found when W and z are known at some point along the beam waveguide. Positive z is usually defined in the direction of beam propagation with negative z values from the narrowest beam waist, W, toward the excitation source.

In the example case programmed the beam-waist radius used to start the computation is the radius of the approximate circle on the first curved mirror where the feed (horn) E-field pattern falls to $1/e$ or -8.686 dB from the pattern peak. Because there can be many beam-guiding mirrors along the BWG system, new values and positions of W arise for each curved mirror. The succeeding mirrors (or lenses) along the beam path are characterized by a focal length calculated from the thin-lens approximation formula:

$$1/F_n = 1/R_a + 1/R_b$$

where R_a is the distance from the center of the mirror to one GO focal point and R_b is the distance to another. When the mirror is a hyperboloid then the distance R_b (behind the reflector) not in the beam is considered negative. For the first ellipsoidal mirror $R_a = A_0$ and $R_b = B_0$.

These equations are programmed into CHU-BW.BAS. First the focal lengths of mirrors or other elements in the BWG system are calculated by the program. Then using descriptive prompts the program computes beam-waist radii and positions along the system and also the radii of curvature of phase fronts arriving and departing a beam waveguide element. By following the computer prompts the waist positions and radii along the BWG can be found. The unfolded beam data for the example layout of Figure 8.8 is given in Figure 8.10. This example can be recalculated using CHU-BW, and then a more complicated beam waveguide system found in [11], for example, can be entered. In [11] Chu derives important equations for optimizing the bandwidth of BWG systems. His method of configuration for wide bandwidth is an important contribution to multielement BWG feeding systems.

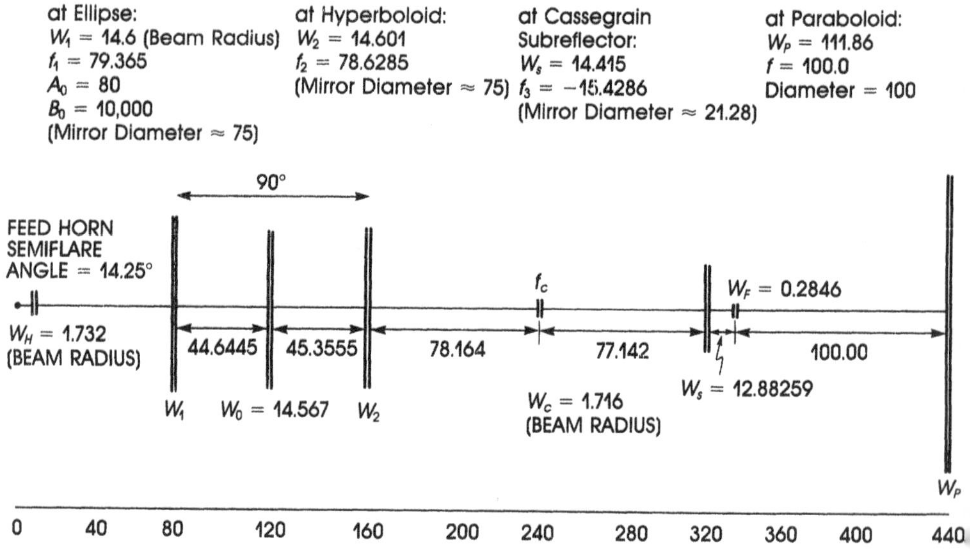

Figure 8.10 Unfolded BWG Antenna Feeding System Diagram for the Example Case

The programs BWAVEG, BWG-H, and CHU-BW are sufficient to begin the physical layout of a BWG system. The mirrors must be made large enough so that spillover, as computed using SPILL.BAS, is small for the horn illumination and the waist illumination on subsequent elements. Mechanical tolerances on the curved and flat surfaces should be $\lambda/32$ or better for good performance. An important consideration in antenna feeding is the level of cross polarization introduced by the BWG system. This design parameter can be addressed using Equation (8.7), and programs POLARB.BAS and SUM-G2.BAS, which are discussed next.

8.4.3 Cross-Polarization Estimation in BWGs

The initial w_0 beam waist is estimated from the near-field pattern of a conical corrugated horn or other launcher. We assume the feed pattern is free from cross-polarization components and can be represented by a linearly polarized Huygens source.

The first curved mirror with the horn-feed axis tilted nearly 90° from the ellipsoid axis can be expected to cause rather high cross polarization of the aperture fields. When the eccentricity e is nearly one, as is often the case for BWG mirrors, the program POLARB.BAS written for the offset paraboloid can be used to calculate the aperture cross polarization for an ellipsoid or a hyperboloid (with $e \sim 1$). As may be guessed from our experience with dual-reflector systems it is possible to orient two BWG mirrors such that in the optical limit the cross polarization in the fields reflected from the second mirror vanishes. One orientation presented in Gans [9, pp. 302, 304] analyses two BWG mirrors facing each other (very similar to the case shown in Figure 8.8 where the curved mirrors are very nearly paraboloids). In the example problem for POLARB.BAS the paraboloid focal lengths are both $F = 40$ with the offset midpoints $Y_c = 80$ for the 90° tilt angles. The feed horn directed at Y_c edited into POLARB .BAS has a semiflare angle of 14.25° and a feed pattern approximated by $\cos^N\theta$ where $N = 125.25$.

Running POLARB.BAS gives important information about the cross-polarized aperture field of the 90° tilted offset paraboloid. The aperture distribution for cross polarization in dB can be seen as different X points at $Y = Y_c = 80$ and are entered into the program. For $X = 0$, $Y = 80$ the X-polarized cross components are zero. We are at the peak of the zero-th order Gaussian beam. For all values of Y with $X = 0$ the cross polarization is zero. This means that in the y-z plane, called the plane of bending, there are no cross-polarization maxima. By entering other values of X at $Y = 80$ we can find the cross-polarization peak which lies in the plane orthogonal to the plane of bending. We see that the peak is at about $X = 10$. The level is -22.3 dB from the peak of the zero-th order mode. POLARB prints out the E-field vector components, L_2, M_2, N_2 of the incident ray for the selected aperture point, X, Y, computed from the Y-polarized Huygens source. Also printed are the field components, L_4, M_4, N_4, reflected from the facing paraboloid. These components are seen to be identical except for sign changes indicating direction of energy flow. Thus, for this ray-optics model of the facing BWG paraboloids, cross-polarization components reflected from the second mirror are zero. If one paraboloid is rotated 180°, then the cross-polarization aperture fields would increase by 6 dB after reflecting from the inverted second mirror. This demonstration software allows the designer to map the distributions of the cross polarization (higher mode) and the zero-th mode fields (which also contain the small higher-mode co-planar, asymmetric fields). It also suggests a geometry for reducing the BWG cross-polarization levels on a two-mirror system (based on geometric-optics criteria). (More general situations including frequency dependence are treated in [13].) A physical explanation for the cross-polarization cancellation at the second mirror is that the zero-th order mode field excites cross-polarization components on the facing dish that just cancel the incident-propagating cross-polarized field generated by the first paraboloid.

Using the higher-order Gaussian mode equation (8.7) cross-polarization field distributions propagating along the BWG system can be examined. Much attention is given in [9,13] to calculating the cross-polarization coupling coefficients when a directive horn pattern is incident on an offset paraboloid or ellipsoid. One of the expressions derived [9, p. 1331] for finding the coupling coefficient is

$$C_a = \theta_{-10\text{dB}} \sqrt{10} \tan(\theta_0/2)/\sqrt{(e \ln 10)}$$

For the $\cos^N\theta$ feed pattern in our example case the -10 dB patterns radius is at $\theta = 10.95°$. θ_0 is the feed tilt angle which is here 90)°. Substituting numbers $C_a = -22.3$ dB, the same value as found by POLARB.BAS at a radius of about $X = 10$. To compare fields of the two Gaussian modes a coefficient must be entered in the higher-order mode equation (8.7b). The value 0.076389 is used for K_1 in the program SUM-G2.BAS discussed below.

In comparing the zero-th and first-order mode equations, (8.7a) and (8.7b), the phase terms, arctan() and 2arctan(), are seen to have the same argument when $W_0 = W_1$. The factor of 2 means there is a relative phase shift between the two modes as a function of Z, the distance from the beam waists, W_0 and W_1 (these minimum waists are colocated). Note that Z is (defined to be) negative to the left (l) of the beam waist, W_0, and positive to the right (r) in Figure 8.10. To find the relative phase shift, $\Delta\Phi$, between the two modes for a distance Δz, we write

$$\Delta\Phi = \Phi_1 - \Phi_0 = \arctan(z_r/kW_0^2) - \arctan(-z_1/kW_0^2) \tag{8.9}$$

This phase shift between modes means that the cross-polarization cancellations for correctly oriented mirrors in the geometric-optics (GO) limit will not necessarily produce zero cross polarization at microwave frequencies. For the example case z_l (left) and z_r (right) are small compared to kW_0^2 so that $\Delta\Phi$ is small between the two facing mirrors and the cross-polarization cancellation is nearly complete. For many beam waveguide systems these conditions are not met and the cross-polarization level may be even 6 dB higher for the beam after leaving the second mirror. Because the beam waist for an offset paraboloid is at the aperture the cross polarization produced by the reflector will always be circularly polarized in the far-field of the antenna.

Program SUM-G2.BAS evaluates the phase terms for the propagation paths shown in Figure 8.10 up to the antenna subreflector. This program also sums the E-field phasors that indicate how the two modes combine at given cross sections along the beam. For the facing ellipsoid and hyperboloid with eccentricities near one in our example the small phase-shift effect that reduces the cancellation is given by K_2, cross-polarized coupling coefficient, at the hyperboloid

$$K_2 = K_1[(1 + \cos(\pi + V))^2 + \sin^2 V]^{1/2}$$

where $V = \Delta\Phi$ in (8.9) evaluated from the ellipsoid to the hyperboloid.

The higher Gaussian mode also contains co-polar field components that phasors add to the zero-th mode fields causing slight beam asymmetry. These co-polar fields are maximum in the plane of bending and, as a study of Equation (8.7) indicates, they have the same relative magnitude (compared to the fundamental mode fields) as the cross-polarized fields. SUM-G2.BAS calculates only the co-polar fields along the y-axis for $Y > Y_c$. As the co-polar higher-order mode fields have the same general form as the cross-polarized fields, however, this program serves to illustrate the superposition of the two modes and allows the BWG fields to be computed for any cross section along the BWG line. The beam-waist radii and waist positions as computed by CHU-BW.BAS are needed as inputs in SUM-G2.BAS. These can be taken from Figure 8.10 for the example case. Amplitude profiles at the first mirror after reflection of the horn pattern is first printed out. The Y-cut pattern at the subreflector location is computed and displayed with the cross-polarization levels and profiles at both cuts. These results are approximate.

These simple computer tools are sufficient for layout of BWG systems and estimations of beam waveguide performance. More rigorous diffraction methods [15,16] are needed accurately to compute the fields throughout the BWG system. GTD or other near-field diffraction methods can be used. Two Gaussian modes can give only an approximation to the fields in BWGs and more Gaussian modes can be introduced to improve accuracy. Even with "perfect" conic section mirrors some beam flattening is observed on patterns computed using rigorous formulations for near-field diffraction [16]. When there are structural deviations from quartic surfaces, especially when additional mirror edges are involved, GTD near-field methods are perhaps the most convenient to use. Shaping of BWG mirrors to alter the launcher phase or amplitude patterns or to synthesize a desired antenna aperture illumination can be done using the shaping techniques already developed in this chapter.

8.5 IMPROVEMENT TECHNIQUES

For generation of high-gain multibeams patterns over a limited scan sector, offset dual-reflector (Cassegrain) antennas have important advantages. Generally, only one horn feed for each beam is needed, and these feeds are capable of handling high power. Even when cluster feeds or focal region phased arrays are used the number of radiating elements is much less than for phased-array antennas. The reflector antennas have lower losses, especially at higher frequencies, and also wider bandwidths and lower construction costs than comparable phased-array antennas. How far the Cassegrain antenna can be pushed to radiate narrow beams with low sidelobes at wide angles depends on the resourcefulness of the designer. Several schemes for improvements beyond those developed in software are sketched in this chapter.

In Section 8.3 Gregorian subreflector shaping to improve single beam pattern performance has concentrated on methods for controlling both near-in and far-out sidelobes with high aperture efficiency. The subreflector edge is identified as the key factor in the control of both near-in and far-out lobes caused by high fields on the main reflector edge region. SHAPH.FOR is developed to correct aperture phase distributions due to horn-phase errors. Reduction of far-out sidelobes is accomplished by shaping the subreflector edges and adding a shaped skirt based on subreflector contours computed by SHAPM.FOR. The aperture amplitude distribution can be made near to the "ideal" distributions of Chapter 1 by introducing the desired radial functions into SHAPM.FOR, but too much amplitude shaping causes aperture phase errors, and shaping of the main reflector to correct these can be expensive.

Historically, the main emphasis in shaping dual-reflector antennas has been to extract the highest possible antenna gain and aperture efficiency. The control of far-out lobes that raise antenna noise temperatures and increase interference has received less attention. Far-out lobes often do not appear on the principal azimuth plane pattern cuts on offset antennas.

The reader will find the beam waveguide software useful for configuring millimeter-wave feed systems.

References

[1] Sletten, C.J., "Numerical Technique for Shaping Offset Subreflectors to Increase Multibeam Antenna Coverage," URSI Symp. Digest, Boston, MA, p. 13, June 25–29, 1984.

[2] Sletten, C.J., and S.E. Carrillo, "Scanning Communication Antennas," IEEE AP-S Symp. Digest, Vol. 1, pp. 474–477, Boston, MA, June 25–29, 1984.

[3] Sletten, C.J., W. Payne, and W.P. Shillue, "Offset Dual Reflector Antennas for Very Low Sidelobes," Microwave Journal, Vol. 29, No. 5, pp. 221–240, May 1986.

[4] Press, W.H., *et al., Numerical Recipes,* "The Art of Scientific Computation," Cambridge Univ. Press, 1986, pp. 569–572.

[5] Makino, S., Y. Koboyashi, and T. Katogi, "Front-fed Offset Cassegrain Type Antenna," IEEE AP-S Symp. Digest, pp. 341–344, June 17–21, 1985.

[6] Sletten, C.J., and W.G. Mavroides, "A Method of Sidelobe Reduction," Proc. Sidelobe Conference, Naval Res. Lab. Rep. 4043, pp. 1–12, April 1952.

[7] Wood, P.J., "Field Correlation Diffraction Theory of the Symmetrical Cassegrain Antenna," IEEE Trans., Vol. AP-19, pp. 191–197, March 1971.

[8] Goubau, G., and F. Schwering, "On the Guided Propagation of Electromagnetic Wave Beams," IRE Trans., Vol. AP-9, pp. 248–256, May 1961.

[9] Gans, M.J., and R.A. Semplak, "Some Far-Field Studies of an Offset Launcher," BSTJ, pp. 1319–1340, Sept. 1975.

[10] Chu, T.S., "Geometrical Representation of Gaussian Beam Propagation," BSTJ, pp. 287–290, Feb. 1966.

[11] Chu, T.S., "An Imaging Beam Waveguide Feed," IEEE Trans., Vol. AP-31, No. 4, pp. 614–619, July 1983.
[12] Mizusawa, M., and T. Kitsuregawa, "A Beam Waveguide Feed Having a Symmetric Beam for Cassegrain Antennas," IEEE Trans., Vol. AP-21, pp. 884–886, Nov. 1973.
[13] Gans, M.J., "Cross Polarization in Reflector-type Beam Waveguides and Antennas," BSTJ, pp. 289–315, March 1976.
[14] Kogelnik, H., and T. Li, "Laser Beams and Resonators," Proc. IEEE, Vol. 54, No. 10, pp. 1312–1329, Oct. 1966.
[15] Galindo-Isreal, V., T. Veruttipong, and W. Imbriale, "GTD, Physical Optics, and Jacobi-Bessel Diffraction Analysis of Beam Waveguide Ellipsoids," IEEE AP-S Symp. Digest, Vol. 1, Philadelphia, PA, pp. 643–646, June 8–13, 1986.
[16] Yoh, V.P., "Beam Waveguide Efficiency Analysis," IEEE AP-S Symp. Digest, Vol. 1, Philadelphia, PA, pp. 787–789, June 8–13, 1986.

Appendix 8A
List of Chapter 8 Computer Programs and Source Code Listings of Selected Programs

In this appendix the names of computer programs developed especially for Chapter 8 are listed. The source code for some of the simpler programs is listed. These tutorial programs are discussed in the text. All programs are written in Microsoft BASIC and FORTRAN for IBM PC and compatible computers. The source code for all programs with instructions for operating them is available on 5¼-inch diskettes in the optional software package and user's manual with example booklet.

The computer program names are given in the first few lines of the source code. The suffix .FOR indicates a FORTRAN program, .BAS a BASIC program, and .DAT a data file.

SOFTWARE LIST, CHAPTER 8

PROGRAM NAME:	ASSOCIATED PROGRAMS, FILES AND SUBROUTINES:
SHAPEC.FOR;	SHAP2.DAT, MNEWT, LUDCMP, LUBKSP, YUP.DAT, YDP.DAT
TRACXX.BAS;	SHAPEC.FOR, YUP.DAT
CLCCS.BAS;	TRACXX.BAS
SHAPH.FOR;	HAT.DAT, YCCD.DAT, YCCU.DAT
SHAPM.FOR;	MAT.DAT, YCUTT.DAT, YCUTB.DAT
TERMF.FOR;	COEF.DAT, ZBRAK, ZBRENT, TFM.DAT, SHAPM.FOR
EQFEED.BAS	
BWAVEG.BAS	
BWG-H.BAS	
*CHU-BW.BAS	

*POLARB.BAS
SUM-G2.BAS

* Source code for these programs is printed in this appendix.

The programs MNEWT, LUDCMP, LUBKSP, ZBRAK, and ZBRENT are *Numerical Recipes* subroutines in FORTRAN, and they must be linked to the programs that they follow.

```
10 REM Propagation programs for beam waveguides.
20 PRINT "CHU-BW.BAS"
30 K=1
40 PRINT "To compute focal lengths based on thin lens formula"
50 INPUT "ENTER A0,B0 ", A0,B0
60 TT=1/A0 +1/B0
70 FF=1/TT
80 PRINT "Fn= ";FF
90 INPUT "ENTER (1) TO CONTINUE, TO COMPUTE MORE FOCAL LENGTHS ENTER (2) ", K
100 IF K=2 THEN GOTO 50
110 PI=3.141592654# : FR=30   : U=30/FR
120 PRINT " W is -8.686 dB level diameter on mirror surface,"
130 PRINT " and R is (phase) radius of curvature of mirror"
140 PRINT " in direction of propagation."
150 REM R is usually positive.
160 INPUT "W,R= ";W,R : PRINT
170 W0=W/SQR(1+(PI*W*W/U/R)^2)
180 Z=R/(1+(U*R/PI/W/W)^2)
190 PRINT "W0,NEXT BEAM WAIST= ";W0
200 PRINT "Z, DISTANCE FROM MIRROR TO WAIST= ";Z
210 PRINT
220 INPUT "ZN,DISTANCE FROM WAIST TO NEXT MIRROR ",Z
230 W3=W0*SQR(1+(U*Z/PI/W0/W0)^2)
240 R3=Z*(1+(PI*W0*W0/U/Z)^2)
250 PRINT "W3,WAIST DIAMETER ON NEXT MIRROR ";W3
260 PRINT "R3, RADIUS OF WAVE INCIDENT ON NEXT MIRROR "R3
270 PRINT
280 INPUT "ENTER F2, OF MIRROR AND R3, COMPUTED ABOVE=";F2,R3
290 Y=1/F2-1/R3
300 R4=1/Y
310 PRINT "R4, RADIUS OF WAVE DEPARTING NEXT MIRROR ";R4
320 PRINT
330 PRINT "W3 AND R4 ARE NOW CYCLED TO GET NEXT WAIST AND ITS LOCATION"
340 PRINT "ENTER THEM AS W,R"
350 GOTO 160
360 END
```

```
10 REM *THIS IS A DEMONSTRATION PROGRAM TO SHOW*
20 REM *THE CROSS POLARIZATION CAUSED IN BWGs*
30 PRINT "POLARB.BAS"
40 PI=3.14159265359#
50 F=40 : EP=.00001 : LC=.4342944819# : MM=125.25/2
60 INPUT "Tilt angle of feed in DEGS.=",A0
70 A0=A0*PI/180
80 REM The Feed is a Y-polarized Huygen's Source.
90 PRINT : INPUT "Enter x,y: ",X,Y
100 X=X+EP
110 Z=F-(X*X+Y*Y)/4/F
120 R1=SQR(X*X+Y*Y+4*F*F)
130 L=X/R1 : M=Y/R1 : N=2*F/R1
140 REM Rotation of coordinates to tilt, Z1-axis
150 Y1=Y*COS(A0)-Z*SIN(A0)
160 Z1=Y*SIN(A0)+Z*COS(A0)
170 X1=X
180 REM PRINT "X1=";X1;"Y1=";Y1;"Z1=";Z1
190 REM PRINT "L,M,N=";L;M;N
200 R=SQR(X*X+Y*Y+Z*Z)
210 T=Z1/R
220 A=ATN(SQR(1-T*T)/T)
230 B=ATN(Y1/X1)
232 IF X1<=0 THEN B=B+PI
235 MG=COS(A)^MM
240 REM Huygen source polarization
250 L1=-(1-COS(A))*SIN(B)*COS(B)
260 M1=1-SIN(B)^2*(1-COS(A))
270 N1=-SIN(A)*SIN(B)
280 L2=L1
290 REM Rotation of polarization vectors back to Z-axis.
300 M2=M1*COS(A0)+N1*SIN(A0)
310 N2=N1*COS(A0)-M1*SIN(A0)
320 B2=2*(L*L2+M*M2+N*N2)
330 L3=B2*L-L2
340 M3=B2*M-M2
350 N3=B2*N-N2
360 REM PRINT "A,B=";A*180/PI;B*180/PI
370 REM PRINT "L1,M1,N1=";L1;M1;N1
380 PRINT "L2,M2,N2=";L2;M2;N2
390 PRINT "EAX,EAY,EAZ=";MG*L3;MG*M3;MG*N3
400 P0=10*LC*LOG(M3*M3*MG*MG)
410 P1=10*LC*LOG(L3*L3*MG*MG+EP)
420 PRINT "Po=";P0;"Px=";P1
430 REM Facing Paraboloid Reflection
440 L=-L : L3=-L3
450 REM Snell's Law again with S1=ray from 1st Mirror
460 B4=2*(L*L3+M*M3+N*N3)
470 L4=B4*L-L3
480 M4=B4*M-M3
490 N4=B4*N-N3
500     PRINT "L4,M4,N4=";L4;M4;N4
505 GOTO 90
510 END
```

Glossary of Programs

Program Name	*Location**	*Data Files, Subroutines, and Associated Programs***
4XSTUP.BAS	2S3.12.1	
APRX.BAS	2S3.2.2	
ASYMM.BAS	2S3.4	
BWAVEG.BAS	2S8.4.1	
BWG-H.BAS	2S8.4.1	
CASS.BAS	2S1.4	
CAUSTIC.BAS	2S7.2	
CHRN.FOR	2S1.5	CORH.DAT, SDIM.DAT
CHU-BW.BAS	2S8.4.2	
CIRC.BAS	1S1.2.3	CIRC (DAT), LINEPL.BAS
CLCCG.BAS	1S7.4	CLCC.BAS, CIRC.BAS, LINEPL.BAS
CLCCS.BAS	2S8.2.3	TRACXX.BAS, SHAPEC.FOR
CLCGG.BAS	2S7.4	CLCG.BAS, CIRC.BAS, LINEPL.BAS
CROSS.BAS	2S1.6.2	
CSC.FOR	2S3.8.1	CSC.DAT, QTRAP, TRAPZD
DCCH.FOR	3S2.10	BDCCH.FOR, ANALCCH.RES, SYNTCCH.RES
DCPH.FOR	3S2.10	BDCPH.FOR, ANALCPH.RES, SYNTCPH.RES
DIPOL.BAS	2S3.3	
DRANE.FOR	2S1.3.2	BESSI1
DRAWCGA.BAS	3S2.11,12	DRAWEX.DAT
DRAWHGC.BAS	3S2.11,12	DRAWEX.DAT
DSCH.FOR	3S2.10	BDSCH.FOR, ANALSCH.RES, SYNTSCH.RES
DSPH.FOR	3S2.10	BDSPH.FOR, ANALSPH.RES, SYNTSPH.RES
EQFEED.BAS	2S8.3.2	
FFOC.BAS	2S1.4	

Program Name	Location*	Data Files, Subroutines, and Associated Programs**
GREG.BAS	2S1.4	
HANSEN.FOR	2S1.3.1	BESSJ1, BESSI0, BESSI1, ZBRENT
HOLTG.BAS	2S5.5.2	HOLT6.BAS, SP1F6(DAT), SP2F6(DAT)
HOLTT.BAS	2S5.5	SP1FIT(DAT), SP2FIT(DAT)
HORN.FOR	3S2.10	BDATA.FOR, SWPH.FOR, CPH.FOR, CCH.FOR, MSCH.FOR
HORN1.FOR	2S1.5	F1N.FOR, ASLEGS.FOR, ZBRENT.FOR
HORN2.FOR	2S1.5	F1N.FOR, ASLEGS.FOR, H1.DAT
HORN3.FOR	2S1.5	SPHBES.FOR, H2.DAT
HORN4.FOR	2S1.5	F1N.FOR, ASLEGS.FOR, SPHBES.FOR, H3.DAT
HORN5.FOR	2S1.5	SVDFIT, SVDCMP, SVBKSB, H4.DAT
HORNFO.BAS	2S3.12	
ICAS.FOR	2S3.9	QTRAP, TRAPZD
KOUZN. BAS	2S5.3	SP1.DAT, SP2.DAT
LENAMF.BAS	2S5.3	SP1.DAT, SP2.DAT
LENAMP.BAS	2S5.2.3	
LENP.FOR	2S5.8	SOLV1.FOR, SOLV2.FOR, ZBRAK, ZBRENT, LEN2.DAT, SP1FIT.DAT, SP2FIT.DAT, ALAM.DAT, PHAMP.DAT, PATT.FOR, CJS.DAT, PLPATT.BAS
LENS4G.BAS	2S5.5.2	LENS4.BAS, STSET(DAT)
LENS4S.BAS	2S5.5.2	STEP.DAT, STSET(DAT)
LENTO4.FOR	2S5.8	SOLV.FOR, ZBRAK, ZBRENT, LEN.DAT, STSET.DAT, PHAMP.DAT, ALAM.DAT, PATT.FOR, CJS.DAT, PLPATT.BAS
LINEPL.BAS	1S1.2.3	GRAPHICS.COM, CIRC.BAS, PLPATT.BAS, PLMET.BAS, PLREF.BAS
LINSOD.BAS	2S7.9.2	
LINSOU.BAS	2S7.9.2	
LOCI.BAS	2S7.2	
LPTST.BAS	1S1.2.3	LINEPL.BAS
MEASE.BAS	2S3.13.1	
MENISG.BAS	2S5.5.2	MENISC.BAS
MKPL.BAS	2S4.4	PLREF.BAS, LINEPL.BAS, REF.FOR
MULTE.FOR	2S7.4.1	CLCGG.BAS, DATINE.DAT, PHAMP.DAT, ALAM.DAT, PATT.FOR, CJS.DAT, PLPATT.BAS

Program Name	*Location**	*Data Files, Subroutines, and Associated Programs***
MULTH.FOR	2S7.4.1	CLCCG.BAS, DATINH.DAT, PHAMP.DAT, ALAM.DAT, PATT.FOR, CJS.DAT, PLPATT.BAS
NEARF.FOR	3S2.11	NEARF0.DAT, NEARF0.RES, RES0.DRA
OFFSET.FOR	2S3.3	DISH1.DAT, PHAMP.DAT, ALAM.DAT, PATT.FOR, CJS.DAT, PLPATT.BAS
PARAP.FOR	2S7.3	MNEWT, LUDCMP, LUBKSB, PARD.DAT, PHAMP.DAT, ALAM.DAT, PATT.FOR, CJS.DAT, PLPATT.BAS
PARBF0.BAS	2S3.12.2	
PARBUP.BAS	2S3.12.2	
PATT.FOR	2S1.2.3	PHAMP.DAT, ALAM.DAT
PFCAS.BAS	2S3.5	CJS.DAT, PLPATT.BAS
PHACEN.FOR	3S2.12	PHACEN0.DAT, RESO.DRA, RES1.DRA, CPHEX00.PHC
PLPATT.BAS	1S1.2.3	CJS.DAT, PATT.FOR, LINEPL.BAS
PLREF.BAS	4S4.4	REF.FOR, LINEPL.BAS, MKPL.BAS
POLARB.BAS	2S8.4.3	
POLGEN.BAS	2S5.4	
PRIMEG.BAS	2S7.2	PRIME.BAS, CIRC.BAS,
PRISM.BAS	2S5.4	LINEPL.BAS
RAYRAE.BAS	2S3.11	
RAYRAH.BAS	2S3.11.1	
RAYRSH.BAS	2S3.11	
REF.FOR	4S4.4	XQ: XQ2.DAT, LINEPL.BAS, OUTXQ2.DAT, PLREF.BAS
RESIDS.BAS	2S3.7	H3.DAT
SGREG.BAS	2S7.6	
SHAPE.FOR	2S7.5	MNEWT, LUDCMP, LUBKSB, CLCCS.BAS, SHAP1.DAT, TIP.DAT,
SHAPEC.FOR	2S8.2.3	MNEWT, LUDCMP, LUBKSB, SHAP2.DAT, YUP.DAT, YDP.DAT, TRACXX.BAS
SHAPH.FOR	2S8.3.1	HAT.DAT, YCCD.DAT, YCCU.DAT
SHAPM.FOR	2S8.3.2	MAT.DAT, COEF.DAT, TERMF.FOR, YCUTT.DAT, YCUTB.DAT
SLCOD.FOR	2S5.7,8	POLCOD, POLIND, SP1.DAT, SP2.DAT, SP1FIT.DAT, SP2FIT.DAT
SLOTS.BAS	2S3.3	
SPHAP.FOR	2S7.6	MNEWT, LUDCMP, LUBKSB, SPH.DAT, PATT.FOR,

Program Name	Location*	Data Files, Subroutines, and Associated Programs**
		ALAM.DAT, PHAMP.DAT, PATT.FOR, CJS.DAT, PLPATT.BAS
SPHPOL.BAS	2S7.6	
SPILL.BAS	2S1.6.1	
SPL.FOR	2S5.7	SPL.IN, SP1.DAT, SP2.DAT
SPLFIT.FOR	2S5.7	GAUSSS, SP1.DAT, SP2.DAT, SP1FIT.DAT, SP2FIT.DAT
STCOD.FOR	2S5.6	FRIED.DAT, POLCOD, POLIND, STSET.DAT
STCODS.FOR	2S5.6	FRIED.DAT, STEP1.DAT, POLCOD, POLIND, STSET.DAT
STEPF.BAS	2S5.6	
STOCH.BAS	2S1.6.4	
SUBPAT.FOR	2S3.5	DISH2.DAT, PHAMP.DAT, ALAM.DAT, PATT.FOR, CJS.DAT, PLPATT.BAS
SUM-G2.BAS	2S8.4.3	
SYMPF.FOR	2S3.2.2	DISH.DAT, PHAMP.DAT ALAM.DAT, PATT.FOR, CJS.DAT, PLPATT.BAS
SYMPF2.FOR	2S3.2.2	DISH3.DAT, PHAMP.DAT, ALAM.DAT, PATT.FOR, CJS.DAT, PLPATT.BAS
SYMPFH.FOR	1S3.2.2	DISH.DAT, H4.DAT, PHAMP.DA, ALAM, PATT.FOR, CJS.DAT, PLPATT.BAS
SYNTH.BAS	2S5.3	
SYNTHS.BAS	2S5.3	
TERMF.FOR	2S8.3.2	ZBRAK, ZBRENT, SHAPM.FOR, COEF.DAT, TFM.DAT
TILT.BAS	2S7.9.3	
TORAP.FOR	2S7.7	TORD.DAT, MNEWT, LUDCMP, LUBKSB, ALAM.DAT, PHAMP.DAT, PATT.FOR, CJS.DAT, PLPATT.BAS
TORCA.FOR	2S7.8	MNEWT, LUDCMP, LUBKSB, SHAP.DAT, YD.DAT, YU.DAT
TORPOL.BAS	2S7.7	
TORRAY.BAS	2S7.7	
TRACE.BAS	2S3.13	
TRACXX.BAS	2S8.2.3	SHAPEC.FOR, CLCCS.BAS, YUP.DAT

**Location*—The first digit indicates the Diskette Number 1, 2, 3, and 4. N indicates an output file generated by programs and that no sample data are provided on the diskettes. S signifies section and the first digit following S is the chapter in both the software manual and text where the program is discussed or listed. The section in the text is specified by the numbers after S.
** The files listed without suffixes, ZBRAK, for example, are Numerical Recipes FORTRAN subroutines available as object modules in Diskette #1 in the subdirectory \SUB.

Glossary of Subroutines and Data Files

File	*Location**	*Related Programs***
ALAM.DAT	NS1.2.3	SYMPFH.FOR (and 11 other pattern programs)
ANALCCH.RES	4S2.10	DCCH.FOR
ANALCPH.RES	4S2.10	DCPH.FOR
ANALSCH.RES	4S2.10	DSCH.FOR
ANALSPH.RES	4S2.10	DSPH.FOR
ASLEGS.FOR	1S1.5	HORN1.FOR, HORN2.FOR, HORN4.FOR
BDATA.FOR	3S2.10	HORN.FOR
BDCCH.FOR	3S2.10	DCCH.FOR
BDCPH.FOR	3S2.10	DCPH.FOR
BDSCH.FOR	3S2.10	DSCH.FOR
BDSPH.FOR	3S2.10	DSPH.FOR
BESSI0.OBJ	1S1.3.1	HANSEN.FOR
BESSI1.OBJ	1S1.3.2	DRANE.FOR
BESSJ1.OBJ	1S1.3.1	HANSEN.FOR
CCH.FOR	3S2.10	HORN.FOR
CIRC	1S1.2.3	CIRC.BAS, LEN4SG.BAS, MENISG.BAS, CLCGG.BAS, CLCCG.BAS, HOLTG.BAS, PRIMEG.BAS
CJS.DAT	NS7.4.1	MULTE.FOR (and many pattern programs)
COEF.DAT	1S8.3.2	TERMF.FOR, SHAPM.FOR
CORH.DAT	1S1.5	CHRN.FOR
CPH.FOR	3S2.10	HORN.FOR
CPHEX.DAT	4S2.11,12	DRAWCGA.BAS
CPHEX.FED	4S2.11,12	DRAWCGA.BAS (Chapter 4)
CPHEX.RES	4S2.11,12	DRAWCGA.BAS
CPHEX00.DRC	4S2.11,12	DRAWCGA.BAS
CPHEX00.DRF	4S2.11,12	DRAWCGA.BAS
CPHEX00.DRN	4S2.11,12	DRAWCGA.BAS
CPHEX00.PHC	4S2.12	PHACEN.FOR

File	Location*	Related Programs**
CPHEX45.DRC	4S2.11,12	DRAWCGA.BAS
CPHEX45.DRX	4S2.11,12	DRAWCGA.BAS
CPHEX90.DRC	4S2.11,12	DRAWCGA.BAS
CPHEX90.DRF	4S2.11,12	DRAWCGA.BAS
CSC.DAT	1S3.8.1	CSC.FOR
DATINE.DAT	1S7.4.1	MULTE.FOR
DATINH.DAT	1S7.4.1	MULTH.FOR
DISH.DAT	1S3.2.2	SYMPF.FOR, SYMPFH.FOR
DISH1.DAT	1S3.3	OFFSET.FOR
DISH2.DAT	1S3.5	SUBPAT.FOR
DISH3.DAT	1S3.2.2	SYMPF2.FOR
DRAW.DAT	4S2.11,12	DRAWCGA.BAS
DRAWEX.DAT	4S2.11,12	DRAWCGA.BAS, DRAWHGC.BAS
DRAWEXF.DAT	4S2.11,12	DRAWCGA.BAS
DRAWNEAR.DAT	4S2.11,12	DRAWCGA.BAS
FIN.FOR	1S1.5	HORN1.FOR, HORN2.FOR, HORN4.FOR
FRIED.DAT	1S5.6	STCOD.FOR, STCODS.FOR
H4.DAT	1S3.2.2	SYMPFH.FOR, HORN4.FOR (H1, H2, H3, H5 are also used)
HAT.DAT	1S8.3.1	SHAPH.FOR
LEN.DAT	1S5.8	LENTO4.FOR
LEN2.DAT	1S5.8	LENP.FOR
LUBKSB.OBJ	1S7.3	PARAP.FOR, SHAPE, SHAPEC, SPHAP, TORAP, TORCA
LUDCMP.OBJ	1S7.3	PARAP.FOR, SHAPE, SHAPEC, SPHAP, TORAP, TORCA
MAT.DAT	1S8.3.2	SHAPM.FOR
MNEWT.OBJ	1S7.3	PARAP.FOR, SHAPE, SHAPEC, SPHAP, TORAP, TORCA
MSCH.FOR	3S2.10	HORN.FOR
NEARF.RES	4S2.11	NEARF.FOR
NEARF0.DAT	4S2.11	NEARF.FOR
NEARF0.RES	4S2.11	NEARF.FOR
NEARF1.DAT	4S2.11	NEARF.FOR
PARD.DAT	1S7.3	PARAP.FOR
PHACEN0.DAT	4S2.12	PHACEN.FOR
PHACEN1.DAT	4S2.12	PHACEN.FOR
PHAMP.DAT	NS1.2.3	SYMPFH.FOR (and 11 other pattern programs)
PIHORN.DAT	4S2.11,12	DRAWCGA.BAS
POLCOD.OBJ	1S5.7,8	SLCOD.FOR, STCOD.FOR, STCODS.FOR
POLIND.OBJ	1S5.7,8	SLCOD.FOR, STCOD.FOR, STCODS.FOR

File	Location*	Related Programs**
QTRAP.OBJ	1S3.9	ICAS.FOR,CSC.FOR
RES0.DRA	4S2.11	NEARF.FOR
RES1.DRA	4S2.12	PHACEN.FOR
SHAP.DAT	1S7.8	TORCA.FOR
SHAP1.DAT	1S7.5	SHAPE.FOR
SHAP2.DAT	1S8.2.3	SHAPEC.FOR
SOLV.FOR	1S5.8	LENTO4.FOR
SOLV1.FOR	1S5.8	LENP.FOR
SOLV2.FOR	1S5.8	LENP.FOR
SP1.DAT	NS5.3	KOUZN.BAS, LENAMF.BAS, SPL.FOR, SLCOD.FOR
SP1F6	1S5.5.2	HOLTG.BAS, HOLT6.BAS, SLCOD.FOR
SP1FIT	1S5.5	HOLTT.BAS, SPL.FOR, SLCOD.FOR
SP1FIT.DAT	NS5.8	LENP.FOR, SPL.FOR, SLCOD.FOR
SP2.DAT	NS5.3	KOUZN.BAS, LENAMF.BAS, SPL.FOR, SLCOD.FOR
SP2F6	1S5.5.2	HOLTG.BAS, HOLT6.BAS, SPL.FOR, SLCOD.FOR
SP2FIT	1S5.5	HOLTT.BAS, SPL.FOR, SLCOD.FOR
SP2FIT.DAT	NS5.D	LENP.FOR, SPL.FOR, SLCOD.FOR
SPH.DAT	1S7.6	SPHAP.FOR
SPHBES.FOR	1S1.5	HORN3.FOR, HORN4.FOR
SPL.IN	1S5.7	SPL.FOR
SQ.DAT	4S4.4	REF.FOR
STEP.DAT	1S5.5.2	LENS4S.BAS
STEP1.DAT	1S5.6	STCODS.FOR
STSET	1S5.5.2	LENS4G.BAS, LENS4.BAS, LENS4S.BAS
STSET.DAT	NS5.8	
SVBKSB.OBJ	1S1.5	LENTO4.FOR, STCOD.FOR
SVDCMP.OBJ	1S1.5	HORN5.FOR
SVDFIT.OBJ	1S1.5	HORN5.FOR
SWCOHO.DAT	4S2.11,12	HORN5.FOR
SWPH.FOR	3S2.10	DRAWCGA.BAS
SYNTCCH.RES	4S2.10	HORN.FOR
SYNTCPH.RES	4S2.10	DCCH.FOR
SYNTSCH.RES	42S2.10	DCPH.FOR
SYNTSPH.RES	4S2.10	DSCH.FOR
		DSPH.FOR

File	*Location**	*Related Programs***
TFM.DAT	NS8.3.2	TERMF.FOR
TIP.DAT	NS7.5	SHAPE.FOR
TORD.DAT	1S7.7	TORAP.FOR
TRAPZD.OBJ	1S3.9	ICAS.FOR,CSC.FOR
YCCD.DAT	NS8.3.1	SHAPH.FOR
YCCU.DAT	NS8.3.1	SHAPH.FOR
YCUTB.DAT	NS8.3.2	SHAPM.FOR
YCUTT.DAT	NS8.3.2	SHAPM.FOR
YD.DAT	NS7.7	TORCA.FOR
YDP.DAT	NS8.2.3	SHAPEC.FOR
YU.DAT	NS7.7	TORCA.FOR
YUP.DAT	1S8.2.3	SHAPEC.FOR
ZBRAK.OBJ	1S5.8	LENP.FOR, LENTO4.FOR, TERMF.FOR
ZBRENT.OBJ	1S1.3.1	HANSEN.FOR, HORN1.FOR, LENP.FOR, LENTO4.FOR, TERMF.FOR

**Location*—The first digit indicates the Diskette Number 1, 2, 3, and 4. N indicates an output file generated by programs and that no sample data are provided on the diskettes. S signifies section and the first digit following S is the chapter in both the software manual and text where the program is discussed or listed. The section in the text is specified by the numbers after S.
** *Related Programs* are usually the programs using the file for input data or to generate the file as output.

Index

www.ingramcontent.com/pod-product-compliance
Lightning Source LLC
LaVergne TN
LVHW101454200926
848483LV00001B/127
9780890062401